Introductory Statistical Inference

STATISTICS: Textbooks and Monographs

Recent Titles

Statistical Process Monitoring and Optimization, *edited by Sung H. Park and G. Geoffrey Vining*

Statistics for the 21st Century: Methodologies for Applications of the Future, *edited by C. R. Rao and Gábor J. Székely*

Probability and Statistical Inference, *Nitis Mukhopadhyay*

Handbook of Stochastic Analysis and Applications, *edited by D. Kannan and V. Lakshmikantham*

Testing for Normality, *Henry C. Thode, Jr.*

Handbook of Applied Econometrics and Statistical Inference, *edited by Aman Ullah, Alan T. K. Wan, and Anoop Chaturvedi*

Visualizing Statistical Models and Concepts, *R. W. Farebrother and Michaël Schyns*

Financial and Actuarial Statistics: An Introduction, *Dale S. Borowiak*

Nonparametric Statistical Inference, Fourth Edition, Revised and Expanded, *Jean Dickinson Gibbons and Subhabrata Chakraborti*

Computer-Aided Econometrics, *edited by David E.A. Giles*

The EM Algorithm and Related Statistical Models, *edited by Michiko Watanabe and Kazunori Yamaguchi*

Multivariate Statistical Analysis, Second Edition, Revised and Expanded, *Narayan C. Giri*

Computational Methods in Statistics and Econometrics, *Hisashi Tanizaki*

Applied Sequential Methodologies: Real-World Examples with Data Analysis, *edited by Nitis Mukhopadhyay, Sujay Datta, and Saibal Chattopadhyay*

Handbook of Beta Distribution and Its Applications, *edited by Arjun K. Gupta and Saralees Nadarajah*

Item Response Theory: Parameter Estimation Techniques, Second Edition, *edited by Frank B. Baker and Seock-Ho Kim*

Statistical Methods in Computer Security, *edited by William W. S. Chen*

Elementary Statistical Quality Control, Second Edition, *John T. Burr*

Data Analysis of Asymmetric Structures, *Takayuki Saito and Hiroshi Yadohisa*

Mathematical Statistics with Applications, *Asha Seth Kapadia, Wenyaw Chan, and Lemuel Moyé*

Advances on Models, Characterizations and Applications, *N. Balakrishnan, I. G. Bairamov, and O. L. Gebizlioglu*

Survey Sampling: Theory and Methods, Second Edition, *Arijit Chaudhuri and Horst Stenger*

Statistical Design of Experiments with Engineering Applications, *Kamel Rekab and Muzaffar Shaikh*

Quality by Experimental Design, Third Edition, *Thomas B. Barker*

Handbook of Parallel Computing and Statistics, *Erricos John Kontoghiorghes*

Statistical Inference Based on Divergence Measures, *Leandro Pardo*

A Kalman Filter Primer, *Randy Eubank*

Introductory Statistical Inference, *Nitis Mukhopadhyay*

Introductory Statistical Inference

Nitis Mukhopadhyay

University of Connecticut

Storrs, Connecticut

CRC Press
Taylor & Francis Group
Boca Raton London New York

CRC Press is an imprint of the
Taylor & Francis Group, an **informa** business

A CHAPMAN & HALL BOOK

CRC Press
Taylor & Francis Group
6000 Broken Sound Parkway NW, Suite 300
Boca Raton, FL 33487-2742

First issued in paperback 2019

ISBN-13: 978-1-57444-613-5 (hbk)
ISBN-13: 978-0-367-39115-7 (pbk)

Library of Congress Cataloging-in-Publication Data

Mukhopadhyay, Nitis, 1950-
   Introductory statistical inference / Nitis Mukhopadhyay.
      p. cm. -- (Statistics, textbooks and monographs ; v. 187)
   Includes bibliographical references and indexes.
   ISBN 1-57444-613-4 (acid-free paper)
      1. Mathematical statistics--Textbooks. 2. Probabilities--Textbooks. I. Title. II. Series.

QA276.M79 2006
519.5--dc22
                                      2005044880

Visit the Taylor & Francis Web site at
http://www.taylorandfrancis.com

and the CRC Press Web site at
http://www.crcpress.com

Dedicated to My Parents-in-Law

Late Professor Atindra Mojumder

Mrs. Manika Majumder

Preface

My department at the University of Connecticut-Storrs offers a standard *one-year* sequence on probability theory (fall semester) and statistical inference (spring semester) for our own first-year graduate students. Through this process, strong students tend to blossom and they move forward to take more advanced statistics courses.

Some years ago, a colleague from another program requested my department to create a *one-semester*-long course at a Ph.D. level, covering *both* probability theory and statistical inference. An ideal course will (i) *preserve* the integrity and depth of the original one-year's coverage, and also (ii) *emphasize* the key concepts, principles, as well as mathematical techniques.

The request to offer such a service course gained momentum as similar requests started arriving from a number of graduate programs. Those programs felt the serious need for such a heavy-duty course for their Ph.D. students. The students successfully completing it would be ready to take advanced courses at a Ph.D. level in their substantive fields quickly. Also, they could be paired with appropriate research projects earlier than usual.

Nearly four years ago, I was given the charge to create this course and try it out. I must admit that I was skeptical at first. However, I decided to accept this challenge, and what a challenge it has been! I have enjoyed every bit of it and I earnestly thank Professor Dipak K. Dey, my colleague and the department head, for giving me the opportunity to show some extra creativity this way.

I looked for an appropriate textbook as I began preparing the material for the course. Several textbooks, including my own (*Probability and Statistical Inference*: Marcel Dekker, Inc., 2000), came to mind, but these were meant for a *one-year sequence* for first-year statistics students aspiring to do Ph.D. work. Some books were available for a Masters level sequence, but a majority of them lacked the necessary depth, breadth, and detail that I surely needed. These books were also meant for a *one-year* sequence! I felt strongly that it would be very hard for my students to focus on the required material if they were asked to follow one of those lengthy textbooks.

I decided to teach from my own notes in the spring of 2003. I taught the course again with subsequent revisions in the spring semesters of 2004 and 2005. By that time, I felt convinced that such a service course could be offered at any school if there was an appropriate textbook to follow at a Ph.D. level.

I thank all those students who bravely took the new course from me in the spring semesters of 2003, 2004, and 2005. Their urgent need and serious devotion to learn this heavy-duty material inspired me to write this book. Each student has helped me in more ways than he or she will ever know in the last three years, and I thank them all profusely.

The prerequisite is three semesters' worth of calculus (including several variables) and at least one junior- or senior-level statistics course. These may suffice for generally understanding a major part of this book. There are sections for which some familiarity with linear algebra will be beneficial. I have reviewed some important mathematical tools (for example, the gamma and beta functions) in Section 1.6.3. Section 4.7 provides a review of matrices and vectors.

The material is presented at the level of a standard one-year sequence for the first-year statistics graduate students. The level is similar to what one finds in my other textbook (*Probability and Statistical Inference*: Marcel Dekker, Inc., 2000). However, the present textbook is less than one-half the size of standard books that may be followed in a one-year sequence. That in itself will help the students and instructors alike to focus on the required material. It is a textbook for a semester-long service course and it is not going to compete with anything else, because there is no other book like it.

The first four chapters introduce some basic concepts in probability theory, including conditional probability, random variables, probability distributions, moments, moment generating functions, multivariate random variables, exponential family of distributions, transformations, and sampling distributions. Chapter 5 introduces convergence in probability and distribution as well as the central limit theorem. Chapter 1 is by far the longest (37 pages), but it is mostly review material, except perhaps Section 1.3.

The remainder of the book develops fundamental concepts of statistical inference. I introduce sufficiency, Neyman factorization, information, minimal sufficiency, completeness, and ancillarity in Chapter 6. Basu's theorem and the location, scale, and location-scale families of distributions are also explained.

Maximum likelihood estimators, Rao-Blackwell Theorem, Cramér-Rao Inequality, Lehmann-Scheffé Theorems, and uniformly minimum variance unbiased estimators are developed in Chapter 7. Chapter 8 provides the foundation of statistical tests of hypotheses by developing the Neyman-Pearson theory. Confidence interval methods are developed in Chapter 9 through a pivotal approach. Chapter 10 is devoted to Bayesian methods. Likelihood ratio and other tests are developed in Chapter 11. Chapter 12 includes some basic large-sample confidence interval and test procedures, including the variance stabilizing transformations.

Chapter 13 presents (i) a list of abbreviations and notations, (ii) brief historical notes, and (iii) standard statistical tables. An extensive list of references is then given that is followed by the answers for some selected exercises. The author index and a very extensive subject index wrap up the book.

This textbook is meant primarily for a one-semester-long course at the Ph.D. level in a number of programs including Economics, Agricultural and Resource Economics, Operations and Information Management, Financial Mathematics, or Engineering. The textbook may *also* be adopted for a one-year sequence, either at the Masters or at the Ph.D. level, for the programs that have been mentioned and others. Additionally, I believe that this book may help immensely as a supplementary text in higher level courses, for example, in Linear Models, Decision Theory, and Advanced Statistical Inference.

Comments from several readers and reviewers have been helpful and I thank them. Everyone within Chapman & Hall/CRC (Taylor & Francis Group) has helped immensely to move this project forward. I express my heartfelt gratitude to all of them. I especially mention my indebtedness to Ms. Linda Manis, the Project Editor. Thank you, Linda.

My dear friend and a former Ph.D. student, Greg Cicconetti, gave valuable pointers on a number of chapters. My two sons, Shankha and Ranjan, looked over the material diligently. Ranjan did so a number of times! I remain grateful to Greg, Shankha, and Ranjan for their invaluable help. In the end, however, I take full responsibility for any remaining mistakes and typographical errors. I will appreciate it very much if the readers will be kind enough to forward to me any mistakes or typographical errors. I hope to correct these in a next printing.

Shankha and Ranjan were encouraging as I was being swayed away by this project. My wife, Mahua, stood by me graciously throughout this labor of love. Without her steadfast support, I could never have finished this book. Thanks to Mahua, Shankha, and Ranjan for their patience and understanding.

Enjoy and Celebrate Statistics!

Nitis Mukhopadhyay
Department of Statistics
University of Connecticut, Storrs

Contents

1

Probability and Distributions

1.1   Introduction

In studying the subject of *probability*, we first imagine an appropriate *random experiment*. A random experiment has three important components which are:

> a)   multiplicity of outcomes,
> b)   uncertainty regarding the outcomes, and
> c)   repeatability of the experiment in identical fashions.

Suppose that one tosses a regular coin up in the air. The coin has two sides, namely the head (H) and tail (T). Let us assume that the tossed coin will land on either H or T. Every time one tosses the coin, there is the possibility of the coin landing on its head or tail (*multiplicity of outcomes*). But, no one can say with absolute certainty whether the coin would land on its head, or for that matter, on its tail (*uncertainty regarding the outcomes*). One may toss this coin as many times as one likes under identical conditions (*repeatability*) provided the coin is not damaged in any way in the process of tossing it successively.

A random experiment provides in a natural fashion a list of *all* possible outcomes, also referred to as *simple events*. These simple events act like "atoms" in the sense that the experimenter is going to observe only one of these simple events as a possible outcome when the particular random experiment is performed. A *sample space* is a set, denoted by **S**, which enumerates every possible simple event. Then, a probability scheme is generated on subsets of **S**, including **S** itself, in a way which mimics the nature of the random experiment itself. Throughout, we will write $P(A)$ for the probability of a statement $A(\subseteq \mathbf{S})$. First, let us consider two simple examples to illustrate the terminologies.

Example 1.1.1 Suppose that we toss a *fair coin* three times and record outcomes from the first, second, and third toss, respectively. Then, the possible simple events are $HHH, HHT, HTH, HTT, THH, THT, TTH$, or TTT. Thus, the sample space is given by:

$$\mathbf{S} = \{HHH, HHT, HTH, HTT, THH, THT, TTH, TTT\}$$

Since the coin is assumed to be fair, this particular random experiment generates the following probability scheme: $P(HHH) = P(HHT) = P(HTH)$ $= P(HTT) = P(THH) = P(THT) = P(TTH) = P(TTT) = \frac{1}{8}$. ▲

Example 1.1.2 Suppose that we roll two *fair* dice, one red and the other yellow, at the same time and record the scores on their faces landing upward. Then, each simple event would constitute, for example, a pair (i, j) where i is the number of dots on the face of the red dice that lands up and j is the number of dots on the face of the yellow dice that lands up. The sample space is then given by:

$$\mathbf{S} = \{(1,1), (1,2), ..., (1,6), (2,1), ..., (2,6), ..., (6,1), ..., (6,6)\}$$

consisting of exactly 36 possible simple events. Since both dice are assumed to be fair, this particular random experiment generates the following probability scheme: $P\{(i, j)\} = \frac{1}{36}$ for all $i, j = 1, ..., 6$. ▲

Some elementary notions of set operations are reviewed in Section 1.2. Section 1.3 describes the setup for developing a formal *theory of probability*. Section 1.4 introduces the concept of *conditional probability* followed by notions such as the *additive* and *multiplicative rules*. Sections 1.5 and 1.6 respectively introduce the *discrete* and *continuous random variables*, and the associated notions of a *probability mass function* (pmf), *probability density function* (pdf), and the *distribution function* (df). Section 1.7 summarizes some probability distributions which are frequently used in statistics.

1.2   About Sets

The union and intersection operations satisfy the following laws: For any subsets A, B, C of $\mathbf{S}$, we have

Commutative Law:	$A \cup B = B \cup A, A \cap B = B \cap A$
Associative Law:	$(A \cup B) \cup C = A \cup (B \cup C)$
	$(A \cap B) \cap C = A \cap (B \cap C)$ (1.2.1)
Distributive Law:	$(A \cup B) \cap C = (A \cap C) \cup (B \cap C)$
	$A \cup (B \cap C) = (A \cup B) \cap (A \cup C)$

We say that A and B are *disjoint* if and only if there is no common element between A and B, that is, if and only if $A \cap B = \varnothing$, an empty set. Two disjoint sets A and B are also referred to as being *mutually exclusive*.

Now, consider $\{A_i; i \in I\}$, a collection of subsets of **S**. This collection may be finite, countably infinite, or uncountably infinite. We define

Union: $\cup_{i \in I} A_i = \{x : x \in A_i \text{ for at least one } i \in I\}$

Intersection: $\cap_{i \in I} A_i = \{x: x \in A_i \text{ for all } i \in I\}$ (1.2.2)

There is another kind of set operation that one may find useful. It is called a *symmetric difference*.

Symmetric Difference: $A_i \Delta A_j = \{x: x \in (A_i \cap A_j^c) \cup (A_j \cap A_i^c)\}$,
$i, j \in I$
(1.2.3)

The Equation (1.2.2) lays down the set operations involving the union and intersection among an arbitrary number of sets. When we specialize $I = \{1, 2, 3, ...\}$ in the definition given by Equation (1.2.2), we can combine the notions of a countably infinite number of unions and intersections to come up with some interesting sets. Let us denote

$$B = \cap_{j=1}^{\infty} \cup_{i=j}^{\infty} A_i \text{ and } C = \cup_{j=1}^{\infty} \cap_{i=j}^{\infty} A_i \qquad (1.2.4)$$

Interpretation of the Set B: Here the set B is the intersection of the collection of sets $\cup_{i=j}^{\infty} A_i, j \geq 1$. In other words, an element x will belong to B if and only if x belongs to $\cup_{i=j}^{\infty} A_i$ for each $j \geq 1$, which is equivalent to saying that there exists a sequence of positive integers $i_1 < i_2 < ... < i_k < ...$ such that $x \in A_{i_k}$ for all $k = 1, 2, ...$. That is, the set B corresponds to elements which are hit *infinitely often* and hence B is referred to as the *limit* (as $n \to \infty$) *supremum* of the sequence of sets $A_n, n = 1, 2, ...$.

Interpretation of the Set C: On the other hand, the set C is the union of the collection of sets $\cap_{i=j}^{\infty} A_i, j \geq 1$. In other words, an element x will belong to C if and only if x belongs to $\cap_{i=j}^{\infty} A_i$ for some $j \geq 1$, which is equivalent to saying that x belongs to $A_j, A_{j+1}, ...$ for some $j \geq 1$. That is, the set C corresponds to the elements which are hit *eventually* and hence C is referred to as the *limit* (as $n \to \infty$) *infimum* of the sequence of sets $A_n, n = 1, 2, ...$.

$$\cap_{j=1}^{\infty} \cup_{i=j}^{\infty} A_i = \limsup_{n \to \infty} A_n \text{ and } \cup_{j=1}^{\infty} \cap_{i=j}^{\infty} A_i = \liminf_{n \to \infty} A_n.$$

Theorem 1.2.1 (DeMorgan's Law) *Consider* $\{A_i; i \in I\}$, *a collection of subsets of* **S**. *Then,*

$$\left(\cup_{i \in I} A_i \right)^c = \cap_{i \in I} (A_i^c) \qquad (1.2.5)$$

Definition 1.2.1 *A collection of sets $\{A_i; i \in I\}$ is said to consist of disjoint sets if and only if no two sets in this collection share a common element, that is when $A_i \cap A_j = \varnothing$ for all $i \neq j \in I$. A collection $\{A_i; i \in I\}$ is called a partition of* **S** *if and only if*

(*i*) $\{A_i; i \in I\}$ *consists of disjoint sets only, and*

(*ii*) $\{A_i; i \in I\}$ *spans the whole space* **S**, *that is $\cup_{i \in I} A_i = $* **S**.

Example 1.2.1 Let **S** $=(0, 1]$ and define the collection of sets $\{A_i; i \in I\}$ where $A_i = (\frac{1}{2^i}, \frac{1}{2^{i-1}}], i \in I = \{1, 2, 3, ...\}$. One should check that the given collection of intervals form a partition of $(0, 1]$. ▲

1.3   Axiomatic Development of Probability

The axiomatic theory of probability was developed by Kolmogorov in his 1933 monograph, originally written in German. Its English translation is cited as Kolmogorov (1950b). Before we describe this approach, we need to fix some ideas first.

Definition 1.3.1 *A sample space is a set, denoted by* **S**, *which enumerates each and every possible outcome or simple event.*

In general an *event* is an *appropriate* subset of the sample space **S**, including the empty set $\varnothing$ and the whole set **S**. In what follows we make this notion more precise.

Definition 1.3.2 *Suppose that $\mathcal{B} = \{A_i : A_i \subseteq$* **S**$, i \in I\}$ *is a collection of subsets of* **S**. *Then, $\mathcal{B}$ is called a Borel sigma-field or Borel sigma-algebra if the following conditions hold:*

(*i*)   *The empty set $\varnothing \in \mathcal{B}$;*

(*ii*)   *If $A \in \mathcal{B}$, then $A^c \in \mathcal{B}$;*

(*iii*)   *If $A_i \in \mathcal{B}$ for $i = 1, 2, ...$, then $\cup_{i=1}^{\infty} A_i \in \mathcal{B}$.*

In other words, the Borel sigma-field $\mathcal{B}$ is closed under the operations of complement and countable union of its members. It is obvious that the whole space **S** belongs to the Borel sigma-field $\mathcal{B}$ since we can write **S** $= \varnothing^c \in \mathcal{B}$, by the requirements (i) and (ii) in the Definition 1.3.2. Also if $A_i \in \mathcal{B}$ for $i = 1, 2, ..., k$, then $\cup_{i=1}^{k} A_i \in \mathcal{B}$, since with $A_i = \varnothing$ for $i = k + 1, k + 2, ...$, we can express $\cup_{i=1}^{k} A_i$ as $\cup_{i=1}^{\infty} A_i$ which belongs to $\mathcal{B}$ in view of (iii) in the Definition 1.3.2. That is, $\mathcal{B}$ is obviously closed under the operation of finite unions of its members. See Exercise 1.3.1 in this context.

Definition 1.3.3 *Suppose that a fixed collection of subsets $\mathcal{B} = \{A_i : A_i \subseteq$* **S**$, i \in I\}$ *is a Borel sigma-field. Then, a subset A of* **S** *is called a Borel set or an event if and only if $A \in \mathcal{B}$.*

Having started with a fixed Borel sigma-field $\mathcal{B}$ of subsets of **S**, a *probability scheme* is simply a way to assign numbers between zero and one to every event. Such assignment of numbers must satisfy some conditions given next.

Definition 1.3.4 *A probability scheme assigns a unique number to a set $A \in \mathcal{B}$, denoted by $P(A)$, for every set $A \in \mathcal{B}$ in such a way that the following conditions hold:*

$$
\begin{aligned}
&(i) \quad && P(\mathbf{S}) = 1; \\
&(ii) \quad && P(A) \geq 0 \text{ for every } A \in \mathcal{B}; \\
&(iii) \quad && P(\cup_{i=1}^{\infty} A_i) = \Sigma_{i=1}^{\infty} P(A_i) \text{ for all } A_i \in \mathcal{B} \text{ such that} \\
& && A_i\text{'s are all pairwise disjoint, } i = 1, 2, \ldots
\end{aligned}
\tag{1.3.1}
$$

Theorem 1.3.1 *Suppose A, B are any two events. Suppose also that a sequence of events $\{B_i; i \geq 1\}$ forms a partition of $\mathbf{S}$. Then,*

(i) $P(\varnothing) = 0$ and $P(A) \leq 1$;

(ii) $P(A^c) = 1 - P(A)$;

(iii) $P(B \cap A^c) = P(B) - P(B \cap A)$;

(iv) $P(A \cup B) = P(A) + P(B) - P(A \cap B)$;

(v) If $A \subseteq B$, then $P(A) \leq P(B)$;

(vi) $P(A) = \Sigma_{i=1}^{\infty} P(A \cap B_i)$.

Proof (i) Observe that $\varnothing \cup \varnothing^c = \mathbf{S}$ and also $\varnothing, \varnothing^c$ are disjoint events. Hence, by part (iii) in the Definition 1.3.4, we have $1 = P(\mathbf{S}) = P(\varnothing \cup \varnothing^c) = P(\varnothing) + P(\varnothing^c)$. Thus, $P(\varnothing) = 1 - P(\varnothing^c) = 1 - P(\mathbf{S}) = 1 - 1 = 0$. The second part follows from part (ii). ♦

(ii) Observe that $A \cup A^c = \mathbf{S}$ and then proceed as before. Observe that A and A^c are disjoint events. ♦

(iii) Notice that $B = (B \cap A) \cup (B \cap A^c)$ where $B \cap A$ and $B \cap A^c$ are disjoint events. Hence, by part (iii) in the Definition 1.3.4, we claim that

$$
P(B) = P\{(B \cap A) \cup (B \cap A^c)\} = P(B \cap A) + P(B \cap A^c)
$$

Now, the result is immediate. ♦

(iv) It is easy to verify that $A \cup B = (A \cap B^c) \cup (B \cap A^c) \cup (A \cap B)$ where the three events $A \cap B^c, B \cap A^c, A \cap B$ are also disjoint. Thus, we have

$$
\begin{aligned}
&P(A \cup B) \\
&= P(A \cap B^c) + P(B \cap A^c) + P(A \cap B) \\
&\qquad \text{in view of part (iii), Definition 1.3.4} \\
&= \{P(A) - P(A \cap B)\} + \{P(B) - P(A \cap B)\} + P(A \cap B)
\end{aligned}
$$

which leads to the desired result. ◆

(v) We leave it out as Exercise 1.3.2. ◆

(vi) Since the sequence of events $\{B_i; i \geq 1\}$ forms a partition of the sample space $\mathbf{S}$, we can write $A = A \cap \mathbf{S} = A \cap \left(\cup_{i=1}^{\infty} B_i\right) = \cup_{i=1}^{\infty}\left(A \cap B_i\right)$ where the events $A \cap B_i, i = 1, 2, \ldots$ are also disjoint. Now, the result follows from part (iii) in the Definition 1.3.4. ∎

Example 1.3.1 A fair coin is tossed three times. Let us define three events as follows:

$$
\begin{array}{ll}
A: & \text{We observe two heads} \\
B: & \text{We observe at least one tail} \\
C: & \text{We observe no heads}
\end{array}
\qquad (1.3.2)
$$

How can we obtain the probabilities of these events? First notice that as subsets of $\mathbf{S}$, we can rewrite these events as $A = \{HHT, HTH, THH\}, B = \{HHT, HTH, HTT, THH, THT, TTH, TTT\}$, and $C = \{TTT\}$. Now, it becomes obvious that $P(A) = \frac{3}{8}, P(B) = \frac{7}{8}$, and $P(C) = \frac{1}{8}$. One can also see that $A \cap B = \{HHT, HTH, THH\}$ so that $P(A \cap B) = \frac{3}{8}$, whereas $A \cup C = \{HHT, HTH, THH, TTT\}$ so that $P(A \cup C) = \frac{1}{2}$. ▲

Example 1.3.2 We roll a red die and a yellow die. Both dice are fair. Consider the following events:

$$
\begin{array}{ll}
D: & \text{The total from the red and yellow dice is 8} \\
E: & \text{The red die comes up with 2 more than the yellow die}
\end{array}
\qquad (1.3.3)
$$

Now, we express these events as $D = \{(2,6),(3,5),(4,4),(5,3),(6,2)\}$ and $E = \{(3,1),(4,2),(5,3),(6,4)\}$ as subsets of $\mathbf{S}$. It is now obvious that $P(D) = \frac{5}{36}$ and $P(E) = \frac{1}{9}$. ▲

> Having a sample space $\mathbf{S}$ and appropriate events from a Borel sigma-field $\mathcal{B}$ of subsets of $\mathbf{S}$, and a probability scheme satisfying Equation (1.3.1), one can evaluate the probability of the legitimate events only. The members of $\mathcal{B}$ are the only legitimate events.

1.4 Conditional Probability and Independent Events

Let us reconsider Example 1.3.2. Suppose that the two fair dice, one red and another yellow, are rolled. After the toss, the experimenter announces that the event D has been observed. Recall that $P(E)$ was $\frac{1}{9}$ to begin with, but we know now that D has happened, and so the probability of the event E should be appropriately updated.

When we are told that the event D has been observed, then D should replace the "sample space" since $\mathbf{S}$ is irrelevant at this point. This is the fundamental idea behind the concept of *conditioning*.

Definition 1.4.1 *Let $\mathbf{S}$ and $\mathcal{B}$ be respectively the sample space and the Borel sigma-field. Suppose that A, B are arbitrary events. Then, the conditional probability of A given B, denoted by $P(A \mid B)$, is defined as:*

$$P(A \mid B) = \frac{P(A \cap B)}{P(B)} \quad provided\ that\ P(B) > 0 \qquad (1.4.1)$$

Definition 1.4.2 *Arbitrary events A, B are independent (dependent) if and only if $P(A \mid B) = (\neq)P(A)$, that is, having the knowledge that the event B has been observed does not (does) affect the probability of the event A, provided that $P(B) > 0$.*

Example 1.4.1 (Example 1.3.2 Continued) Recall that $P(D \cap E) = \frac{1}{36}$ and $P(D) = \frac{5}{36}$, so that we have $P(E \mid D) = P(D \cap E)/P(D) = \frac{1}{5}$. But, $P(E) = \frac{1}{9}$ which is different from $P(E \mid D)$. In other words, D and E are dependent events. ▲

The proofs of the following theorems are left as Exercise 1.4.1.

Theorem 1.4.1 *The events A and B are independent if and only if $P(A \cap B) = P(A)P(B)$.*

Theorem 1.4.2 *Suppose A, B are events. Then, the following statements are equivalent:*

(i)   *A and B are independent;*
(ii)   *A^c and B are independent;*
(iii) *A and B^c are independent;*
(iv)   *A^c and B^c are independent.*

Definition 1.4.3 *A collection of events $A_1, ..., A_n$ is called mutually independent if and only if every subcollection consists of independent events, that is,*

$$P(\cap_{j=1}^k A_{i_j}) = \Pi_{j=1}^k P(A_{i_j})$$

for all $1 \leq i_1 < i_2 < ... < i_k \leq n$ and $2 \leq k \leq n$.

> A collection of events $A_1, ..., A_n$ may be *pairwise independent*, that is, any two events are independent according to the Definition 1.4.2, but the whole collection of sets may not be mutually independent. See Example 1.4.2.

Example 1.4.2 Consider the random experiment of tossing a fair coin twice. Let us define the following events:

A_1 :   Observe a head (H) on the first toss
A_2 :   Observe a head (H) on the second toss
A_3 :   Observe the same outcome on both tosses

The sample space is given by $\mathbf{S} = \{HH, HT, TH, TT\}$ with each outcome being equally likely. Now, rewrite $A_1 = \{HH, HT\}$, $A_2 = \{HH, TH\}$, $A_3 = \{HH, TT\}$. Thus, we have $P(A_1) = P(A_2) = P(A_3) = \frac{1}{2}$. Now, $P(A_1 \cap A_2) = P(HH) = \frac{1}{4} = P(A_1)P(A_2)$, that is, the two events A_1, A_2 are independent. Similarly, one should verify that A_1, A_3 are independent, and so are also A_2, A_3. But, observe that $P(A_1 \cap A_2 \cap A_3) = P(HH) = \frac{1}{4}$ and it is not the same as $P(A_1)P(A_2)P(A_3)$. That is, A_1, A_2, A_3 are not mutually independent, but they are pairwise independent. ▲

1.4.1   Calculus of Probability

Suppose that A and B are two arbitrary events. Here we summarize some standard rules involving probabilities.

Additive Rule:

$$P(A \cup B) = P(A) + P(B) - P(A \cap B) \tag{1.4.2}$$

Conditional Probability Rule:

$$P(A \mid B) = P(A \cap B)/P(B) \text{ provided that } P(B) > 0 \tag{1.4.3}$$

Multiplicative Rule:

$$P(A \cap B) = P(A)P(B \mid A) \text{ provided that } P(A) > 0 \tag{1.4.4}$$

Example 1.4.3 A company publishes two magazines, M1 and M2. Based on its record of subscriptions in a suburb, the company found that 60 percent of the households subscribed only for M1, 45 percent subscribed only for M2, while only 20 percent subscribed for both M1 and M2. If a household is picked at random from this suburb, the publisher would like to address the following questions: What is the probability that the randomly selected household is a subscriber for (i) at least one of the magazines M1,M2, (ii) none of those magazines M1,M2, (iii) magazine M2 given that the same household subscribes for M1? Let us now define

A : The selected household subscribes for M1
B : The selected household subscribes for M2

We are told that $P(A) = 0.60, P(B) = 0.45$, and $P(A \cap B) = 0.20$. Then, $P(A \cup B) = P(A) + P(B) - P(A \cap B) = 0.85$, that is, there is an 85% chance that a randomly chosen household is a subscriber for at least one of the magazines M1,M2. In order to answer question (ii), we need to evaluate $P(A^c \cap B^c)$ which can simply be written as $1 - P(A \cup B) = 0.15$, that is, there is a 15% chance that the randomly selected household subscribes for none of the magazines M1,M2. Next, to answer question (iii), we obtain $P(B \mid A) = P(A \cap B)/P(A) = \frac{1}{3}$, that is, there is a one-in-three chance that the randomly selected household subscribes for the magazine M2 given that this household already receives the magazine M1. ▲

> In the Example 1.4.3, $P(A \cap B)$ was given to us, and so we could use Equation (1.4.3) to find the conditional probability $P(B \mid A)$.

Example 1.4.4 An urn contains eight green and twelve blue marbles, all of equal size and weight. The marbles are mixed and then one marble is picked at random. The first drawn marble is not returned to the urn. The remaining marbles inside the urn are again mixed and one marble is picked at random. This selection process is called *sampling without replacement*. What is the probability that (i) both the first and second drawn marbles are green, and (ii) the second drawn marble is green? Let us define the two events

A: The first selected marble is green
B: The second selected marble is green

Obviously, $P(A) = \frac{8}{20} = 0.4$ and $P(B \mid A) = \frac{7}{19}$. One observes that the experimental setup itself dictates the value of $P(B \mid A)$. A result such as Equation (1.4.3) does not help here. In order to answer question (i), we proceed by using Equation (1.4.4) to evaluate $P(A \cap B) = P(A)P(B \mid A) = \frac{8}{20}\frac{7}{19} = \frac{14}{95}$. Obviously, $\{A, A^c\}$ forms a partition of the sample space. Now, in order to answer question (ii), using Theorem 1.3.1, part (vi), we write:

$$P(B) = P(A \cap B) + P(A^c \cap B) \qquad (1.4.5)$$

But, as before we have $P(A^c \cap B) = P(A^c)P(B \mid A^c) = \frac{12}{20}\frac{8}{19} = \frac{24}{95}$. Thus, from Equation (1.4.5), we have $P(B) = \frac{14}{95} + \frac{24}{95} = \frac{38}{95} = \frac{2}{5} = 0.4$. Here, note that $P(B \mid A) \neq P(B)$ and so by Definition 1.4.2, two events A, B are dependent. The reader should check that $P(A)$ would equal $P(B)$ whatever the number of green or blue marbles in the urn. ▲

> In Example 1.4.4, $P(B \mid A)$ was known to us, and so we could use Equation (1.4.4) to find the joint probability $P(A \cap B)$.

1.4.2 Bayes' Theorem

Example 1.4.5 An experimenter has urns #1 and #2 at his disposal. The urn #1 (#2) has eight (ten) green and twelve (eight) blue marbles, all of the same size and weight. The experimenter selects one urn with equal probability and picks a marble at random. It is announced that the selected marble turned out blue. What is the probability that the blue marble came from urn #2? We will answer it shortly. ▲

Theorem 1.4.3 (Bayes' Theorem) *Suppose the events $A_1, ..., A_k$ form a partition of* **S**, *and B is another event. Then,*

$$P(A_j \mid B) = \frac{P(A_j)P(B \mid A_j)}{\Sigma_{i=1}^{k} P(A_i)P(B \mid A_i)}, \text{ for fixed } j = 1, ..., k$$

Proof Since $\{A_1, ..., A_k\}$ form a partition of **S**, in view of the Theorem 1.3.1, part (vi) we can immediately write:

$$P(B) = \Sigma_{i=1}^{k} P(B \cap A_i) = \Sigma_{i=1}^{k} P(A_i)P(B \mid A_i) \qquad (1.4.6)$$

by using Equation (1.4.4). Next, using Equation (1.4.3), let us write:

$$P(A_j \mid B) = P(A_j \cap B)/P(B) \qquad (1.4.7)$$

The required result follows by combining Equations (1.4.6) and (1.4.7). ■

In the statement of Theorem 1.4.3, note that the conditioning events on the *right-hand side* (rhs) are $A_1, ..., A_k$, but on the *left-hand side* (lhs) one has conditioning event B instead. $P(A_i), i = 1, ..., k$ are often referred to as *prior* probabilities, whereas $P(A_j \mid B)$ is referred to as the *posterior probability*.

Example 1.4.6 (Example 1.4.5 Continued) Define events

A_i: The urn #i is selected, $i = 1, 2$

B :   The marble picked from the selected urn is blue

It is clear that $P(A_i) = \frac{1}{2}$ for $i = 1, 2$, whereas we have $P(B \mid A_1) = \frac{12}{20}$ and $P(B \mid A_2) = \frac{8}{18}$. Now, applying Bayes' Theorem, we have

$$P(A_2 \mid B)$$
$$= P(A_2)P(B \mid A_2)/\{P(A_1)P(B \mid A_1) + P(A_2)P(B \mid A_2)\}$$
$$= (\tfrac{1}{2})(\tfrac{8}{18})/\{(\tfrac{1}{2})(\tfrac{12}{20}) + (\tfrac{1}{2})(\tfrac{8}{18})\}$$

which simplifies to $\frac{20}{47}$. ▲

> The Bayes' Theorem helps in finding the conditional
> probabilities when the original conditioning events
> $A_1, ..., A_k$ and the event B reverse their roles.

Example 1.4.7 Suppose that 40% of the individuals in a population have a particular disease. The diagnosis of the presence or absence of this disease in an individual is reached by blood test. But, like most clinical tests, this particular test is not perfect. The manufacturer of the blood-test kit made the accompanying information available. If an individual has the disease, the test indicates the absence (*false negative*) of the disease 10% of the time, whereas if an individual does not have the disease, the test indicates the presence (*false positive*) of the disease 20% of the time. From this population an individual is selected at random and blood tested. The health professional found that the test indicated presence of the disease. What is the probability that this individual does indeed have the disease? Define events:

A_1:   The individual has the disease
A_1^c: The individual does not have the disease
B :   The blood test indicates the presence of the disease

We are given the following information: $P(A_1) = 0.4, P(A_1^c) = 0.6, P(B \mid A_1) = 0.9$, and $P(B \mid A_1^c) = 0.2$. We need to find the conditional probability of A_1 given B. Denote $A_2 = A_1^c$ and use Bayes' Theorem:

$$P(A_1 \mid B) = \frac{P(A_1)P(B \mid A_1)}{P(A_1)P(B \mid A_1) + P(A_2)P(B \mid A_2)}$$
$$= (0.4)(0.9)/\{(0.4)(0.9) + (0.6)(0.2)\} = 0.75.$$

So, there is a 75% chance that a tested individual has the disease. ▲

1.4.3   Selected Counting Rules

A Fundamental Rule of Counting: Suppose that there are k different tasks and the ith task can be completed in n_i ways, $i = 1, ..., k$. Then, the total number of ways these k tasks can be completed is given by $\Pi_{i=1}^k n_i$.

Permutations: The word *permutation* refers to *arrangements* of objects taken from a collection of *distinct* objects. Suppose that we have n distinct objects. The number of ways we can arrange k of these objects, denoted by the symbol ^nP_k, is given by

$$^nP_k = n(n-1)(n-2)...(n-k+1) \text{ or } \Pi_{i=1}^k(n-i+1) \qquad (1.4.8)$$

The number of ways we can arrange all n objects is given by ^nP_n, which is denoted by

$$n! = n \times (n-1) \times \ldots \times 3 \times 2 \times 1 \qquad (1.4.9)$$

Combinations: The word *combination* refers to *selection* of some objects from a set of *distinct* objects without regard to order. Suppose that we have n distinct objects. The number of ways we can select k of these objects, denoted by the symbol $\binom{n}{k}$ or ^nC_k, is given by

$$\binom{n}{k} = \frac{n!}{k!(n-k)!} \qquad (1.4.10)$$

Observe that $n! = n(n-1)\ldots(n-k+1)(n-k)!$ and hence we can rewrite Equation (1.4.10) as

$$\binom{n}{k} = \frac{n(n-1)(n-2)\ldots(n-k+1)}{k!} = \frac{^nP_k}{k!} \qquad (1.4.11)$$

Theorem 1.4.4 (Binomial Theorem) *For any two real numbers a, b and a positive integer n, one has:*

$$(a+b)^n = \Sigma_{k=0}^n \binom{n}{k} a^k b^{n-k} \qquad (1.4.12)$$

Example 1.4.8 Suppose that a fair coin is tossed five times. Now the outcomes would look like $HHTHT, THTTH$ and so on. By the fundamental rule of counting, $\mathbf{S}$ will consist of 2^5 outcomes each of which has five components, and these are equally likely. How many of these outcomes would include two heads? Imagine five distinct positions in a row and each position will be filled by the letter H or T. Out of these five positions, choose two positions and fill them both with the letter H while the remaining three positions are filled with the letter T. This can be done in $\binom{5}{2}$ or 10 ways. So, $P(\text{Two Heads}) = \binom{5}{2}/2^5 = 5/16.$ ▲

Example 1.4.9 There are ten students in a class. In how many ways can a teacher form a committee of four students? In this selection process, naturally the order of selection is *not* pertinent. A committee of four can be chosen in $\binom{10}{4}$ or 210 ways and $\mathbf{S}$ would consist of 210 equally likely outcomes. ▲

Example 1.4.10 (Example 1.4.9 Continued) Suppose that there are six men and four women in the small class. Then, what is the probability that a randomly selected committee of four students would consist of two men and two women? Two men and two women can be chosen in $\binom{6}{2}\binom{4}{2}$

ways, that is, in 90 ways. In other words, P(Two men and two women are selected) $= \binom{6}{2}\binom{4}{2}/\binom{10}{4} = 3/7.$ ▲

Example 1.4.11 John, Sue, Rob, Dan, and Molly have gone to see a movie. Inside the theatre, they picked a row where there were exactly five empty seats next to each other. These five friends can sit in exactly 5! ways, that is, in 120 ways and arrangement is naturally pertinent. What is the probability that John and Molly would sit next to each other? John and Molly may occupy the first two seats and the other three friends may permute in 3! ways in the remaining chairs. But then John and Molly may occupy the second and third or the third and fourth or the fourth and fifth seats while the other three friends would take the remaining three seats in any order. One can also permute John and Molly in 2! ways. That is, we have P(John and Molly sit next to each other) $= (2!)(4)(3!)/5! = 2/5.$ ▲

1.5 Discrete Random Variables

Loosely speaking, a *discrete random variable* takes a finite or countably infinite number of possible values with associated probabilities. A more formal definition will follow shortly. In a collection agency, for example, the manager may look at the pattern of the number (X) of delinquent accounts. In a packaging plant, one may be interested in studying the pattern of the number (X) of the defective packages. One may ask: how many times (X) must one drill in a field in order to hit oil? These are some examples of *discrete* random variables.

In order to illustrate further, let us go back to the Example 1.3.2. Suppose that X is the total score from the red and yellow dice. The possible values of X would be any number from the set consisting of $2, 3, ..., 11, 12$. We had discussed earlier how one could proceed to evaluate $P(X = x), x = 2, ..., 12$. In this case, the event $[X = 2]$ corresponds to the subset $\{(1, 1)\}$, the event $[X = 3]$ corresponds to the subset $\{(1, 2), (2, 1)\}$, the event $[X = 4]$ corresponds to the subset $\{(1, 3), (2, 2), (3, 1)\}$, and so on. Thus, $P(X = 2) = \frac{1}{36}, P(X = 3) = \frac{2}{36}, P(X = 4) = \frac{3}{36}$, and so on. The reader should easily verify the following entries:

x:	2	3	4	5	6	7	8	9	10	11	12
$P(X = x)$:	$\frac{1}{36}$	$\frac{2}{36}$	$\frac{3}{36}$	$\frac{4}{36}$	$\frac{5}{36}$	$\frac{6}{36}$	$\frac{5}{36}$	$\frac{4}{36}$	$\frac{3}{36}$	$\frac{2}{36}$	$\frac{1}{36}$

$$(1.5.1)$$

Here, the set of the possible values of X is *finite*.

On the other hand, when tossing a fair coin, let Y be the number of independent tosses required to observe the first head (H) come up. Then,

$P(Y = 1) = P(H) = \frac{1}{2}$, and $P(Y = 2) = P(TH) = (\frac{1}{2})(\frac{1}{2}) = \frac{1}{4}$. Similarly, $P(Y = 3) = P(TTH) = \frac{1}{8}, ...,$ that is,

$$P(Y = y) = (\tfrac{1}{2})^y, y = 1, 2, ...$$

Here, the set of the possible values for the random variable Y is *countably infinite*.

1.5.1 Probability Mass and Distribution Functions

A random variable X is a mapping (that is, a function) from **S** to a subset $\mathcal{X}$ of the real line $\Re$ which amounts to saying that the random variable X induces events ($\in \mathcal{B}$) in the context of **S**. We may express this by writing $X : \mathbf{S} \to \mathcal{X}$. Then, X takes possible values $x_1, x_2, x_3, ...$ with respective probabilities $p_i = P(X = x_i), i = 1, 2, ...$, evaluated as follows:

$$p_i = P(X = x_i) = \sum_{s \in \mathbf{S}: X(s) = x_i} P(s), i = 1, 2, ... \qquad (1.5.2)$$

In Equation (1.5.1), we found $P(X = i)$ for $i = 2, 3, 4$ whereas the space $\mathcal{X} = \{2, 3, ..., 12\}$ and $\mathbf{S} = \{(1, 1), ..., (1, 6), (2, 1), ..., (6, 1), ..., (6, 6)\}$.

While assigning or evaluating these probabilities, one makes sure that the following conditions are met:

$$
\begin{aligned}
(i) \quad & p_i \geq 0 \text{ for all } i = 1, 2, ..., \text{ and} \\
(ii) \quad & \Sigma_{i=1}^{\infty} p_i = 1
\end{aligned}
\qquad (1.5.3)
$$

When both these conditions are met, we call an assignment such as

$$
\begin{array}{llllll}
X \text{ values :} & x_1 & x_2 & \cdot \ \cdot \ \cdot & x_i & \cdot \ \cdot \ \cdot \\
\text{Probabilities :} & p_1 & p_2 & \cdot \ \cdot \ \cdot & p_i & \cdot \ \cdot \ \cdot
\end{array}
\qquad (1.5.4)
$$

a *discrete distribution* or a *discrete probability distribution* of X.

> The function $f(x) = P(X = x)$ for $x \in \mathcal{X} = \{x_1, x_2, ..., x_i, ...\}$ is customarily known as the *probability mass function* (pmf) of X.

We also define another useful function associated with X as follows:

$$F(x) = P\{X \leq x\} = P\{s : s \in \mathbf{S} \text{ such that } X(s) \leq x\}, \ x \in \Re \qquad (1.5.5)$$

which is customarily called a *distribution function* (df) or a *cumulative distribution function* (cdf) of X. Sometimes we may write $F_X(x)$ for the df of X.

> A *distribution function* or a *cumulative distribution function* $F(x)$ or $F_X(x)$ for X is defined for all real numbers x.

Once a pmf $f(x)$ is given as in the case of Equation (1.5.4), one can find the probabilities of events which are defined through the random variable X. If we denote a set $A(\subseteq \Re)$, then:

$$P(X \in A) = \sum_{i:x_i \in A \cap \mathcal{X}} P(X = x_i) \qquad (1.5.6)$$

Example 1.5.1 Suppose that X is a discrete random variable having the following probability distribution:

$$
\begin{array}{llll}
X \text{ values}: & 1 & 2 & 4 \\
\text{Probabilities}: & 0.2 & 0.4 & 0.4
\end{array}
\qquad (1.5.7)
$$

The associated df is then given by

$$
F(x) = \begin{cases}
0 & \text{if } x < 1 \\
0.2 & \text{if } 1 \leq x < 2 \\
0.6 & \text{if } 2 \leq x < 4 \\
1.0 & \text{if } 4 \leq x
\end{cases}
\qquad (1.5.8)
$$

This df is the step function shown in Figure 1.5.1. From Figure 1.5.1, it becomes clear that the jump in the value of $F(x)$ at the points $x = 1, 2, 4$ respectively amounts to $0.2, 0.4$ and 0.4. These jumps obviously correspond to the assigned values of $P(X = 1), P(X = 2)$ and $P(X = 4)$. Also, the df is nondecreasing in x and it is discontinuous at the points $x = 1, 2, 4$, namely at those points where the probability distribution from Equation (1.5.7) assigns positive masses. ▲

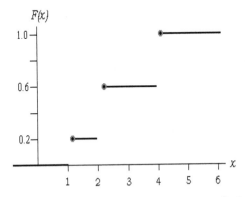

Figure 1.5.1. Plot of $F(x)$ from Equation (1.5.8).

Suppose that we write

$$F(x-) = \lim F(h) \text{ as } h \uparrow x, \text{ for } x \in \Re \qquad (1.5.9)$$

This $F(x-)$ is the limit of $F(h)$ as h converges to x from the lhs of x. In Example 1.5.1, one can verify that $F(1-) = 0, F(2-) = 0.2$, and $F(4-) = 0.6$. In other words, the *jump* of the df or cdf $F(x)$ at the point

$$
\begin{aligned}
&x = 1 \text{ equals } F(1) - F(1-) \text{ which is } P(X = 1) = 0.2 \\
&x = 2 \text{ equals } F(2) - F(2-) \text{ which is } P(X = 2) = 0.4 \qquad (1.5.10) \\
&x = 4 \text{ equals } F(4) - F(4-) \text{ which is } P(X = 4) = 0.4
\end{aligned}
$$

There is a natural correspondence between the jumps of a df, the left-hand limit of a df, and the associated pmf.

> For a discrete random variable X, one can obtain $P(X = x)$ as $F(x)-$ $F(x-)$ where the left-hand limit $F(x-)$ comes from Equation (1.5.9).

Example 1.5.2 (Example 1.3.2 Continued) For the random variable X, the total from the two dice, the probability that it will be smaller than 4 or larger than 9 can be found as follows: Let us denote an event $A = \{X < 4 \cup X > 9\}$ and we exploit Equation (1.5.6) to write $P(A) = P(X < 4) + P(X > 9) = P(X = 2, 3) + P(X = 10, 11, 12) = \frac{1}{4}$. What is the probability that the total from the two dice will differ from 8 by at least 2? Let us denote an event $B = \{|X - 8| \geq 2\}$ and we again exploit Equation (1.5.6) to write $P(B) = P(X \leq 6) + P(X \geq 10) = P(X = 2, 3, 4, 5, 6) + P(X = 10, 11, 12) = \frac{7}{12}$. ▲

Example 1.5.3 (Example 1.5.2 Continued) Reconsider two events A, B. One can obtain $P(A), P(B)$ alternatively by using the expression of the associated df $F(x)$. Now, $P(A) = F(3) + \{1 - F(9)\} = \frac{3}{36} + \{1 - \frac{30}{36}\} = \frac{1}{4}$. Also, $P(B) = F(6) + \{1 - F(9)\} = \frac{15}{36} + \{1 - \frac{30}{36}\} = \frac{7}{12}$. ▲

Theorem 1.5.1 *Consider the df $F(x)$ with $x \in \Re$, defined by Equation (1.5.5), for a discrete random variable X. Then, the following properties hold:*

(i)   *$F(x)$ is nondecreasing, that is, $F(x) \leq F(y)$ for all $x \leq y$ where $x, y \in \Re$;*

(ii)   *$\lim_{x \to -\infty} F(x) = 0$, $\lim_{x \to \infty} F(x) = 1$;*

(iii)   *$F(x)$ is right continuous, that is, $F(x + h) \downarrow F(x)$ as $h \downarrow 0$, for all $x \in \Re$.*

1.6 Continuous Random Variables

A *continuous random variable* takes values in subintervals or within subsets generated by subintervals of the real line, $\Re$. At a particular location in a lake, for example, the manager of a local park and recreation department may study the depth (X) of the water level. In a high-rise office building, one may study the waiting time (X) for an elevator. These are some examples of *continuous* random variables.

1.6.1 *Probability Density and Distribution Functions*

Assume that **S** is a subinterval of $\Re$. Now, a continuous real valued random variable X is a mapping of **S** into $\Re$. We no longer talk about $P(X = x)$ because this probability will be zero regardless of what specific value $x \in \Re$. The waiting time at a bus stop would often be postulated as a continuous random variable, and thus the probability of one's waiting *exactly* five minutes or *exactly* seven minutes for the next bus is zero.

In modeling a continuous random phenomenon, we start with a function $f(x)$ associated with $x \in \Re$ satisfying the following properties:

$$
\begin{array}{ll}
(i) & f(x) \geq 0 \text{ for all } x \in \Re, \text{ and} \\
(ii) & \int_{\Re} f(x)dx = 1.
\end{array}
\qquad (1.6.1)
$$

Note that the Equations (1.5.3) and (1.6.1) are similar in spirit. Any $f(x)$ satisfying Equation (1.6.1) is called a *probability density function* (pdf).

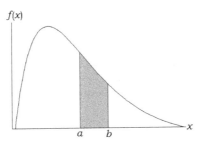

Figure 1.6.1. Shaded area under $f(x)$ is $P(a < X < b)$.

Once a pdf $f(x)$ is specified, we can find probabilities of events defined in terms of the random variable X. If we denote a set $A(\subseteq \Re)$, then

$$
P(X \in A) = \int_A f(x)dx
\qquad (1.6.2)
$$

where the convention is that we would integrate the function $f(x)$ only on that part of the set A wherever $f(x)$ is positive.

In other words, $P(X \in A)$ is given by the area under the curve $\{(x, f(x));$ for all $x \in A$ wherever $f(x) > 0\}$. In Figure 1.6.1, we let A be the interval (a, b) and the shaded area represents the corresponding probability, $P(a < X < b)$.

We define the *distribution function* (df) of a continuous random variable X by modifying the discrete analog from Equation (1.5.5). We let

$$F(x) = P\{X \le x\} = \int_{-\infty}^{x} f(y)dy, x \in \Re \qquad (1.6.3)$$

which also goes by the name *cumulative distribution function* (cdf). Again, note that $F(x)$ is defined for all real values x.

> Theorem 1.5.1 also holds for the distribution function $F(x)$ of a continuous random variable X. Theorem 1.5.1 and its converse hold for all random variables.

$$(1.6.4)$$

Now, we state an important characteristic of a df without supplying its proof. One will find its proof in Rao (1973, p. 85), among other places.

Theorem 1.6.1 *Suppose that* $F(x), x \in \Re$, *is the df of an arbitrary random variable* X. *Then, the set of points of discontinuity of* $F(x)$ *is finite or countably infinite.*

Example 1.6.1 Consider the following discrete random variable X first:

$$\begin{array}{lccc} X \text{ value:} & -1 & 2 & 5 \\ \text{Probability:} & 0.1 & 0.3 & 0.6 \end{array} \qquad (1.6.5)$$

So, $F_X(x)$ is discontinuous at $x = -1, 2$ and 5 only. ▲

Example 1.6.2 Look at the next random variable Y. Suppose that

$$P(Y = y) = \left(\tfrac{1}{2}\right)^y, y = 1, 2, 3, \dots \qquad (1.6.6)$$

Here, $F_Y(y)$ is discontinuous at a countably infinite number of points, $y = 1, 2, 3, \dots$. ▲

Example 1.6.3 Suppose that a random variable U has an associated *nonnegative* function $f(u)$ given by

$$f(u) = \begin{cases} \left(\tfrac{1}{2}\right)^{u+1} & \text{if } u = 1, 2, 3, \dots \\ u & \text{if } 0 < u < 1 \\ 0 & \text{elsewhere} \end{cases}$$

Observe that

$$\textstyle\sum_{u=1}^{\infty}\left(\tfrac{1}{2}\right)^{u+1} + \int_{u=0}^{\infty} f(u)du = \tfrac{1}{2} + \tfrac{1}{2} = 1$$

and thus $f(u)$ happens to be the probability distribution of U. This random variable U is *neither* discrete *nor* continuous. One may check that $F_U(u)$ is discontinuous at a countably infinite number of points, $u = 1, 2, 3, \ldots$. ▲

If X is continuous, then $F(x)$ turns out to be continuous at every point $x \in \Re$. However, the df $F(x)$ may not be differentiable at a number of points. Consider the following examples.

Example 1.6.4 Suppose that W has its pdf

$$f(w) = \begin{cases} \frac{3}{7}w^2 & \text{if } 1 < w < 2 \\ 0 & \text{if } w \leq 1 \text{ or } w \geq 2 \end{cases} \tag{1.6.7}$$

The associated df $F_W(w)$ is given by

$$F_W(w) = \begin{cases} 0 & \text{if } w \leq 1 \\ \frac{1}{7}(w^3 - 1) & \text{if } 1 < w < 2 \\ 1 & \text{if } w \geq 2 \end{cases} \tag{1.6.8}$$

The expression for $F_W(w)$ is clear for $w \leq 1$ since in this case we integrate the zero density. The expression for $F_W(w)$ is also clear for $w \geq 2$ because the associated integral can be written as:

$$\int_{-\infty}^1 f(v)dv + \int_1^2 f(v)dv + \int_2^w f(v)dv = 1$$

For $1 < w < 2$, the expression for $F_W(w)$ can be written as:

$$\int_{-\infty}^1 f(v)dv + \int_1^w f(v)dv = \int_1^w \frac{3}{7}v^2 dv = \frac{1}{7}(w^3 - 1)$$

It should be obvious that $F_W(w)$ is continuous at all points $w \in \Re$, but $F_W(w)$ is *not differentiable* at $w = 1, 2$. ▲

Example 1.6.5 Consider a function $f(x)$ defined as follows:

$$f(x) = \begin{cases} 7(4)^{-i} & \text{whenever } x \in \left(\frac{1}{2^i}, \frac{1}{2^{i-1}}\right], i = 1, 2, 3, \ldots \\ 0 & \text{whenever } x \leq 0 \text{ or } x > 1 \end{cases} \tag{1.6.9}$$

One may verify that (i) $f(x)$ is a genuine pdf, (ii) the associated df $F(x)$ is continuous at all points $x \in \Re$, and (iii) $F(x)$ is not differentiable at the points x belonging to the set $\{1, \frac{1}{2}, \frac{1}{2^2}, \frac{1}{2^3}, \ldots\}$ which is countably infinite. In Example 1.6.2, we had the point masses at a countably infinite number of points. But, the present example is little different in its construction. ▲

At all points $x \in \Re$ wherever the df $F(x)$ is differentiable, one must have the following result:

$$dF(x)/dx = f(x), \text{ the pdf of } X \text{ at the point } x \qquad (1.6.10)$$

This is a restatement of the fundamental theorem of integral calculus.

Example 1.6.6 (Example 1.6.4 Continued) Consider $F_W(w)$ from Equation (1.6.8). $F_W(w)$ is not differentiable at $w = 1, 2$. Except at $w = 1, 2$, the pdf $f(w)$ of the random variable W can be obtained from Equation (1.6.8) as follows: $f(w) = \frac{d}{dw} F_W(w)$, which will coincide with zero when $-\infty < w < 1$ or $2 < w < \infty$, whereas for $1 < w < 2$ we would have

$$\frac{d}{dw}\{\tfrac{1}{7}(w^3 - 1)\} = \tfrac{3}{7}w^2$$

This agrees with the pdf from Equation (1.6.7) except when $w = 1, 2$. ▲

1.6.2 Median of a Distribution

Often a pdf has interesting features including the position of its *median*. When comparing two income distributions, for example, one may consider median incomes from these distributions. For a *continuous* random variable X with df $F(x)$, we say that x_m is the median if and only if $F(x_m) = \frac{1}{2}$, that is, $P(X \le x_m) = P(X \ge x_m) = \frac{1}{2}$.

Example 1.6.7 (Example 1.6.4 Continued) Reconsider $F(w)$ from Equation (1.6.8). The median w_m is a solution of the equation $F(w_m) = \frac{1}{2}$. We obtain $w_m = (4.5)^{1/3} \approx 1.651$. ▲

1.6.3 Selected Reviews from Mathematics

Finite Sums of Powers of Positive Integers:

$$\begin{aligned}
1 + 2 + \ldots + n &= \tfrac{1}{2}n(n+1) \\
1^2 + 2^2 + \ldots + n^2 &= \tfrac{1}{6}n(n+1)(2n+1) \\
1^3 + 2^3 + \ldots + n^3 &= \{\tfrac{1}{2}n(n+1)\}^2 \\
1^4 + 2^4 + \ldots + n^4 &= \tfrac{1}{30}n(n+1)(2n+1)(3n^2 + 3n - 1)
\end{aligned} \qquad (1.6.11)$$

Infinite Sums of Reciprocal Powers of Positive Integers: Let us write

$$\zeta(p) = \Sigma_{n=1}^{\infty} n^{-p} \qquad (1.6.12)$$

Then, $\zeta(p) = \infty$ if $p \le 1$, but $\zeta(p)$ is finite if $p > 1$. It is known that $\zeta(2) = \frac{1}{6}\pi^2$, $\zeta(4) = \frac{1}{90}\pi^4$, and $\zeta(3) \approx 1.2020569$. Refer to Abramowitz and Stegun (1972, pp. 807-811) for related tables.

Results on Limits:

$$\lim_{n\to\infty} n^{1/n} = 1; \quad \lim_{n\to\infty} \left\{1 + \tfrac{a}{n}\right\} = 1, \text{ for fixed } a \in \Re$$

$$\lim_{n\to\infty} \left\{1 + \tfrac{a}{n}\right\}^n = e^a, \text{ for fixed } a \in \Re$$

$(1.6.13)$

Taylor Expansion: Let $f(.)$ be a real valued function having finite n^{th} derivative $\frac{d^n}{dx^n}f(x)$, denoted by $f^{(n)}(.)$, everywhere in an open interval $(a, b) \subseteq \Re$ and assume that $f^{(n-1)}(.)$ is continuous on the closed interval $[a, b]$. Let $c \in [a, b]$. Then, for every $x \in [a, b], x \ne c$, there exists a real number ξ between numbers x and c such that

$$f(x) = f(c) + \Sigma_{i=1}^{n-1} \frac{(x-c)^i}{i!} f^{(i)}(c) + \frac{(x-c)^n}{n!} f^{(n)}(\xi) \qquad (1.6.14)$$

Some Infinite Series Expansions:

$$e^x \quad = 1 + \tfrac{x}{1!} + \tfrac{x^2}{2!} + \tfrac{x^3}{3!} + ... \text{ for all } x \in \Re$$

$$(1-x)^{-m} \quad = 1 + mx + \tfrac{m(m+1)}{2!}x^2 + \tfrac{m(m+1)(m+2)}{3!}x^3 + ... \qquad (1.6.15)$$
$$\text{for all } x \in (-1, 1), m > 0$$

$$\log(1+x) \quad = x - \tfrac{1}{2}x^2 + \tfrac{1}{3}x^3 - ... \text{ for all } x > -1$$

Differentiation under Integration (Leibnitz's Rule): Suppose that $f(x, \theta), a(\theta)$, and $b(\theta)$ are differentiable functions with respect to θ for $x \in \Re, \theta \in \Re$. Then,

$$\frac{d}{d\theta}\left[\int_{a(\theta)}^{b(\theta)} f(x, \theta)dx\right] = f(b(\theta), \theta)\left(\tfrac{d}{d\theta}b(\theta)\right) - f(a(\theta), \theta)\left(\tfrac{d}{d\theta}a(\theta)\right)$$
$$+ \int_{a(\theta)}^{b(\theta)} \left(\tfrac{\partial}{\partial\theta}f(x, \theta)\right)dx$$

$(1.6.16)$

Differentiation under Integration: Suppose that $f(x, \theta)$ is a differentiable function in θ, for $x \in \Re, \theta \in \Re$. Let there be another function $g(x, \theta)$ such that (i) $\left|\frac{\partial}{\partial\theta}f(x, \theta)\right|_{\theta=\theta_0}\right| \le g(x, \theta)$ for all θ_0 belonging to some interval $(\theta - \varepsilon, \theta + \varepsilon)$, and (ii) $\int_{-\infty}^{\infty} g(x, \theta)dx < \infty$. Then,

$$\frac{d}{d\theta}\left[\int_{-\infty}^{\infty} f(x, \theta)dx\right] = \int_{-\infty}^{\infty} \left(\tfrac{\partial}{\partial\theta}f(x, \theta)\right)dx \qquad (1.6.17)$$

Monotone Function of a Single Real Variable: Suppose that $f(x)$ is a real valued function of $x \in (a, b) \subseteq \Re$. Let us assume that $\frac{d}{dx}f(x)$ exists at each $x \in (a, b)$ and that $f(x)$ is continuous at $x = a, b$. Then,

(i) $f(x)$ is strictly increasing in (a, b) if $\frac{d}{dx}f(x) > 0$ for $x \in (a, b)$

(ii) $f(x)$ is strictly decreasing in (a, b) if $\frac{d}{dx}f(x) < 0$ for $x \in (a, b)$.

$$(1.6.18)$$

Gamma Function and Gamma Integral: The expression $\Gamma(\alpha)$, known as the *gamma function* evaluated at α, is defined as

$$\Gamma(\alpha) = \int_0^\infty e^{-x}x^{\alpha-1}dx \text{ for } \alpha > 0 \qquad (1.6.19)$$

The representation given in the rhs of Equation (1.6.19) is referred to as the *gamma integral*. The gamma function has interesting properties including the following:

$$\Gamma(\alpha + 1) = \alpha\Gamma(\alpha) \text{ with arbitrary } \alpha > 0; \Gamma(\tfrac{1}{2}) = \sqrt{\pi}$$
$$\text{and } \Gamma(n) = (n-1)! \text{ for arbitrary } n = 1, 2, ... \qquad (1.6.20)$$

Stirling's Approximation: Recall $\Gamma(\alpha), \alpha > 0$. Now,

$$\Gamma(\alpha) \backsim \sqrt{2\pi}e^{-\alpha}\alpha^{\alpha-\frac{1}{2}} \text{ for large values of } \alpha \qquad (1.6.21)$$

Writing $\alpha = n+1$ where n is a positive integer, one can immediately claim that

$$n! \backsim \sqrt{2\pi}e^{-n}n^{n+\frac{1}{2}} \text{ for large values of } n \qquad (1.6.22)$$

The approximation for $n!$ given by Equation (1.6.22) works well even for n as small as five or six.

Ratios of Gamma Functions: Recall $\Gamma(\alpha), \alpha > 0$. Then, one has

$$\Gamma(c\alpha + d) \backsim \sqrt{2\pi}e^{-c\alpha}(c\alpha)^{c\alpha+d-\frac{1}{2}} \text{ for large values of } \alpha \qquad (1.6.23)$$

with fixed numbers $c(> 0)$ and d, assuming that the gamma function itself is defined. Also, one has

$$\alpha^{d-c}\frac{\Gamma(\alpha + c)}{\Gamma(\alpha + d)} \backsim 1 + \tfrac{1}{2\alpha}(c-d)(c+d-1) \text{ for large values of } \alpha \quad (1.6.24)$$

with fixed numbers c and d, assuming that the gamma functions involved are defined.

Beta Function and Beta Integral: The expression $b(\alpha, \beta)$, known as the *beta function* evaluated at α and β in that order, is defined as

$$b(\alpha, \beta) = \int_0^1 x^{\alpha-1}(1-x)^{\beta-1}dx \text{ for } \alpha > 0, \beta > 0 \qquad (1.6.25)$$

The representation given in the rhs of Equation (1.6.25) is referred to as the *beta integral*. Note that $b(\alpha, \beta)$ can alternatively be expressed as:

$$b(\alpha, \beta) = \{\Gamma(\alpha)\Gamma(\beta)\}/\{\Gamma(\alpha + \beta)\} \text{ for } 0 < \alpha, \beta < \infty \qquad (1.6.26)$$

Maximum and Minimum of a Function of One Real Variable: For some integer $n \geq 1$, suppose that $f(x)$ is a real valued function of a single real variable $x \in (a, b) \subseteq R$, having a continuous n^{th} derivative $\frac{d^n}{dx^n} f(x)$, denoted by $f^{(n)}(x)$, everywhere in the open interval (a, b). Suppose also that for some point $\xi \in (a, b)$, one has $f^{(1)}(\xi) = f^{(2)}(\xi) = ... = f^{(n-1)}(\xi) = 0$, but $f^{(n)}(\xi) \neq 0$. Then,

(i)   for n *even*, $f(x)$ has a local minimum at $x = \xi$ if $f^{(n)}(\xi) > 0$

(ii)   for n *odd*, $f(x)$ has a local maximum at $x = \xi$ if $f^{(n)}(\xi) < 0$.

$$(1.6.27)$$

Maximum and Minimum of a Function of Two Real Variables:

Suppose that $f(\mathbf{x})$ is a real valued function of a two-dimensional variable $\mathbf{x} = (x_1, x_2) \in (a_1, b_1) \times (a_2, b_2) \subseteq \Re^2$. The process of finding where this function $f(\mathbf{x})$ attains its maximum or minimum requires knowledge of matrices and vectors. We briefly review notions involving matrices and vectors in Section 4.7. Hence, we defer to state this particular result from calculus in Section 4.7.

Integration by Parts: Consider two real valued functions $f(x), g(x)$ where $x \in (a, b)$, an open subinterval of $\Re$. Let us denote $\frac{d}{dx} f(x)$ by $f'(x)$ and the indefinite integral $\int g(x)dx$ by $h(x)$. Then,

$$\int_a^b f(x)g(x)dx = [f(x)h(x)]_{x=a}^{x=b} - \int_a^b f'(x)h(x)dx \qquad (1.6.28)$$

assuming that all integrals and $f'(x)$ are finite.

L'Hôpital's Rule: Suppose that $f(x)$ and $g(x)$ are two *differentiable* real valued functions. Assume that $\lim_{x \to a} f(x) = 0$ and $\lim_{x \to a} g(x) = 0$ where a is a fixed real number, $-\infty$ or $+\infty$. Then,

$$\lim_{x \to a} \{f(x)/g(x)\} = \lim_{x \to a} \{f'(x)/g'(x)\} \qquad (1.6.29)$$

where $f'(x) = df(x)/dx, g'(x) = dg(x)/dx$.

Triangular Inequality: For any two real numbers a and b, the following holds:

$$|a + b| \le |a| + |b| \qquad (1.6.30)$$

It follows that $|a - b| \ge |a| - |b|$.

1.7 Some Useful Distributions

> As a convention, we often write down the pmf or pdf $f(x)$ only for those $x \in \mathcal{X}$ where $f(x)$ is positive.

1.7.1 Discrete Distributions

Bernoulli Distribution: A random variable X has Bernoulli(p) distribution if its pmf is

$$f(x) = P(X = x) = p^x (1 - p)^{1-x} \text{ for } x = 0, 1 \qquad (1.7.1)$$

where $0 < p < 1$ is a *parameter*. In applications, one may collect dichotomous data, for example, record whether an item is defective ($x = 0$) or nondefective ($x = 1$) or whether an individual is married ($x = 0$) or unmarried ($x = 1$). In each situation, p stands for $P(X = 1)$.

Binomial Distribution: A random variable X has Binomial(n, p) distribution if its pmf is

$$f(x) = P(X = x) = \binom{n}{x} p^x (1 - p)^{n-x} \text{ for } x = 0, 1, ..., n \qquad (1.7.2)$$

where $0 < p < 1$ is a *parameter*.

The Binomial(n, p) distribution arises as follows. Repeat a Bernoulli experiment independently n times and each time one observes the outcome (0 or 1) and $p = P(X = 1)$. Clearly, $f(x) \ge 0$ for all x. But, also recall part (ii) in Equation (1.5.3) which demands that all probabilities given by Equation (1.7.2) must add up to one. In order to verify this, we write:

$$\Sigma_{x=0}^{n} f(x) = \Sigma_{x=0}^{n} \binom{n}{x} p^x (1 - p)^{n-x} = [p + (1 - p)]^n = (1)^n = 1 \qquad (1.7.3)$$

Example 1.7.1 In a multiple-choice quiz, suppose that there are ten questions, each with five suggested answers. Each question has exactly one correct answer given. A student guessed all answers in a quiz. Suppose that each correct (wrong) answer to a question carries one (zero) point. The student's score (X) has Binomial$(n = 10, p = \frac{1}{5})$ distribution. Then, $P(X = 0) = \binom{10}{0}(\frac{1}{5})^0(\frac{4}{5})^{10} \approx 0.10737$. Also, $P(X \geq 8) = \binom{10}{8}(\frac{1}{5})^8(\frac{4}{5})^2 + \binom{10}{9}(\frac{1}{5})^9(\frac{4}{5})^1 + \binom{10}{10}(\frac{1}{5})^{10}(\frac{4}{5})^0 \approx 7.7926 \times 10^{-5}$. ▲

Poisson Distribution: A random variable X has a Poisson(λ) distribution if its pmf is

$$f(x) = P(X = x) = e^{-\lambda}\lambda^x/x! \text{ for } x = 0, 1, 2, ... \qquad (1.7.4)$$

where $0 < \lambda < \infty$ is a *parameter*.

Clearly, $f(x) \geq 0$ for all x. But, in order to verify directly that all probabilities add up to one, one may use the infinite series expansion of e^x:

$$\Sigma_{x=0}^{\infty}f(x) = e^{-\lambda}\left\{1 + \frac{\lambda}{1!} + \frac{\lambda^2}{2!} + ...\right\} = e^{-\lambda}e^{\lambda} = 1 \qquad (1.7.5)$$

A Poisson distribution may arise as follows. Reconsider a binomial distribution but pretend having a situation like this: Make $n \to \infty$ and $p \to 0$ in such a way that np remains a constant, say, $\lambda(> 0)$. Now then, rewrite the binomial probability for any fixed $x = 0, 1, 2, ...$ as follows:

$$
\begin{aligned}
f(x) &= \binom{n}{x}p^x(1-p)^{n-x} \\
&= \frac{n(n-1)...(n-x+1)}{x!}(\frac{\lambda}{n})^x(1-\frac{\lambda}{n})^{n-x} \\
&= \frac{1}{x!}\lambda^x(1-\frac{1}{n})(1-\frac{2}{n})...(1-\frac{x-1}{n})(1-\frac{\lambda}{n})^{-x}(1-\frac{\lambda}{n})^n
\end{aligned} \qquad (1.7.6)
$$

Observe that $(1 - \frac{k}{n}) \to 1$ as $n \to \infty$ for all fixed $k = 1, 2, ..., x - 1, \lambda$. Also, $(1 - \frac{\lambda}{n})^n \to e^{-\lambda}$ as $n \to \infty$. Now, from Equation (1.7.6) we can conclude that $\binom{n}{x}p^x(1-p)^{n-x} \to e^{-\lambda}\lambda^x/x!$ as $n \to \infty$.

Example 1.7.2 We are inspecting a particular brand of concrete slab specimens for visible cracks. Suppose that the number (X) of cracks per concrete slab has a Poisson distribution with $\lambda = 2.5$. What is the probability that a randomly selected slab will have at least two cracks? Now, $P(X \geq 2) = 1 - P(X \leq 1) = 1 - [\{e^{-2.5}(2.5)^0/0!\} + \{e^{-2.5}(2.5)^1/1!\}] \approx 0.7127$. ▲

Geometric Distribution: A random variable X has a Geometric(p) distribution if its pmf is

$$f(x) = P(X = x) = p(1-p)^{x-1} \text{ for } x = 1, 2, 3, ... \qquad (1.7.7)$$

where $0 < p < 1$ with p as a *parameter*.

The Geometric(p) distribution arises as follows. Consider repeating a Bernoulli experiment independently until we observe the value $x = 1$ for the first time.

Example 1.7.3 A geological exploration may indicate that a well drilled for oil in a region in Texas would strike oil with probability 0.3. Assuming that oil strikes are independent from one drill to another, what is the probability that the first oil strike will occur on the sixth drill? Let X be the number of drills. Then, X is distributed as Geometric($p = 0.3$). So, $P(X = 6) = (0.3)(0.7)^5 \approx 0.050421$. ▲

Negative Binomial Distribution: A random variable X is said to have a negative binomial distribution, denoted by NB(μ, k), if its pmf is

$$f(x) = P(X = x) = \binom{k+x-1}{k-1} \left(\frac{\mu}{\mu+k}\right)^x \left(\frac{k}{\mu+k}\right)^k \text{ for } x = 0, 1, 2, ... \quad (1.7.8)$$

where $0 < \mu, k < \infty$ are *parameters*. The parameterization given in Equation (1.7.8) is due to Anscombe (1949), which is widely used in areas such as entomology, plant science, and soil science.

Uniform Distribution: Suppose a random variable X takes the only possible values $x_1, ..., x_k$ each with the same probability $\frac{1}{k}$. Such X is said to have a discrete uniform distribution with the pmf:

$$f(x) = \frac{1}{k} \text{ for } x = x_1, ..., x_k, \text{ with a fixed positive integer } k \quad (1.7.9)$$

1.7.2 Continuous Distributions

Uniform Distribution: A random variable X has a uniform distribution on an interval (a, b), denoted by Uniform(a, b), if its pdf is

$$f(x) = (b - a)^{-1} \text{ for } a < x < b \quad (1.7.10)$$

where $-\infty < a < b < \infty$. Here, a, b are *parameters*.

The function $f(x)$ is nonnegative for all $x \in \Re$. Next, write

$$\int_{-\infty}^{\infty} f(x)dx = (b-a)^{-1} \int_a^b dx = (b-a)^{-1} [x]_{x=a}^{x=b} = (b-a)^{-1}(b-a) = 1$$

since $b \neq a$. In other words, Equation (1.7.10) defines a genuine pdf.

Example 1.7.4 The waiting time X at a bus stop, measured in minutes, may be uniformly distributed between zero and five. What is the probability that one would wait more than 3.8 minutes for a bus? $P(X > 3.8) = \int_{3.8}^5 \frac{1}{5} dx = \frac{1}{5} [x]_{x=3.8}^{x=5} = 0.24$. ▲

Normal Distribution: A random variable X has the normal distribution with *parameters* μ and σ^2, denoted by $N(\mu, \sigma^2)$, if its pdf is:

$$f(x) = \{\sigma\sqrt{2\pi}\}^{-1} \exp\{-(x-\mu)^2/(2\sigma^2)\} \text{ for } x \in \Re \qquad (1.7.11)$$

with $-\infty < \mu < \infty$ and $0 < \sigma < \infty$. A normal distribution is widely used in statistics.

C. F. Gauss, the celebrated German mathematician of the eighteenth century, had discovered this distribution while analyzing measurement errors in astronomy. Hence, the normal distribution is also called a *Gaussian* distribution.

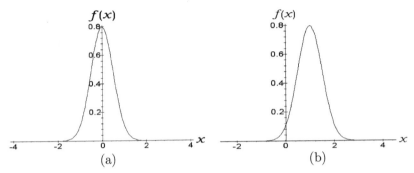

Figure 1.7.1. Normal densities: (a) $N(0, 0.25)$; (b) $N(1, 0.25)$.

The function $f(x)$ is obviously positive for all $x \in \Re$. Next, let us verify directly that $\int_{-\infty}^{\infty} f(x)dx = 1$. Substitute $u = (x-\mu)/\sigma, v = -u, w = \frac{1}{2}v^2$ successively, and rewrite the full integral as:

$\int_{-\infty}^{\infty} f(x)dx$

$$
\begin{aligned}
&= \{\sqrt{2\pi}\}^{-1}\left[\int_{-\infty}^{0} \exp\{-\tfrac{1}{2}u^2\}du + \int_{0}^{\infty} \exp\{-\tfrac{1}{2}u^2\}du\right] \\
&= \{\sqrt{2\pi}\}^{-1}\left[\int_{\infty}^{0} \exp\{-\tfrac{1}{2}v^2\}(-dv) + \int_{0}^{\infty} \exp\{-\tfrac{1}{2}u^2\}du\right] \\
&= \{\sqrt{2\pi}\}^{-1}\left[\int_{0}^{\infty} \exp\{-\tfrac{1}{2}v^2\}dv + \int_{0}^{\infty} \exp\{-\tfrac{1}{2}u^2\}du\right]
\end{aligned} \qquad (1.7.12)
$$

In the last step in Equation (1.7.12) since the two integrals are the same, we can claim that $\int_{-\infty}^{\infty} f(x)dx$ is:

$$
\begin{aligned}
2\{\sqrt{2\pi}\}^{-1}\int_{0}^{\infty} \exp\{-\tfrac{1}{2}v^2\}dv &= \{\sqrt{\pi}\}^{-1}\int_{0}^{\infty} \exp\{-w\}w^{-1/2}dw \\
&= \{\sqrt{\pi}\}^{-1}\Gamma(\tfrac{1}{2}) = 1, \text{ since } \Gamma(\tfrac{1}{2}) = \sqrt{\pi}
\end{aligned}
$$

An alternative way to prove the same result using polar coordinates has been indicated in Exercise 1.7.8.

The pdf given by Equation (1.7.11) is symmetric around $x = \mu$, that is, we have $f(x - \mu) = f(x + \mu)$ for all fixed $x \in \Re$. In other words, once the curve $(x, f(x))$ is plotted, if we pretend to fold the curve around the vertical line $x = \mu$, then the two sides of the curve will lie exactly on one another. See Figure 1.7.1.

Standard Normal Distribution: A normal distribution with $\mu = 0, \sigma = 1$ is called *standard normal*. A standard normal random variable is commonly denoted by Z. The standard normal pdf and the df are respectively denoted by

$$\phi(z) = \{\sqrt{2\pi}\}^{-1} exp(-z^2/2) \text{ for } z \in \Re$$
$$\Phi(z) = P(Z \le z) = \int_{-\infty}^{z} \phi(u)du \text{ for } z \in \Re \tag{1.7.13}$$

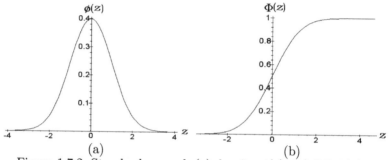

(a)   (b)
Figure 1.7.2. Standard normal: (a) density $\phi(z)$; (b) DF $\Phi(z)$.

Unfortunately, there is no available simple analytical expression for the df $\Phi(z)$. The standard normal table, namely Table 13.3.1, will facilitate finding various probabilities associated with Z. One should easily verify the following:

$$\Phi(-a) = 1 - \Phi(a) \text{ for all fixed } a > 0 \tag{1.7.14}$$

How is the df of $N(\mu, \sigma^2)$ related to the df of a standard normal random variable Z? With any fixed $x \in \Re$, $v = (u - \mu)/\sigma$, observe that we can proceed as in Equation (1.7.13) to write $P\{X \le x\}$ as

$$\{\sigma\sqrt{2\pi}\}^{-1} \int_{-\infty}^{x} exp\{-(u - \mu)^2/(2\sigma^2)\}du$$
$$= \{\sqrt{2\pi}\}^{-1} \int_{-\infty}^{(x-\mu)/\sigma} exp\{-\tfrac{1}{2}v^2\}dv = \Phi\left(\tfrac{x-\mu}{\sigma}\right) \tag{1.7.15}$$

In Figure 1.7.1, we plotted the pdfs corresponding to the $N(0, 0.25)$ and $N(1, 0.25)$ distributions. By comparing the two plots in Figure 1.7.1 we see that the shapes of the pdfs are exactly the same whereas in (b) the curve's point of symmetry $(x = 1)$ has moved by a unit on the right-hand side.

By comparing the plots in Figures 1.7.1(a) and 1.7.2(a), we see that the $N(0, 0.25)$ pdf is more concentrated than the standard normal pdf around their points of symmetry $x = 0$.

We added earlier that the parameter μ indicated the point of symmetry of the pdf. When σ is held fixed, this pdf's shape remains intact whatever the value of μ. But when μ is held fixed, the pdf becomes more (less) concentrated around the fixed center μ as σ becomes smaller (larger).

Example 1.7.5 Suppose that the scores of female students on the recent Mathematics Scholastic Aptitude Test were normally distributed with $\mu = 520$ and $\sigma = 100$ points. Find the proportion of female students taking this exam who scored (i) between 500 and 620, and (ii) more than 650. To answer the first part, we find $P(500 < X < 620) = \Phi(\frac{620-520}{100}) - \Phi(\frac{500-520}{100}) = \Phi(1) - \Phi(-0.2) = \Phi(1) + \Phi(.2) - 1$, using Equation (1.7.14). Thus, reading the entries from the standard normal table (see Chapter 13), we find that $P(500 < X < 620) = 0.84134 + 0.57926 - 1 = 0.4206$. We again use Equation (1.7.14) and the standard normal table to write: $P(X > 650) = 1 - \Phi(\frac{650-520}{100}) = 1 - \Phi(1.3) = 1 - 0.90320 = 0.0968$. ▲

Theorem 1.7.1 *Suppose that X has a $N(\mu, \sigma^2)$ distribution. Then, $Z = (X - \mu)/\sigma$ has a standard normal distribution.*

Proof Let us first find the df of Z. For any fixed $z \in \Re$, in view of Equation (1.7.15), we have:

$$P(Z \le z) = P(X \le \mu + \sigma z) = \Phi\left(\frac{(\mu+\sigma z)-\mu}{\sigma}\right) = \Phi(z) \qquad (1.7.16)$$

Since $\Phi(z)$ is differentiable for all $z \in \Re$, using Equation (1.6.10) we can claim that the pdf of Z would be $d\Phi(z)/dz = \phi(z)$, which is the pdf of a standard normal random variable. ■

Gamma Distribution: A random variable X has a gamma distribution involving α and β, denoted by Gamma(α, β), if its pdf is:

$$f(x) = \{\beta^\alpha \Gamma(\alpha)\}^{-1} e^{-x/\beta} x^{\alpha-1} \text{ for } 0 < x < \infty \qquad (1.7.17)$$

where $0 < \alpha, \beta < \infty$ are *parameters*. By varying the values of α and β, one can generate interesting shapes for the associated pdf. This distribution appears frequently as a statistical model for data obtained from reliability and survival experiments as well as in clinical trials.

How can one directly check that $f(x)$ given by Equation (1.7.17) is indeed a pdf? The function is nonnegative for all $x \in \Re$. Next, we need to verify directly that $\int_0^\infty f(x)dx = 1$. Let us substitute $u = x/\beta$ which is one-to-one and rewrite the whole integral as:

$$\{\beta^\alpha \Gamma(\alpha)\}^{-1} \int_0^\infty e^{-x/\beta} x^{\alpha-1} dx = \{\beta^\alpha \Gamma(\alpha)\}^{-1} \int_0^\infty e^{-u}(\beta u)^{\alpha-1}(\beta du)$$
$$= \{\Gamma(\alpha)\}^{-1} \int_0^\infty e^{-u} u^{\alpha-1} du = 1$$

(1.7.18)

since $\Gamma(\alpha) = \int_0^\infty e^{-u} u^{\alpha-1} du$.

Exponential Distribution: This is the same as the Gamma$(1, \beta)$ distribution where $\beta(> 0)$ is a *parameter*. When $\beta = 1$, we refer to it as the *standard exponential distribution*. The pdf is:

$$f(x) = \beta^{-1} e^{-x/\beta} \text{ for } 0 < x < \infty \qquad (1.7.19)$$

This distribution is also skewed to the right. An exponential distribution is widely used in reliability and survival analyses.

One can easily check that the corresponding df is given by

$$F(x) = \begin{cases} 0 & \text{if } -\infty < x \le 0 \\ 1 - exp(-x/\beta) & \text{if } x > 0 \end{cases} \qquad (1.7.20)$$

Suppose that X represents the life of a component in a machine and X has an exponential distribution defined by Equation (1.7.19). With any two fixed positive numbers a and b, one has

$$P\{X > a + b \mid X > a\} = exp\,(-b/\beta) = P\{X > b\} \qquad (1.7.21)$$

which does not involve a. This conveys the following message: given that a component has lasted until time a, the conditional probability of its survival beyond a time $a + b$ is the same as $P(X > b)$, regardless of the magnitude of a. In other words, the life of a component ignores the aging process regardless of its own age. This is referred to as *memoryless property*. The edited volume of Balakrishnan and Basu (1995) provides a synthesis of gamma, exponential, and other related distributions.

Chi-square Distribution: We say that a positive random variable X has a Chi-square distribution with ν degrees of freedom, denoted by χ_ν^2 with $\nu = 1, 2, 3, ...$, if X has a Gamma$(\frac{1}{2}\nu, 2)$ distribution. Here, the *parameter* is the degree of freedom, ν.

Lognormal Distribution: A positive random variable X has a lognormal distribution if its pdf is:

$$f(x) = \{\sigma\sqrt{2\pi}\}^{-1} x^{-1} exp[-\{log(x) - \mu\}^2/(2\sigma^2)] \text{ for } 0 < x < \infty \quad (1.7.22)$$

with $-\infty < \mu < \infty$ and $0 < \sigma < \infty$ as *parameters*. We can verify that

$$P(X \le x) = \Phi(\{log(x) - \mu\}/\sigma), \text{ for all } x > 0 \qquad (1.7.23)$$

and

$$P\{log(X) \le x\} = P\{X \le e^x\} = \Phi(\{x - \mu\}/\sigma), \text{ for all } x \in \Re \quad (1.7.24)$$

That is, the pdf of $log(X)$ must coincide with that of a $N(\mu, \sigma^2)$ random variable. Thus, the name "lognormal" appears natural.

Student's t Distribution: The pdf of a *Student's t* random variable with ν degrees of freedom, denoted by t_ν, is:

$$f(x) = a(1 + \tfrac{1}{\nu}x^2)^{-\frac{1}{2}(\nu+1)} \quad \text{for } -\infty < x < \infty \quad (1.7.25)$$

with $a \equiv a(\nu) = \{\sqrt{\nu\pi}\}^{-1}\Gamma(\tfrac{1}{2}(\nu+1))\{\Gamma(\tfrac{1}{2}\nu)\}^{-1}, \nu = 1, 2, 3, ...$ where ν is a *parameter*. One can easily verify that this distribution is symmetric about $x = 0$. This distribution plays a key role in statistics. W. S. Gosset introduced this distribution in his 1908 publication, but he used the pseudonym "Student" as its author. As the degree of freedom ν increases, one can check easily that the pdf has less spread around the point of symmetry $x = 0$. It is true, however, that the tail of a t distribution is "heavier" than that of a standard normal distribution.

Cauchy Distribution: This corresponds to the pdf of a Student's t_ν random variable with $\nu = 1$. The pdf from Equation (1.7.25) simplifies to

$$f(x) = \pi^{-1}(1 + x^2)^{-1} \quad \text{for } -\infty < x < \infty \quad (1.7.26)$$

One may verify that the df of a Cauchy distribution is:

$$F(x) = P(X \le x) = \tfrac{1}{\pi} \arctan(x) + \tfrac{1}{2} \text{ for all } x \in \Re \quad (1.7.27)$$

The Cauchy distribution has a heavier tail compared with that of a standard normal distribution. From Equation (1.7.26) it is clear that the Cauchy pdf $f(x) \approx \tfrac{1}{\pi}x^{-2}$ for large values of $|x|$. It can be argued then that $\tfrac{1}{\pi}x^{-2} \to 0$ slowly compared with $(\sqrt{2\pi})^{-1}e^{-x^2/2}$ when $|x| \to \infty$.

F Distribution: A positive random variable X has an F distribution with ν_1, ν_2 degrees of freedom, in that order and denoted by F_{ν_1,ν_2}, if its pdf is

$$f(x) = kx^{\frac{1}{2}(\nu_1-2)}\{1 + (\nu_1/\nu_2)x\}^{-\frac{1}{2}(\nu_1+\nu_2)} \quad \text{for } 0 < x < \infty \quad (1.7.28)$$

with $k \equiv k(\nu_1, \nu_2) = (\nu_1/\nu_2)^{\frac{1}{2}\nu_1}\Gamma((\nu_1 + \nu_2)/2)\{\Gamma(\nu_1/2)\Gamma(\nu_2/2)\}^{-1}$ where ν_1 and ν_2 are the *parameters*.

Beta Distribution: A random variable X, defined on the interval $(0,1)$, has a beta distribution with *parameters* α and β, denoted by Beta(α, β), if its pdf is

$$f(x) = \{b(\alpha, \beta)\}^{-1}x^{\alpha-1}(1-x)^{\beta-1} \text{ for } 0 < x < 1 \quad (1.7.29)$$

where $0 < \alpha, \beta < \infty$. It is a simple matter to verify that a random variable distributed as Beta$(1, 1)$ is equivalent to a Uniform$(0, 1)$ random variable.

Negative Exponential Distribution: A random variable X has the negative exponential distribution involving γ and β, if its pdf is

$$f(x) = \beta^{-1} e^{-(x-\gamma)/\beta} \text{ for } \gamma < x < \infty \qquad (1.7.30)$$

with $-\infty < \gamma < \infty, 0 < \beta < \infty$ as *parameters*. This distribution is widely used in modeling data arising from experiments in reliability and life tests. When the minimum *threshold* parameter γ is assumed zero, one then goes back to an exponential distribution.

Weibull Distribution: A positive random variable X has a Weibull distribution if its pdf is

$$f(x) = \alpha \beta^{-\alpha} x^{\alpha-1} exp\left(-[x/\beta]^{\alpha}\right) I(x > 0) \qquad (1.7.31)$$

with $\alpha(> 0)$ and $\beta(> 0)$ as *parameters*.

Rayleigh Distribution: A positive random variable X has a Rayleigh distribution if its pdf is

$$f(x) = 2\theta^{-1} x exp(-x^2/\theta) \text{ for } 0 < x < \infty \qquad (1.7.32)$$

with $\theta(> 0)$ as a *parameter*. In reliability studies and related areas, this distribution is used frequently.

1.8   Exercises and Complements

1.1.1 (Example 1.1.2 Continued) Find the probability of observing (i) a difference of one between the scores and (ii) the red die scoring higher than the yellow die.

1.1.2 Five equally qualified individuals consisting of four men and one woman apply for two identical positions in a company. Two positions are filled by selecting two from this applicant pool at random.
   (*i*)   Write down the sample space **S**;
   (*ii*)   Assign probabilities to the simple events in **S**;
   (*iii*) Find the probability that the woman applicant is selected.

1.2.1 Suppose that A, B, C are subsets of **S**. Show that

$$A \triangle C \subseteq (A \triangle B) \cup (B \triangle C)$$

where $\triangle$ denotes symmetric difference.

1.3.1 Show that the Borel sigma-field $\mathcal{B}$ is closed under the operation of (i) finite intersection and (ii) countably infinite intersection of its members. {*Hint*: Can DeMorgan's Law (Theorem 1.2.1) be used here?}

1.3.2 Prove part (v) in Theorem 1.3.1.

1.3.3 Consider a sample space $\mathbf{S}$ and suppose that $\mathcal{B}$ is the Borel sigma-field of subsets of $\mathbf{S}$. Let A, B, C be events, that is, these belong to $\mathcal{B}$. Now, prove the following statements:

(i) $P(A \cup B) \le P(A) + P(B)$;

(ii) $P(A \cup B \cup C) = P(A) + P(B) + P(C) - P(A \cap B)$
        $-P(A \cap C) - P(B \cap C) + P(A \cap B \cap C)$.

1.3.4 (Exercise 1.3.3 Continued) Suppose $A_1, ..., A_n$ are Borel sets, that is, they belong to $\mathcal{B}$. Show that

$$P(A_1 \cup ... \cup A_n) = \Sigma_{i=1}^n P(A_i) - \Sigma\Sigma_{j>i=1}^n P(A_i \cap A_j)$$

$$+\Sigma\Sigma\Sigma_{k>j>i=1}^n P(A_i \cap A_j \cap A_k) - ... + (-1)^{n-1} P(A_1 \cap ... \cap A_n).$$

1.3.5 Suppose $A_1, ..., A_n$ are Borel sets, that is, they belong to $\mathcal{B}$. Show that

$$P(A_1 \cap ... \cap A_n) \ge \Sigma_{i=1}^n P(A_i) - (n-1)$$

This is commonly called the *Bonferroni Inequality*.

1.3.6 Let A_1, A_2 be Borel sets, that is, they belong to $\mathcal{B}$. Show that

$$P(A_1 \triangle A_2) = P(A_1) + P(A_2) - 2P(A_1 \cap A_2)$$

1.3.7 Let $A_1, ..., A_n$ be Borel sets, that is, they belong to $\mathcal{B}$. Define: $B_1 = A_1, B_2 = A_2 \cap A_1^c, B_3 = A_3 \cap (A_1 \cup A_2)^c, ..., B_n = A_n \cap (A_1 \cup ... \cup A_{n-1})^c$. Show that

(i) $B_1, ..., B_n$ are Borel sets;

(ii) $B_1, ..., B_n$ are disjoint sets;

(iii) $\cup_{i=1}^n A_i = \cup_{i=1}^n B_i$.

1.4.1 Prove Theorems 1.4.1 and 1.4.2.

1.4.2 Five seniors and eight first-year graduate students are available to fill vacancies of local news reporters at the radio station in a college campus. If four students are to be randomly selected from this pool for interviews, find the probability that at least two first-year graduate students are among the chosen group.

1.4.3 Four cards are drawn at random from a standard playing pack of fifty-two cards. Find the probability that the random draw will yield

(i) an ace and three kings;

(ii) the ace, king, queen, and jack of clubs;

(iii) the ace, king, queen, and jack from the same suit;

(iv) the four queens.

1.4.4 In the context of Example 1.1.2, let us define the three events:

E : The sum of the scores from the two dice exceeds 8

F : The score on the red die is twice that of the yellow die

G : The red die scores one point more than the yellow die

Are events E, F independent? Are events E, G independent?

1.4.5 (Example 1.4.7 Continued) Assume that the prevalence of the disease was $100p\%$ in the population whereas the diagnostic blood test had the $100\alpha\%$ false negative rate and $100\beta\%$ false positive rate, $0 < p, \alpha, \beta < 1$. Now, from this population an individual is selected at random and his blood tested. The health professional is informed that the test indicated the presence of the particular disease. Find the expression of the probability, involving p, α and β, that this individual does indeed have the disease. {*Hint*: Use Bayes' Theorem.}

1.4.6 In a twin-engine plane, we are told that the engines ($\#1, \#2$) function independently. We are also told that the plane flies just fine when at least one of the two engines are working. During a flying mission, individually the engine $\#1$ and $\#2$ respectively may fail with probability 0.001 and 0.01. During a flying mission, what is the probability that the plane would crash? What is the probability the plane would complete its mission?

1.4.7 Let $A_1, ..., A_k$ be disjoint events and B be another event. Show that

$$P\{\cup_{i=1}^{k} A_i \mid B\} = \Sigma_{i=1}^{k} P\{A_i \mid B\}$$

1.4.8 Let A_1, A_2 be events. Show that

$$P(A_1) = P(A_2) \text{ if and only if } P(A_1 \cap A_2^c) = P(A_1^c \cap A_2)$$

1.4.9 Let A_1, A_2 be disjoint events. Show that

(i) $P(A_1 \mid A_2^c) = P(A_1)/\{1 - P(A_2)\}$ if $P(A_2) \neq 1$;

(ii) $P(A_1 \mid A_1 \cup A_2) = P(A_1)/\{P(A_1) + P(A_2)\}$.

1.5.1 A random variable X has the following pmf:

X values:	-2	0	1	3	8
Probabilities:	0.2	p	0.1	$2p$	0.4

where $p \in (0, 0.1]$. Is it possible to determine p uniquely?

1.5.2 A random variable X has the following pmf:

X values:	-2	0	1	3	8
Probabilities:	0.2	p	0.1	$0.3 - p$	0.4

where $p \in (0, 0.3)$. Is it possible to determine p uniquely?

1.5.3 An envelope has 20 cards of the same size. The numeral 2 is written on 10 cards, the numeral 4 is written on 6 cards, and the numeral 5 is written on the remaining 4 cards. The cards are mixed inside the envelope and we reach in to take out one card at random. Let X be the number written on the selected card.

(i)   Derive the pmf $f(x)$ of X;

(ii)   Derive the df $F(x)$ of X;

(iii) Plot the df $F(x)$ and check its points of discontinuities;

(iv)   Perform the types of calculations as in Equation (1.5.10).

1.6.1 Let c be a positive constant and consider the pdf of X given by

$$f(x) = \begin{cases} c(2 - x) & \text{if } 0 \leq x \leq 1 \\ 0 & \text{elsewhere} \end{cases}$$

Find c. Find the df $F(x)$ and plot it. Does $F(x)$ have any points of discontinuity? Find the set of points, if nonempty, where $F(x)$ is not differentiable. Calculate $P(-0.5 < X \leq 0.8)$.

1.6.2 Let c be a constant and consider the pdf of X given by

$$f(x) = \begin{cases} c + \frac{1}{4}x & \text{if } -2 \leq x \leq 0 \\ c - \frac{1}{4}x & \text{if } 0 \leq x \leq 2 \\ 0 & \text{elsewhere} \end{cases}$$

Find c. Find the df $F(x)$ and plot it. Does $F(x)$ have any points of discontinuity? Find the set of points, if nonempty, where $F(x)$ is not differentiable. Calculate $P(-1.5 < X \leq 1.8)$.

1.7.1 The probability that a patient recovers from a stomach infection is 0.9. Suppose that ten patients are known to have contracted this infection.

(i)   What is the probability that exactly seven will recover?

(ii)   What is the probability that at least five will recover?

(iii) What is the probability that at most seven will recover?

1.7.2 Let X have Binomial(n, p) distribution, $0 < p < 1$. Show that

$$P\{X > 1 \mid X \geq 1\} = \frac{1 - (1 - p)^n - np(1 - p)^{n-1}}{1 - (1 - p)^n}$$

1.7.3 A switchboard rings according to a Poisson distribution on an average five $(= \lambda)$ times in a 10-minute interval. What is the probability that during a 10-minute interval, the service desk will receive

(i)   no more than three calls?

(ii)   at least two calls?

(iii)   exactly five calls?

1.7.4 Expressions of the pdf of different random variables are given. In each case, identify a pdf by its standard name and specify the associated parameters. Find $c(> 0)$ in each case.

(i) $f(x) = c exp(-\pi x)I(x > 0)$;

(ii) $f(x) = c exp(-\pi x^2), x \in \Re$;

(iii) $f(x) = c exp(-x^2 - \frac{1}{4}x), x \in \Re$;

(iv) $f(x) = 4x^c exp(-2x)I(x > 0)$;

(v) $f(x) = \frac{128}{3}x^4 exp(-cx)I(x > 0)$;

(vi) $f(x) = 105x^4(1 - x)^c I(0 < x < 1)$.

1.7.5 Let X have Gamma$(2, 1)$ distribution. Evaluate $P(X < a), P(X > b)$ in their simplest forms where a, b are positive numbers.

1.7.6 Let X have the following pdf involving θ and β:

$$f(x) = \frac{1}{2\beta} exp\{-|x - \theta|/\beta\} \text{ for all } x \in \Re$$

with $\theta \in \Re$ and $\beta \in \Re^+$ as *parameters*. This is the *Laplace* or a *double exponential* distribution. Show that $f(x)$ is symmetric about $x = \theta$. Derive the df explicitly.

1.7.7 Show that

$$f(x) = \{\sigma\sqrt{2\pi}\}^{-1}x^{-1} exp[-\{log(x) - \mu\}^2/(2\sigma^2)]I(x > 0)$$

is a bonafide pdf with $\mu \in \Re, \sigma \in \Re^+$. Verify that

(i) $P(X \leq x) = \Phi(\{log(x) - \mu\}/\sigma)$, for all $x \in \Re^+$;

(ii) $P\{log(X) \leq x\} = \Phi(\{x - \mu\}/\sigma)$, for all $x \in \Re$.

1.7.8 (Polar Transformation) Let $\phi(z) = \{\sqrt{2\pi}\}^{-1}e^{-z^2/2}, z \in \Re$. One may use polar coordinates to show that $\int_{-\infty}^{\infty} \phi(z)dz = 1$ by going through the following steps:

(i)   Show that it is enough to prove $\int_0^\infty e^{-z^2/2}dz = \sqrt{\pi/2}$;

(ii)   Verify that $\left[\int_0^\infty e^{-z^2/2}dz\right]^2 = \int_0^\infty e^{-u^2/2}du \int_0^\infty e^{-v^2/2}dv$. Hence,

show that $\int_{t=0}^\infty \int_{s=0}^\infty e^{-(u^2+v^2)/2}dudv = \frac{1}{2}\pi$;

(iii)   In the double integral found in part (ii), use the substitutions
$u = r\cos(\theta), v = r\sin(\theta), 0 < r < \infty$ and $0 < \theta < 2\pi$. Then,

rewrite $\int_{t=0}^\infty \int_{s=0}^\infty e^{-(u^2+v^2)/2}dudv = \int_{\theta=0}^{2\pi}\left[\int_{r=0}^\infty e^{-r^2/2}rdr\right]d\theta$;

(iv)   Evaluate explicitly to show that $\int_{r=0}^\infty e^{-r^2/2}rdr = 1$;

(v)   Does part (iii) now lead to part (ii)?

1.7.9 A discrete random variable X has *hypergeometric distribution* if
its pmf is:

$$f(x) = \binom{r}{x}\binom{N-r}{n-x}/\binom{N}{n}$$

with $x = 0, 1, 2, ..., n$, but $x \le r$ and $n - x \le N - r$. Show that $f(x)$ is a
genuine pmf.

1.7.10 (Exercise 1.7.9 Continued) A box has 10 CDs of which 3 are
defective and 7 are good. One reaches into the box and grabs 4 CDs at
random. What is the distribution of X, the number of defective CDs among
4 selected CDs? It is hypergeometric with $r = 3, N = 10, n = 4$. Possible
values of X are $0, 1, 2, 3$. Evaluate the associated probabilities.

1.7.11 Show that the function $f(x)$ given in Equation (1.7.31) is a genuine pdf.

1.7.12 Show that the function $f(x)$ given in Equation (1.7.32) is a genuine pdf.

1.7.13 Suppose that X is a random variable having a Rayleigh pdf $f(x)$
given in Equation (1.7.32) with $\theta = 2$. Evaluate the following probabilities.

(i) $P(X > 2)$;
(ii) $P(X > 2 \mid X > 1)$;
(iii) $P(X < 3 \mid X < 4)$;
(iv) $P(\mid X - 2 \mid > 2)$;
(v) $P(X + X^{-1} > 2)$.

1.7.14 Suppose that we write $f(x; \theta) = 2\theta^{-1}x\,exp(-x^2/\theta)I(x > 0)$ for
$\theta > 0$. This is the Rayleigh pdf from Equation (1.7.32). We denote

$$g(x) = \alpha f(x; 2) + (1 - \alpha)f(x; 4) \text{ for } 0 < x < \infty$$

with some fixed number α, $0 < \alpha < 1$.

(i) Show that $g(x)$ is a genuine pdf;

(ii) Evaluate $P(X > 2 \mid X > 1)$ where X has its pdf $g(x)$;

(iii) Evaluate $P(X < 3 \mid X < 4)$ where X has its pdf $g(x)$;

(iv) Evaluate $P(X + X^{-1} > 2)$ where X has its pdf $g(x)$.

1.7.15 Suppose that X is a random variable having a lognormal pdf $f(x)$ given in Equation (1.7.22) with $\mu = 0$ and $\sigma = 2$. Express the following integrals in their simplest forms and evaluate them.

(i) $\int_0^\infty x f(x) dx$;

(ii) $\int_0^\infty x^2 f(x) dx$;

(iii) $\int_0^\infty [log(x)]^2 f(x) dx$.

2

Moments and Generating Functions

2.1 Introduction

In Section 2.2, we summarize the concepts of *expected value* and *variance*. The expected value is interpreted as the "center" of a probability distribution. A *variance* quantifies the average squared deviation from its "center". Section 2.3 introduces the notions of *moments* and a *moment generating function* (mgf) of a random variable. Section 2.4 provides a powerful result which says that a finite mgf determines a probability distribution uniquely. Section 2.5 briefly touches upon *probability generating function* (pgf) which leads to *factorial* moments.

2.2 Expectation and Variance

Consider playing this game: the house will roll a fair die. The player will win $7.20 whenever a six comes up, but the player will pay $0.60 to the house anytime a face other than six comes up. Suppose that from ten successive turns, we observed the following faces on the rolled die to come up: $4, 3, 6, 6, 2, 5, 3, 6, 1, 4$. At this point, the player is ahead by $18(= \$21.60 - \$3.60)$ so that the player's average win per game thus far has been exactly $1.80. But in a long run, what is expected to be the player's win per game? Assume that the player stays in the game k times in succession and by that time six appears n_k times while a number different from six appears $k - n_k$ times. Then, the player's win will amount to $7.2n_k - 0.6(k - n_k)$ so that the player's win per game ($\equiv W_k$, say) should be $\{7.2n_k - 0.6(k - n_k)\}/k$, which may be rewritten as follows:

$$W_k = (7.2)\left(k^{-1}n_k\right) + (-0.6)\{1 - (k^{-1}n_k)\} \qquad (2.2.1)$$

What will W_k amount to when k is "large"? Interpreting probabilities as limiting relative frequencies, we can say that $\lim_{k \to \infty} \frac{n_k}{k}$ should coincide with the probability of seeing the face six in a single roll of a fair die, which is $\frac{1}{6}$. Hence, the player's ultimate win per game is going to be

$$\lim_{k \to \infty} W_k = (7.2)(1/6) + (-0.6)(5/6) = 0.7 \qquad (2.2.2)$$

In other words, the player will win $0.70 or 70 cents on an average per game in the long run. The concept is intrinsically a limiting one.

Let us begin with a random variable X whose *probability mass function* (pmf) or *probability density function* (pdf) is $f(x)$ for $x \in \mathcal{X} \subseteq \Re$. In some examples, the *support* space $\mathcal{X}$ may consist of only finitely many points. In a continuous case in general, the space $\mathcal{X}$ will be the union of subintervals of $\Re$.

Definition 2.2.1 *Start with a random variable X and a summable (or integrable) function $g(X)$. Suppose that $f(x)$ is the pmf or pdf of X where $x \in \mathcal{X}$. Then, the expected value of the random variable $g(X)$, denoted by $E[g(X)]$ or $E(g(X))$, is defined as:*

$$\Sigma_{i:x_i \in \mathcal{X}} \, g(x_i)f(x_i) \quad \text{when } X \text{ is discrete, or}$$
$$\int_{\mathcal{X}} g(x)f(x)dx \quad \text{when } X \text{ is continuous}$$

When we plug in $g(x) = x$, the resulting expectation is called the *expected value* or *average* of X, often denoted by $E[X]$ or $E(X)$ or μ. Next, as we plug in $g(x) = (x - \mu)^2$, the resulting expectation is called the *variance* of a random variable X, often denoted by $V[X]$ or $V(X)$ or σ^2. $V[X]$ is interpreted as the average of the squared distance of X from its center μ. The positive square root of σ^2, namely σ, stands for the *standard deviation* of X.

The two entities μ and σ^2 play important roles in statistics. Intuitively speaking, a large (small) σ^2 indicates that on an average there is a good chance that X may (may not) stray away from the center, μ.

Example 2.2.1 Let us consider the following discrete probability distributions for the random variables X, Y respectively.

X values:	-1	1	3	5	7
Probabilities:	0.1	0.2	0.4	0.2	0.1

(2.2.3)

Y values:	-1	1	3	5	7
Probabilities:	0.2	0.15	0.3	0.15	0.2

Here, we have $\mathcal{X} = \mathcal{Y} = \{-1, 1, 3, 5, 7\}$. One may verify that

$$E(X) = (-1)(0.1) + (1)(0.2) + 3(0.4) + 5(0.2) + 7(0.1) = 3$$
$$E(Y) = (-1)(0.2) + (1)(0.15) + 3(0.3) + 5(0.15) + 7(0.2) = 3$$

(2.2.4)

and summarize by saying that $\mu_X = \mu_Y = 3$. ▲

Example 2.2.2 The technique is conceptually similar in a continuous case. Consider a random variable U whose pdf is given by

$$f(u) = \begin{cases} \frac{3}{8}u^2 & \text{if } 0 < u < 2 \\ 0 & \text{elsewhere} \end{cases}$$

(2.2.5)

Here, we have $\mathcal{U} = (0, 2)$ and write:

$$E(U) = \int_0^2 uf(u)du = \int_0^2 \frac{3}{8}u^3 du = \frac{3}{8}\left[\frac{1}{4}u^4\right]_{u=0}^{u=2}$$

(2.2.6)

so that $\mu_U = \frac{3}{2}$. ▲

Theorem 2.2.1 *Let X be a random variable. Suppose that we have real valued summable (or integrable) functions $g_i(x)$ and constants $a_i, i = 0, 1, ..., k$. Then, we have*

$$E\{a_0 + \Sigma_{i=1}^k a_i g_i(X)\} = a_0 + \Sigma_{i=1}^k a_i E\{g_i(X)\} \qquad (2.2.7)$$

provided that the expectations involved are finite. That is, "expectation" is a linear operation.

Proof We supply a proof assuming that X is a continuous random variable with its pdf $f(x), x \in \mathcal{X}$. In a discrete case, the proof is similar. Let us write:

$$E\{a_0 + \Sigma_{i=1}^k a_i g_i(X)\}$$

$$= \int_{\mathcal{X}} \left[a_0 + \{\Sigma_{i=1}^k a_i g_i(x)\} \right] f(x) dx$$

$$= a_0 \int_{\mathcal{X}} f(x) dx + \Sigma_{i=1}^k \int_{\mathcal{X}} a_i g_i(x) f(x) dx, \text{ since } a_0 \text{ is a}$$

constant and using the property of integration.

Hence, we have:

$$E\{a_0 + \Sigma_{i=1}^k a_i g_i(X)\} = a_0 + \Sigma_{i=1}^k a_i \int_{\mathcal{X}} g_i(x) f(x) dx$$

since $a_1, ..., a_k$ are constants and $\int_{\mathcal{X}} f(x) dx = 1$. Now, the proof of our desired result is complete. ■

> We can immediately express the variance alternatively as follows:
> $$V(X) = E(X^2 - 2\mu X + \mu^2) = E(X^2) - E^2(X).$$

$$(2.2.8)$$

Example 2.2.3 (Example 2.2.1 Continued) Recall that $\mu_X = \mu_Y = 3$. We may ask ourselves: between X and Y, defined via Equation (2.2.3), which one has more "variation"? Note that

$$E(X^2) = (1)(0.1) + (1)(0.2) + 9(0.4) + 25(0.2) + 49(0.1) = 13.8$$
$$E(Y^2) = (1)(0.2) + (1)(0.15) + 9(0.3) + 25(0.15) + 49(0.2) = 16.6$$

Then, using Equation (2.2.8) we have $\sigma_X^2 = 4.8$ and $\sigma_Y^2 = 7.6$. That is, Y is more variable than X. ▲

Example 2.2.4 (Example 2.2.2 Continued) Recall that U had its mean equal 1.5. Note that

$$E(U^2) = \int_0^2 u^2 f(u) du = \frac{3}{8} \int_0^2 u^4 du = \frac{3}{8} \left[\frac{1}{5} u^5 \right]_{u=0}^{u=2} = \frac{12}{5}$$

Thus, $\sigma_X^2 = \frac{12}{5} - (\frac{3}{2})^2 = 0.15.$ ▲

Theorem 2.2.2 *For a random variable X, we have*

$$V(aX + b) = a^2 V(X)$$

where a, b are any fixed real numbers.

Its proof is left out as an exercise.

2.2.1 Bernoulli Distribution

A Bernoulli(p) random variable X takes values 1 and 0 with probability p and $1 - p$, respectively, $0 < p < 1$. Then, we have:

$$\mu = E(X) = (1)(p) + (0)(1 - p) = p$$

$$E(X^2) = (1)^2(p) + (0)^2(1 - p) = p \qquad (2.2.9)$$

$$\sigma^2 = V(X) = E(X^2) - E^2(X) = p - p^2 = p(1 - p)$$

2.2.2 Binomial Distribution

Let X have its pmf $\binom{n}{x}p^x(1 - p)^{n-x}$ where $x = 0, ..., n$, $0 < p < 1$. Note that $x! = x(x - 1)!$, $n! = n(n - 1)!$ for $x \geq 1, n \geq 1$ and so we have:

$$\Sigma_{x=0}^n x\binom{n}{x}p^x(1 - p)^{n-x} = n\sum_{x=1}^n \frac{(n-1)!}{(x-1)!(n-x)!}p^x(1 - p)^{n-x}$$

Thus, using the Binomial Theorem, we obtain:

$$E(X) = np\Sigma_{x=1}^n \binom{n-1}{x-1}p^{x-1}(1 - p)^{n-x} = np\{p + (1 - p)\}^{n-1} = np$$
$$(2.2.10)$$

To evaluate the variance, let us use Theorems 2.2.1 and 2.2.2 and note that

$$V(X) = E\{X(X - 1) + X\} - \mu^2 = E\{X(X - 1)\} + \mu - \mu^2 \qquad (2.2.11)$$

We now proceed along the lines of Equation (2.2.10) by omitting some intermediate steps and write:

$$E\{X(X - 1)\} = n(n - 1)p^2\Sigma_{x=2}^n \binom{n-2}{x-2}p^{x-2}(1 - p)^{n-x} = n(n - 1)p^2$$
$$(2.2.12)$$

In other words, for a Binomial(n, p) variable, one has:

$$\mu = np \text{ and } \sigma^2 = np(1 - p) \qquad (2.2.13)$$

2.2.3 Poisson Distribution

Let X have its pmf $f(x) = e^{-\lambda}\lambda^x/x!$ where $x = 0, 1, ...$, $0 < \lambda < \infty$. Here, an exponential series expansion will help. One should verify:

$$\mu = \lambda \text{ and } \sigma^2 = \lambda \qquad (2.2.14)$$

2.2.4   Normal Distribution

Consider a standard normal variable Z with pdf $\phi(z) = \{\sqrt{2\pi}\}^{-1} exp(-z^2/2)$ for $-\infty < z < \infty$. We make a one-to-one substitution $u = -z$ along the lines of Equation (1.7.12) when $z < 0$, and express $E(Z)$ as

$$\{\sqrt{2\pi}\}^{-1} \left[\int_{-\infty}^{0} z\, exp(-\tfrac{1}{2}z^2)dz + \int_{0}^{\infty} z\, exp(-\tfrac{1}{2}z^2)dz \right]$$
$$= \{\sqrt{2\pi}\}^{-1} \left[-\int_{0}^{\infty} u\, exp(-\tfrac{1}{2}u^2)du + \int_{0}^{\infty} z\, exp(-\tfrac{1}{2}z^2)dz \right] \quad (2.2.15)$$
$$= -I_1 + J_1, \text{ say.}$$

But, the integrals I_1, J_1 are equal and finite. Hence, $E(Z) = 0$. Next,

$$E(Z^2) = \{\sqrt{2\pi}\}^{-1} \left[\int_{-\infty}^{0} z^2\, exp(-\tfrac{1}{2}z^2)dz + \int_{0}^{\infty} z^2\, exp(-\tfrac{1}{2}z^2)dz \right]$$

and we rewrite $E(Z^2)$ as

$$\{\sqrt{2\pi}\}^{-1} \left[\int_{0}^{\infty} u^2\, exp(-\tfrac{1}{2}u^2)du + \int_{0}^{\infty} z^2\, exp(-\tfrac{1}{2}z^2)dz \right]$$
$$= 2\{\sqrt{2\pi}\}^{-1} \int_{0}^{\infty} u^2\, exp(-\tfrac{1}{2}u^2)du \quad (2.2.16)$$

since the integrals are equal and finite. Then, use another one-to-one substitution $v = \tfrac{1}{2}u^2$ when $u > 0$, and proceed as follows:

$$\{\sqrt{2\pi}\}^{-1} \int_{0}^{\infty} u^2\, exp(-\tfrac{1}{2}u^2)du$$
$$= \{\sqrt{\pi}\}^{-1} \int_{0}^{\infty} v^{1/2}\, exp(-v)dv$$
$$= \{\sqrt{\pi}\}^{-1} \Gamma(\tfrac{3}{2}), \text{ with } \Gamma(.) \text{ from Equation (1.6.19)} \quad (2.2.17)$$
$$= \{\sqrt{\pi}\}^{-1} \tfrac{1}{2}\Gamma(\tfrac{1}{2})$$

which reduces to $\tfrac{1}{2}$ since $\Gamma(\tfrac{1}{2}) = \sqrt{\pi}$. Now, combining Equations (2.2.16) and (2.2.17) we see that $E(Z^2) = 1$. Hence, the mean and variance of a standard normal variable Z are given by

$$\mu_Z = 0 \text{ and } \sigma_Z^2 = 1 \quad (2.2.18)$$

Example 2.2.5 How should one evaluate, for example, $E[(Z-2)^2]$ or $E[(Z-2)(2Z+3)]$? Use Theorem 2.2.1 and write $E[(Z-2)^2] = E[Z^2] - 4E[Z] + 4 = 5$. Also, $E[(Z-2)(2Z+3)] = 2E[Z^2] - E[Z] - 6 = -4$. ▲

Next suppose that X has a $N(\mu, \sigma^2)$ distribution with pdf

$$f(x) = \{\sigma\sqrt{2\pi}\}^{-1} exp\{-(x-\mu)^2/(2\sigma^2)\} \text{ for } -\infty < x < \infty$$

Let us sketch a direct method to find the mean and variance of this distribution. We first evaluate $E\{(X-\mu)/\sigma\}$ and $E\{(X-\mu)^2/\sigma^2\}$ by making a one-to-one substitution $w = (x-\mu)/\sigma$ in the respective integrals where $x \in \Re$:

$$E\{(X-\mu)/\sigma\}$$
$$= \{\sigma\sqrt{2\pi}\}^{-1} \int_{-\infty}^{\infty} [(x-\mu)/\sigma] exp\{-(x-\mu)^2/(2\sigma^2)\}dx$$
$$= \{\sqrt{2\pi}\}^{-1} \int_{-\infty}^{\infty} w\, exp\{-w^2/2\}dw \quad (2.2.19)$$
$$= 0, \text{ since } E(Z) = 0 \text{ in view of Equation (2.2.18)}$$

Again, with the same substitution, we look at

$$E\{(X-\mu)^2/\sigma^2\} = \{\sqrt{2\pi}\}^{-1} \int_{-\infty}^{\infty} w^2 exp\{-w^2/2\}dw = 1 \qquad (2.2.20)$$

since $E(Z^2) = 1$ in view of Equation (2.2.18). Now, by appealing to Theorem 2.2.1, one can verify that for X distributed as $N(\mu, \sigma^2)$, one has

$$\mu_X = \mu \text{ and } \sigma_X^2 = \sigma^2 \qquad (2.2.21)$$

2.2.5   Gamma Distribution

Suppose that X has a Gamma(α, β) distribution. Here, we have $(\alpha, \beta) \in \Re^+ \times \Re^+$. We express $E(X)$ as

$$\{\beta^\alpha \Gamma(\alpha)\}^{-1} \int_0^\infty xe^{-x/\beta} x^{\alpha-1} dx = \beta\{\Gamma(\alpha)\}^{-1} \int_0^\infty e^{-u} u^\alpha du$$

which simplifies to

$$\beta\{\Gamma(\alpha)\}^{-1}\Gamma(\alpha+1) = \alpha\beta, \text{ since } \Gamma(\alpha+1) = \alpha\Gamma(\alpha)$$

Then, we express $E(X^2)$ analogously as

$$\{\beta^\alpha \Gamma(\alpha)\}^{-1} \int_0^\infty x^2 e^{-x/\beta} x^{\alpha-1} dx = \beta^2\{\Gamma(\alpha)\}^{-1} \int_0^\infty e^{-u} u^{\alpha+1} du$$

which simplifies to

$$\beta^2\{\Gamma(\alpha)\}^{-1}\Gamma(\alpha+2) = (\alpha+1)\alpha\beta^2, \text{ since}$$
$$\Gamma(\alpha+2) = (\alpha+1)\alpha\Gamma(\alpha), \alpha > 0$$

Hence, for X distributed as Gamma(α, β), we have

$$\mu_X = \alpha\beta \text{ and } \sigma_X^2 = \alpha\beta^2 \qquad (2.2.22)$$

2.3   Moments and Moment Generating Function

We again use Definition 2.2.1 with special choices of functions, namely, $g(x) = x^r$ or $(x-\mu)^r$ for *fixed* $r = 1, 2, \dots$. The rth *moment* of X, denoted by η_r, is given by $E[X^r]$ for $r = 1, 2, \dots$. The first moment η_1 is the mean or the expected value of X. Analogously, the rth *central* moment of X around μ, denoted by μ_r, is $E[(X-\mu)^r]$ with $r = 1, 2, \dots$.

Example 2.3.1 Note that the series $\Sigma_{i=1}^\infty i^{-p}$ converges if $p > 1$, and it diverges if $p \leq 1$. One may refer to Equation (1.6.12). With fixed $p > 1$, let us write $\Sigma_{i=1}^\infty i^{-p} = \zeta(p)$, a positive and finite real number, and define a random variable X which takes the values $i \in \mathcal{X} = \{1, 2, 3, \dots\}$ such that

$P(X = i) = \{\zeta(p)\}^{-1} i^{-p}, i = 1, 2, ...$ with $p = 2$. Since $\zeta(1) = \Sigma_{i=1}^{\infty} i^{-1}$ is not finite, η_1 or $E(X)$ is not finite. This shows that it is quite simple to construct X whose mean is not finite. ▲

> Example 2.3.1 shows that moments of a random variable may not be finite.

Now, for a random variable X, the following conclusions should be obvious:

(i) η_r is finite $\Rightarrow \mu_r$ is finite;
(ii) η_r is finite $\Rightarrow \eta_s$ is finite for all $s = 1, ..., r - 1$;   (2.3.1)
(iii) η_r is finite $\Rightarrow \mu_s$ is finite for all $s = 1, ..., r - 1$.

Part (iii) in Equation (2.3.1) follows immediately from parts (i) and (ii).

> The finiteness of the rth moment η_r does not necessarily imply the finiteness of the sth moment η_s when $s > r$. Look at Example 2.3.2.

Example 2.3.2 (Example 2.3.1 Continued) For X defined in Example 2.3.1, it is easy to verify the claims made in the adjoining table:

Table 2.3.1. Existence of Lower-Order Moments
But Nonexistence of Higher-Order Moments

p	Finite η_r	Infinite η_r
2	none	$r = 1, 2, ...$
3	$r = 1$	$r = 2, 3, ...$
4	$r = 1, 2$	$r = 3, 4, ...$
.	.	.
k	$r = 1, ..., k - 2$	$r = k - 1, k, ...$

Table 2.3.1 shows that it is a simple matter to construct discrete random variables X for which μ is finite but σ^2 is not, or μ, σ^2 are finite but η_3 is not, and so on. ▲

When Definition 2.2.1 is applied with $g(x) = e^{tx}$, one comes up with another special function that is frequently used in statistics.

Definition 2.3.1 *The moment generating function (mgf) of a random variable X, denoted by $M_X(t)$, is defined as*

$$M_X(t) = E[e^{tX}]$$

provided that the expectation is finite for $|t| < a$ with some $a > 0$.

The function $M_X(t)$ bears the name mgf because one can derive all moments from it. Consider the next result.

Theorem 2.3.1 *If a random variable X has a finite mgf $M_X(t)$, for $|t| < a$ with some $a > 0$, then the rth moment η_r of X is the same as $d^r M_X(t)/dt^r$ evaluated at $t = 0$.*

Proof Pretend that X is a continuous random variable so that $M_X(t) = \int_{\mathcal{X}} e^{tx} f(x)dx$. Now, assume that the differentiation operator of $M_X(t)$ with respect to t can be taken inside the integral. We write

$$dM_X(t)/dt = \frac{d}{dt} \int_{\mathcal{X}} e^{tx} f(x)dx = \int_{\mathcal{X}} \frac{\partial}{\partial t} \{e^{tx} f(x)\} dx$$
$$= \int_{\mathcal{X}} x e^{tx} f(x)dx \qquad (2.3.2)$$

which shows that $dM_X(t)/dt$ with $t = 0$ will coincide with $\int_{\mathcal{X}} x f(x)dx = \eta_1(= \mu)$. Similarly, let us use Equation (2.3.2) to claim that

$$d^2 M_X(t)/dt^2 = \int_{\mathcal{X}} \frac{\partial}{\partial t} \{x e^{tx} f(x)\} dx = \int_{\mathcal{X}} x^2 e^{tx} f(x)dx \qquad (2.3.3)$$

Hence, $d^2 M_X(t)/dt^2$ with $t = 0$ will coincide with $\int_{\mathcal{X}} x^2 f(x)dx = \eta_2$. The rest of the proof proceeds similarly. In a discrete scenario, replace integrals with sums. ∎

> A finite mgf $M_X(t)$ determines a unique infinite sequence of moments for a random variable X.

Incidentally, the sequence of moments $\{\eta_r = E(X^r) : r = 1, 2, ...\}$ is sometimes referred to as the sequence of *positive integral moments* of X. Theorem 2.3.1 provides an explicit tool to restore all the positive integral moments of X by successively differentiating its mgf and then letting $t = 0$ in its expression. It is interesting to note that negative integral moments (that is, when we use $r = -1, -2, ...$ in Definition 2.3.1) of X are also hidden inside the same mgf. Negative moments of X can be restored by implementing a process of *successive integration* of the mgf, an operation which is viewed as the opposite of differentiation. Precise regularity conditions under which negative moments can be derived are found in Cressie et al. (1981).

> Positive integral moments of X can be found by successively differentiating its mgf with respect to t and letting $t = 0$ in the final derivative.

The following result is very useful. It is also simple to prove.

Theorem 2.3.2 *A random variable X has its mgf $M_X(t)$, for $|t| < a$ with some $a > 0$. Let $Y = cX + d$ be another random variable where c, d are fixed real numbers. Then, $M_Y(t)$ is related to $M_X(t)$ as follows:*

$$M_Y(t) = e^{td} M_X(tc)$$

2.3.1 Binomial Distribution

Suppose that X has a Binomial(n,p) distribution. Here, for *all* fixed $t \in \Re$ we can express $M_X(t)$ as

$$\Sigma_{x=0}^n \binom{n}{x} (pe^t)^x (1-p)^{n-x}$$
$$= \{(1-p) + pe^t\}^n, \text{ using Binomial Theorem (1.4.12)} \tag{2.3.4}$$

Observe that $log(M_X(t)) = nlog\{(1-p) + pe^t\}$ so that we obtain

$$dM_X(t)/dt = npe^t\{(1-p) + pe^t\}^{-1} M_X(t) \tag{2.3.5}$$

Next, using the chain rule of differentiation, from Equation (2.3.5) we can write

$$d^2 M_X(t)/dt^2 = npe^t\{(1-p) + pe^t\}^{-1} M_X(t) - n(pe^t)^2\{(1-p)$$
$$+pe^t\}^{-2} M_X(t) + npe^t\{(1-p) + pe^t\}^{-1}\{dM_X(t)/dt\} \tag{2.3.6}$$

From Equation (2.3.4) it is obvious that $dM_X(t)/dt$, when evaluated at $t = 0$, reduces to

$$\eta_1 = np\{(1-p) + p\}^{-1} M_X(0) = np \text{ since } M_X(0) = 1 \tag{2.3.7}$$

We found the mean of the distribution in Equation (2.2.13). Here we have another way to check that the mean is np.

Also, we would equate the expression of $E(X^2)$ with $d^2 M_X(t)/dt^2$ evaluated at $t = 0$. From Equation (2.3.7) it is obvious that $d^2 M_X(t)/dt^2$, when evaluated at $t = 0$, should lead to η_2, that is,

$$E(X^2)$$
$$= np\{(1-p) + p\}^{-1} M_X(0) - np^2\{(1-p) + p\}^{-2} M_X(0)$$
$$+ np\{(1-p) + p\}^{-1}\{dM_X(t)/dt\}\mid_{t=0} \tag{2.3.8}$$
$$= np - np^2 + n^2p^2, \text{ since } M_X(0) = 1 \text{ and}$$
$$\{dM_X(t)/dt\}\mid_{t=0} = np$$

Hence, one has $V(X) = E(X^2) - \mu^2 = np - np^2 + n^2p^2 - (np)^2 = np(1-p)$. This matches with the expression of σ^2 given in Equation (2.2.13).

2.3.2 Poisson Distribution

Suppose that X has a Poisson(λ) distribution and $0 < \lambda < \infty$. Here, for *all* fixed $t \in \Re$ we can express $M_X(t)$ as

$$e^{-\lambda}\Sigma_{x=0}^\infty(\lambda e^t)^x\frac{1}{x!} = exp\{-\lambda + \lambda e^t\}, \text{ using the}$$
$$\text{exponential series expansion from Equation (1.6.15)} \tag{2.3.9}$$

One may check that $dM_X(t)/dt$, when evaluated at $t = 0$, reduces to $\eta_1 = \lambda$ and $d^2 M_X(t)/dt^2$, evaluated at $t = 0$, matches with $\eta_2 = \lambda + \lambda^2$. Thus, we have $V(X) = E(X^2) - \mu^2 = \lambda + \lambda^2 - \lambda^2 = \lambda$ and these again match with the expression of μ, σ^2 given in Equation (2.2.14).

2.3.3 Normal Distribution

Let us first suppose that Z has a standard normal distribution. Now, for *all* fixed $t \in \Re$ we can express the mgf $M_Z(t)$ as

$$
\begin{aligned}
&\int_{-\infty}^{\infty} e^{tz}\phi(z)dz \\
&= \{\sqrt{2\pi}\}^{-1} \int_{-\infty}^{\infty} exp(tu - \tfrac{1}{2}u^2)du \\
&= exp(\tfrac{1}{2}t^2) \int_{-\infty}^{\infty} h_t(u)du \quad \text{where} \\
&\quad h_t(u) = \{\sqrt{2\pi}\}^{-1} exp\{-\tfrac{1}{2}(u-t)^2\} \quad \text{for } u \in \Re
\end{aligned}
\tag{2.3.10}
$$

Observe that the function $h_t(u)$ used in the last step in Equation (2.3.10) resembles the pdf of a random variable having a $N(t,1)$ distribution for all fixed $t \in \Re$. So, we must have $\int_{-\infty}^{\infty} h_t(u)du = 1$ for all fixed $t \in \Re$. In other words, Equation (2.3.10) leads to the following conclusion:

$$
M_Z(t) = exp\{\tfrac{1}{2}t^2\} \text{ where } Z \text{ is } N(0,1), \text{ for all } t \in \Re \tag{2.3.11}
$$

Now, suppose that X is distributed as $N(\mu, \sigma^2)$. We can write $X = \sigma Y + \mu$ with $Y = (X - \mu)/\sigma$. We can immediately use Theorem 2.3.2 and claim:

$$
M_X(t) = e^{t\mu} M_Y(t\sigma)
$$

Now, by substituting $y = (x - \mu)/\sigma$ in the integral, the expression of $M_Y(t)$ can be found as follows:

$$
\begin{aligned}
&M_Y(t) \\
&= \{\sigma\sqrt{2\pi}\}^{-1} \int_{-\infty}^{\infty} exp\left\{t\left(\tfrac{x-\mu}{\sigma}\right)\right\} exp\left\{-\tfrac{1}{2}\left(\tfrac{x-\mu}{\sigma}\right)^2\right\} dx \\
&= \{\sqrt{2\pi}\}^{-1} \int_{-\infty}^{\infty} exp(ty) exp\{-\tfrac{1}{2}y^2\}dy
\end{aligned}
$$

which is exactly the same integral that was evaluated in Equation (2.3.10). We can immediately claim that $M_Y(t) = exp(\tfrac{1}{2}t^2)$ and write

$$
M_X(t) = exp\{t\mu + \tfrac{1}{2}t^2\sigma^2\} \text{ where } X \text{ is } N(\mu, \sigma^2), \text{ for all } t \in \Re \tag{2.3.12}
$$

Now, $log(M_X(t)) = t\mu + \tfrac{1}{2}t^2\sigma^2$ so that $dM_X(t)/dt = (\mu + t\sigma^2)M_X(t)$. Hence, $dM_X(t)/dt$, when evaluated at $t = 0$, reduces to $\eta_1 = \mu$. Also, using the chain rule of differentiation, we have $d^2 M_X(t)/dt^2 = \sigma^2 M_X(t) + (\mu + t\sigma^2)^2 M_X(t)$ so that $d^2 M_X(t)/dt^2$, when evaluated at $t = 0$, reduces to $\eta_2 = \sigma^2 + \mu^2$. In view of Theorem 2.3.1, we can say that μ is the mean of X and $V(X) = E(X^2) - \mu^2 = \sigma^2 + \mu^2 - \mu^2 = \sigma^2$. These answers were verified earlier in Equation (2.2.21).

One can also evaluate higher-order derivatives, namely $d^k M_X(t)/dt^k$ with $k = 3$ and 4. When evaluated at $t = 0$, these will reduce to η_3 or $E(X^3)$ and η_4 or $E(X^4)$, respectively. Then, in order to obtain the *third* and *fourth* central moments of X, one should proceed as follows:

$$
\mu_3 = E\{X^3 - 3\mu X^2 + 3\mu^2 X - \mu^3\} = \eta_3 - 3\mu\eta_2 + 2\mu^3 \tag{2.3.13}
$$

by Theorem 2.2.1. Similarly,

$$\mu_4 = E\{X^4 - 4\mu X^3 + 6\mu^2 X^2 - 4\mu^3 X + \mu^4\}$$
$$= \eta_4 - 4\mu\eta_3 + 6\mu^2\eta_2 - 3\mu^4 \qquad (2.3.14)$$

In Exercise 2.3.2, one finds: $\mu_3 = 0$ and $\mu_4 = 3\sigma^4$.

2.3.4 Gamma Distribution

Suppose that a random variable X has a Gamma(α, β) distribution. One should verify the following:

$$M_X(t) = (1 - \beta t)^{-\alpha}, \text{ for all } t < \beta^{-1}. \qquad (2.3.15)$$

In view of Theorem 2.3.1, one can successively obtain the expressions of $E(X^r)$ when r is a positive integer.

But, how should one derive the expression for $E(X^r)$ when r is arbitrary? We no longer restrict r to be a positive integer. Now, the expression of the mgf of X given by Equation (2.3.15) may not be of immediate help. One may pursue a direct approach along the lines of Section 2.2.5 and write

$$E(X^r) = \{\beta^\alpha \Gamma(\alpha)\}^{-1} \int_0^\infty e^{-x/\beta} x^{(\alpha+r)-1} dx \qquad (2.3.16)$$

Observe that the integrand in Equation (2.3.16) is the kernel (that is, the part involving x only) of a gamma pdf provided that $\alpha + r > 0$. In other words, we must have

$$\int_0^\infty e^{-x/\beta} x^{(\alpha+r)-1} dx = \beta^{\alpha+r} \Gamma(\alpha + r) \text{ when } \alpha + r > 0 \qquad (2.3.17)$$

Next, we combine Equations (2.3.16) and (2.3.17) to conclude that

$$E(X^r) = \beta^r \Gamma(\alpha + r)\{\Gamma(\alpha)\}^{-1} \text{ when } \alpha > -r \qquad (2.3.18)$$

> If X is distributed as Gamma(α, β), then $E(X^r)$ is finite if and only if $\alpha > -r$.

Special Case 1: Exponential Distribution

An exponential random variable X with pdf $f(x) = \beta^{-1} e^{-x/\beta} I(x > 0)$ is equivalent to a random variable distributed as Gamma$(1, \beta)$ with $\beta \in \Re^+$. In this situation, we summarize the following results:

$$\text{The mgf is } M_X(t) = (1 - \beta t)^{-1} \text{ for all } t < \beta^{-1}, \text{ and}$$
$$E(X) = \beta, E(X^2) = 2\beta^2, V(X) = \beta^2 \qquad (2.3.19)$$

Special Case 2: Chi-Square Distribution

A Chi-square random variable X with its pdf

$$f(x) = \{2^{\nu/2} \Gamma(\nu/2)\}^{-1} e^{-x/2} x^{(\nu/2)-1} I(x > 0)$$

is equivalent to a random variable distributed as Gamma($\frac{\nu}{2}, 2$) with a positive integral degree of freedom ν. We provide the following results:

$$\text{The mgf of } \chi_\nu^2 \text{ is } (1 - 2t)^{-\nu/2} \text{ for all } t < \tfrac{1}{2}$$

$$E(X) = \nu, E(X^2) = \nu(\nu + 2), V(X) = 2\nu \qquad (2.3.20)$$

2.4 Determination of a Distribution via MGF

Next, we emphasize the role of a finite moment generating function in uniquely determining a probability distribution. We state an important and useful result. Its proof is out of scope.

Theorem 2.4.1 *Let $M(t)$ be a finite mgf for $|t| < a$ with some $a > 0$. Then, $M(t)$ corresponds to the mgf associated with a uniquely determined probability distribution.*

Example 2.4.1 A random variable X takes possible values $0, 1$, and 2 with respective probabilities $\frac{1}{8}, \frac{1}{4}$, and $\frac{5}{8}$. The mgf of X is obviously given by $M_X(t) = \frac{1}{8} + \frac{1}{4}e^t + \frac{5}{8}e^{2t} = \frac{1}{8}(1 + 2e^t + 5e^{2t})$. Observe how the mgf of a discrete random variable is formed and how easy it is to identify the probability distribution by inspecting the appearance of the terms which together build up the function $M_X(t)$. Now, suppose that we have a random variable U whose mgf $M_U(t) = \frac{1}{5}(1 + e^t + 3e^{2t})e^{-t}$. Let us rewrite this mgf as $0.2e^{-t} + 0.2 + 0.6e^t$. We can claim that U takes the possible values $-1, 0$, and 1 with the respective probabilities $0.2, 0.2$, and 0.6. From Theorem 2.4.1, we know that the indicated distribution is unique. ▲

Example 2.4.2 Let U be a random variable with $M_U(t) = \frac{1}{16}(1 + e^t)^4$. Rewrite $M_U(t) = (\frac{1}{2} + \frac{1}{2}e^t)^4$ which agrees with $M_X(t)$ given by Equation (2.3.4) where $n = 4$ and $p = \frac{1}{2}$. Hence, U must be distributed as Binomial$(4, \frac{1}{2})$ by Theorem 2.4.1. ▲

Example 2.4.3 Let U be a random variable such that $M_U(t) = exp\{\pi t^2\}$, which agrees with $M_X(t)$ given by Equation (2.3.12) with $\mu = 0, \sigma^2 = 2\pi$. Hence, U must be distributed as $N(0, 2\pi)$. ▲

| A finite mgf determines a probability distribution uniquely. |

Before we move on, let us attend to one other point involving moments. A finite mgf uniquely determines a probability distribution, but moments alone may not identify a probability distribution associated with those moments.

Example 2.4.4 Rao (1973, p. 152) had discussed the construction of the following two pdfs, originally due to C. C. Heyde. Consider positive continuous random variables X, Y whose pdfs are respectively

$$\begin{aligned} f(x) &= (2\pi)^{-1/2}x^{-1}exp[-\tfrac{1}{2}(log(x))^2] \quad && \text{for } x > 0 \\ g(y) &= f(y)[1 + c\sin(2\pi log(y))] \quad && \text{for } y > 0 \end{aligned} \qquad (2.4.1)$$

where c is a fixed number, $-1 \leq c \leq 1$ and $c \neq 0$. In Exercise 2.4.4, one is asked to show that $E[X^r] = E[Y^r]$ for all $r = 1, 2, \dots$. Certainly X, Y have different probability distributions. See Figure 2.4.1, and also Exercise 2.4.5. ▲

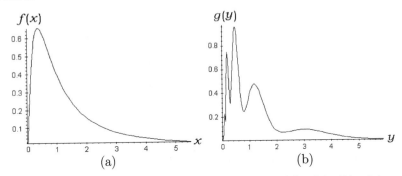

Figure 2.4.1. The pdfs from Equation (2.4.1): (a) $f(x)$; (b) $g(y)$ with $c = \frac{1}{2}$.

> Finite moments alone may not determine a distribution uniquely. Example 2.4.4 highlights this point.

2.5 Probability Generating Function

We have seen that a mgf generates the moments. There is another special function which generates the probabilities. Let X be a nonnegative random variable. A *probability generating function* (pgf) of X is defined by

$$P_X(t) = E[t^X] \text{ with } t > 0 \tag{2.5.1}$$

The explicit form of a pgf is often found when the random variable X is integer-valued. From this point onward, let us include *only nonnegative integer-valued random variables* in this discussion.

Why this function $P_X(t)$ is called a pgf should become clear once we write it out fully as follows:

$$P_X(t) = p_0 + p_1 t + p_2 t^2 + p_3 t^3 + \dots \tag{2.5.2}$$

where $p_i = P(X = i)$ for $i = 1, 2, \dots$. We see immediately that the coefficient of t^i is p_i, the probability that $X = i$ for $i = 1, 2, \dots$.

> In statistics, $E\{X(X-1)\dots(X-k+1)\}$ is referred to as the kth-order factorial moment of X. $\tag{2.5.3}$

In the light of Theorem 2.3.1 one can justify the following result:

$d^k P_X(t)/dt^k$, evaluated at $t = 1$, will reduce to the kth-order factorial moment of X, for $k = 1, 2, \dots$ $\tag{2.5.4}$

One should verify that

$$P_X(t) = M_X(\log(t)) \text{ with } t > 0 \qquad (2.5.5)$$

where $P_X(t)$ is the pgf of X defined in Equation (2.5.1).

> If X has a Binomial(n, p) distribution, one can verify:
> $$E[X(X-1)...(X-k+1)] = p^k \Pi_{i=1}^k (n-i+1), 0 < p < 1.$$

$$(2.5.6)$$

> When X has a Poisson(λ) distribution, one can verify:
> $$E[X(X-1)...(X-k+1)] = \lambda^k, 0 < \lambda < \infty.$$

$$(2.5.7)$$

2.6   Exercises and Complements

2.2.1 An insurance company sells life insurance policies with a face value of $2000 and a yearly premium of $30. If 0.4% of the policyholders are expected to die in a year, what would be the company's expected earnings per policyholder in a year?

2.2.2 Let X have a discrete uniform distribution:

$$P(X = i) = n^{-1} \text{ for } i = 1, ..., n$$

Derive explicit expressions of μ and σ.

2.2.3 Consider an arbitrary random variable Y which may be discrete or continuous. Let A be an arbitrary event (Borel set) defined through the random variable Y. Define a new random variable $X = I(A)$, the indicator variable of the set A, that is:

$$X = \begin{cases} 0 & \text{if } A^c \text{ occurs} \\ 1 & \text{if } A \text{ occurs} \end{cases}$$

Evaluate μ and σ.

2.2.4 Let X have a Geometric(p) distribution with $0 < p < 1$. Derive explicit expressions of μ and σ.

2.2.5 Let c be a positive constant and consider the pdf of X given by

$$f(x) = \begin{cases} c(2-x) & \text{if } 0 \leq x \leq 1 \\ 0 & \text{elsewhere} \end{cases}$$

(i)   Explicitly evaluate c, μ and σ;
(ii)   Evaluate $E\{(1-X)^{-1/2}\}$;
(iii) Evaluate $E\{X^3(1-X)^{3/2}\}$.
{*Hints*: Express expected values as combinations of beta functions.}

2.2.6 Let c be a constant and consider the pdf of a random variable X given by

$$f(x) = \begin{cases} c + \frac{1}{4}x & \text{if } -2 \le x \le 0 \\ c - \frac{1}{4}x & \text{if } 0 \le x \le 2 \\ 0 & \text{elsewhere} \end{cases}$$

Explicitly evaluate μ and σ.

2.2.7 Let X have a Uniform$(0,1)$ distribution. Evaluate

(i) $E[X^{-1/2}(1-X)^{10}]$;

(ii) $E[X^{3/2}(1-X)^{5/2}]$;

(iii) $E[(X^2 + 2X^{1/2} - 3X^{5/2})(1-X)^{10}]$.

{*Hint*: Can these be evaluated using beta integrals?}

2.2.8 A random variable X has its pdf $f(x) = c(x - 2x^2 + x^3)I(0 < x < 1)$ where c is a positive number. Evaluate μ and σ. {*Hint*: Should one match this $f(x)$ with a beta density?}

2.2.9 Let X have a *Rayleigh distribution* with its pdf

$$f(x) = 2\theta^{-1}x\,exp(-x^2/\theta)I(x > 0)$$

where $\theta(>0)$. Evaluate μ and σ.

2.2.10 Let X have a *Weibull distribution* with its pdf

$$f(x) = \alpha\beta^{-\alpha}x^{\alpha-1}\,exp(-[x/\beta]^\alpha)I(x > 0)$$

where $\alpha(>0)$ and $\beta(>0)$. Evaluate μ and σ.

2.2.11 Let X have a pdf $f(x) = \{\sqrt{2\pi}\}^{-1}x^{-1}\,exp[-\{log(x)\}^2/2]I(x > 0)$. Derive expressions of the mean and variance of X.

2.2.12 Let X have a *hypergeometric distribution* with its pmf:

$$f(x) = \binom{r}{x}\binom{N-r}{n-x}/\binom{N}{n}$$

with $x = 0, 1, 2, ..., n$, $x \le r$, and $n - x \le N - r$. Show that $\mu = np$ and $\sigma^2 = np(1-p)\left(\dfrac{N-n}{N-1}\right)$ with $p = r/N$.

2.3.1 Let Z have a standard normal distribution. Show that

(i) $E(Z^r) = 0$ for any odd integer $r > 0$;

(ii) $E(Z^r) = \pi^{-1/2}2^{r/2}\Gamma(\frac{1}{2}r + \frac{1}{2})$ for any even integer $r > 0$;

(iii) $\mu_4 = 3$ using part (ii).

2.3.2 Let X have a $N(\mu,\sigma^2)$ distribution. Show that

(i) $E\{(X-\mu)^r\} = 0$ for all positive odd integers r;

(ii) $E\{(X-\mu)^r\} = \pi^{-1/2}\sigma^r 2^{r/2}\Gamma\left(\frac{1}{2}r + \frac{1}{2}\right)$ for all positive even integers r;

(iii) the fourth central moment μ_4 reduces to $3\sigma^4$.

2.3.3 Let Z have a standard normal distribution. Show that

$$E(|Z|^r) = \pi^{-1/2} 2^{r/2} \Gamma(\tfrac{1}{2}r + \tfrac{1}{2})$$

for any fixed $r > 0$.

2.3.4 Why is it that the answer in Exercise 2.3.1, part (ii) matches with the expression of $E(|Z|^r)$ in Exercise 2.3.3? Is it possible to connect the two pieces?

2.3.5 Let X have a $N(\mu, \sigma^2)$ distribution. Show that $\mu_3 = 0$ and $\mu_4 = 3\sigma^4$ by successively differentiating the mgf of X.

2.3.6 Consider a Cauchy random variable X whose pdf is $f(x) = \pi^{-1}(1 + x^2)^{-1}$, $x \in \Re$. Show that $E(X)$ does not exist.

2.3.7 Give examples of continuous random variables, other than Cauchy, for which the mean μ is infinite. {*Hint*: Try a continuous version of Example 2.3.1. Is $f(x) = x^{-2}I(1 < x < \infty)$ a genuine pdf?}

2.3.8 Give examples of continuous random variables for which

(i) μ is finite, but σ^2 is infinite;

(ii) μ and σ^2 are both finite, but μ_3 is infinite;

(iii) μ_{10} is finite, but μ_{11} is infinite.

2.3.9 Suppose that Z has a standard normal distribution. Derive an expression of the mgf of $|Z|$. Show that $E(e^{|Z|}) = 2\sqrt{e}\Phi(1)$.

2.4.1 For the given expressions of the mgf, (a) identify the probability distribution either by its standard name or by explicitly writing down the pmf or pdf, (b) find μ and σ.

(i) $M_X(t) = e^{5t}$, for $t \in \Re$;

(ii) $M_X(t) = 1$, for $t \in \Re$;

(iii) $M_X(t) = \tfrac{1}{2}(1 + e^t)$, for $t \in \Re$;

(iv) $M_X(t) = \tfrac{1}{3}(e^{-2t} + 1 + e^t)$, for $t \in \Re$;

(v) $M_X(t) = \tfrac{1}{10}(e^{2t} + 3 + 6e^{4t})$, for $t \in \Re$;

(vi) $M_X(t) = \tfrac{1}{2401}(3e^t + 4)^4$, for $t \in \Re$.

2.4.2 For the given expressions of the mgf, (a) identify the probability distribution either by its standard name or by explicitly writing down the pmf or pdf, and (b) find μ and σ.

(i) $E[e^{tX}] = e^{25t^2}$, for $t \in \Re$;

(ii) $E[e^{tX}] = e^{t^2}$, for $t \in \Re$;

(iii) $M_X(t) = (1 - 6t + 9t^2)^{-2}$, for $t < \tfrac{1}{3}$.

2.4.3 Let X have its mgf given by

$$M_X(t) = \frac{1}{5}\left\{\frac{1 + 2e^{2t}}{e^{4t}} + \frac{e^{3t} + e^{6t}}{e^{5t}}\right\} \quad \text{for } t \in \Re$$

Find μ and σ^2. Can the distribution of X be identified?

2.4.4 (Example 2.4.4 Continued) Show that $E(X^r) = E(Y^r)$ for all $r = 1, 2, \dots$.

2.4.5 (Exercise 2.4.4 Continued) Consider the two pdfs $f(x)$ and $g(y)$ with $x > 0, y > 0$, as defined in Exercise 2.4.4. Let $a(x)$ with $x > 0$ be any other pdf which has all its (positive integral) moments finite. For example, $a(x)$ may be the pdf corresponding to a Gamma(α, β) distribution. Consider now nonnegative random variables U and V with respective pdfs $f_0(u)$ and $g_0(v)$ with $u > 0, v > 0$ where

$$f_0(u) = pa(u) + (1-p)f(u)$$

$$g_0(v) = pa(v) + (1-p)g(v) \text{ with fixed } p, 0 < p < 1$$

Show that U and V have the same infinite sequence of moments but different distributions. Plot the two pdfs $f_0(u)$ and $g_0(v)$ with $c = p = \frac{1}{2}, a(x) = \frac{1}{10}e^{-x/10}$, and $x > 0$.

2.4.6 Let X have the following distribution:

$$P(X = i) = n^{-1} \text{ for } i = 1, \dots, n$$

Derive an explicit expression of the mgf of X. Hence, obtain μ, σ.

2.4.7 Let X have its pdf

$$f(x) = c\exp\{-x^2 - dx\} \text{ for } x \in \Re$$

with appropriate constants $c(> 0)$ and d. We are also told that $E(X) = -\frac{3}{2}$.
(i)   Determine c and d;
(ii)   Identify the distribution by its name;
(iii) Evaluate the third and fourth central moments of X.

2.4.8 Let X have its pdf

$$f(x) = c\exp\{-x/3\}x^d \text{ for } x \in \Re^+$$

where $c(> 0)$ and $d(> 0)$ are appropriate constants. We are told that $E(X) = 10.5$ and $V(X) = 31.5$.
(i)   Determine c and d;
(ii)   Identify the distribution by its name;
(iii) Evaluate the third and fourth moments, η_3 and η_4.

2.4.9 A random variable X has its mgf given by

$$M_X(t) = e^t(5 - 4e^t)^{-1} \text{ for } t < 0.223$$

Evaluate $P(X = 4 \text{ or } 5)$.

2.5.1 Let X have a Geometric(p) distribution. Show that the pgf is:

$$P_X(t) = pt\{1 - (1-p)t\}^{-1} \text{ with } t > 0$$

Evaluate the first two factorial moments of X.

2.5.2 Consider the pgf $P_X(t)$ defined by Equation (2.5.1). Suppose that $P_X(t)$ is twice differentiable and assume that the derivatives with respect to t and expectation can be interchanged. Let us denote $P'_X(t) = \frac{d}{dt} P_X(t)$ and $P''_X(t) = \frac{d^2}{dt^2} P_X(t)$. Show that

   (i) $E[X] = P'_X(1)$;
   (ii) $V[X] = P''_X(1) + P'_X(1) - \{P'_X(1)\}^2$.

2.5.3 Let X have a Beta$(3,5)$ distribution. Evaluate the third and fourth factorial moments of X.

3

Multivariate Random Variables

3.1 Introduction

Generally speaking, we may have $k(\geq 2)$ real valued random variables $X_1, ..., X_k$ which vary both individually and jointly. For example, during health awareness week, we may consider college students and record each student's height (X_1), weight (X_2), age (X_3), and blood pressure (X_4). It may be reasonable to assume that these variables jointly follow a four-dimensional probability distribution.

Section 3.2 first introduces *joint, marginal,* and *conditional distributions* in a continuous case. Discrete examples, including the *multinomial distribution,* are then included. *Covariances* and *correlation coefficients* are explored in Section 3.3. Section 3.4 introduces *independent* random variables. A *bivariate normal distribution* is introduced in Section 3.5. Relationships between a zero correlation and independence are explained in Section 3.6. Section 3.7 introduces *exponential families.* Section 3.8 includes useful inequalities.

3.2 Probability Distributions

Notions of the *joint, marginal,* and *conditional distributions* are described here. Section 3.2.1 introduces a bivariate scenario and Section 3.2.2 gives examples in three and higher dimensions. Section 3.2.3 includes discrete examples and the *multinomial distribution.*

3.2.1 Joint, Marginal, and Conditional Distributions

For simplicity, we begin with a bivariate continuous random vector variable (X_1, X_2) taking values $(x_1, x_2) \in \mathcal{X}_1 \times \mathcal{X}_2$, both subintervals of $\Re$. Consider a function $f(x_1, x_2)$. Throughout, we will use the following convention:

> Let us presume that $\mathcal{X}_1 \times \mathcal{X}_2$ is the support of a joint probability distribution in the sense that $f(x_1, x_2) > 0$ for all $(x_1, x_2) \in \mathcal{X}_1 \times \mathcal{X}_2$ and $f(x_1, x_2) = 0$ for all $(x_1, x_2) \in \Re^2 - (\mathcal{X}_1 \times \mathcal{X}_2)$.

A nonnegative function $f(x_1, x_2)$ is a *joint probability density function* (joint pdf) of $\mathbf{X} = (X_1, X_2)$ if

$$\int_{\mathcal{X}_2} \int_{\mathcal{X}_1} f(x_1, x_2) dx_1 dx_2 = 1 \qquad (3.2.1)$$

This integral represents the *total volume* under the surface $z = f(x_1, x_2)$.

The *marginal distribution* of one of the random variables is obtained by integrating the joint pdf with respect to the other variable so that the *marginal* pdf of X_i is

$$f_i(x_i) = \int_{\mathcal{X}_j} f(x_1, x_2) dx_j \text{ for } x_i \in \mathcal{X}_i \text{ with fixed } i = 1, 2, \ j \neq i \quad (3.2.2)$$

Visualizing the notion of a *conditional distribution* in a continuous scenario is a little tricky. In a *discrete case*, $f(x_1, x_2)$ and $f_i(x_i)$ are simply interpreted as $P(X_1 = x_1 \cap X_2 = x_2)$ and $P(X_i = x_i)$, respectively, and hence the conditional probabilities could be found as long as $P(X_i = x_i) > 0, i = 1, 2$. In a continuous case, however, $P(X_i = x_i) = 0$ for all $x_i \in \mathcal{X}_i$ $i = 1, 2$. So, how could one define a *conditional* pdf?

First, we write down the conditional df of X_1 given that $x_2 \leq X_2 \leq x_2 + h$ where $h(> 0)$ is a "small" number. Assuming that $P(x_2 \leq X_2 \leq x_2 + h) > 0$, we have:

$$P\{X_1 \leq x_1 \mid x_2 \leq X_2 \leq x_2 + h\}$$

$$= \frac{P\{X_1 \leq x_1 \cap x_2 \leq X_2 \leq x_2 + h\}}{P\{x_2 \leq X_2 \leq x_2 + h\}}$$

$$= \frac{\int_{t=x_2}^{x_2+h} \int_{s=-\infty}^{x_1} f(s,t) ds dt}{\int_{t=x_2}^{x_2+h} f_2(t) dt} \quad (3.2.3)$$

From the last step in Equation (3.2.3) it is clear that as $h \downarrow 0$, the *limiting* value of this conditional probability takes the form of $0/0$. Thus, by using L'Hôpital's rule, we obtain:

$$\lim_{h \downarrow 0} P\{X_1 \leq x_1 \mid x_2 \leq X_2 \leq x_2 + h\}$$

$$= \lim_{h \downarrow 0} \frac{\frac{d}{dh} \left[\int_{t=x_2}^{x_2+h} \int_{s=-\infty}^{x_1} f(s,t) ds dt \right]}{\frac{d}{dh} \left[\int_{t=x_2}^{x_2+h} f_2(t) dt \right]} \quad (3.2.4)$$

$$= \lim_{h \downarrow 0} \frac{\int_{s=-\infty}^{x_1} f(s, x_2 + h) ds}{f_2(x_2 + h)}$$

Next, by differentiating the last expression in Equation (3.2.4) with respect to x_1, we obtain the expression for the conditional pdf of X_1 given that $X_2 = x_2$ which is:

$$f_{1|2}(x_1) = \frac{f(x_1, x_2)}{f_2(x_2)} \text{ for all } (x_1, x_2) \in \mathcal{X}_1 \times \mathcal{X}_2 \text{ such that } f_2(x_2) > 0$$

$$(3.2.5)$$

The *conditional* pdf of X_2 given $X_1 = x_1$ is analogously given by:

$$f_{2|1}(x_2) = \frac{f(x_1, x_2)}{f_1(x_1)} \text{ for all } (x_1, x_2) \in \mathcal{X}_1 \times \mathcal{X}_2 \text{ such that } f_1(x_1) > 0$$

$$(3.2.6)$$

One may also talk about the *moments* of X_1, X_2 individually (*marginally*) as well as *conditionally* for one of the random variables given the other. For example, we may write:

$$\mu_i = E[X_i] = \int_{\mathcal{X}_i} x_i f_i(x_i) dx_i, i = 1, 2$$
$$\sigma_i^2 = E[X_i^2] - \mu_i^2, i = 1, 2$$
$$\mu_{i|j} = E[X_i \mid X_j = x_j] = \int_{\mathcal{X}_i} x_i f_{i|j}(x_i) dx_i, i \neq j = 1, 2 \qquad (3.2.7)$$
$$E[X_i^2 \mid X_j = x_j] = \int_{\mathcal{X}_i} x_i^2 f_{i|j}(x_i) dx_i, i \neq j = 1, 2$$
$$\sigma_{i|j}^2 = E[X_i^2 \mid X_j = x_j] - \mu_{i|j}^2, i \neq j = 1, 2$$

In general, we can write

$$E[g(X_1, X_2)] = \int_{\mathcal{X}_2} \int_{\mathcal{X}_1} g(x_1, x_2) f(x_1, x_2) dx_1 dx_2$$
$$E[h(X_i) \mid X_j = x_j] = \int_{\mathcal{X}_i} h(x_i) f_{i|j}(x_i) dx_i \text{ with } x_j \in \mathcal{X}_j \qquad (3.2.8)$$
$$\text{whenever } f_j(x_j) > 0 \text{ for } i \neq j = 1, 2$$

Example 3.2.1 Consider X_1, X_2 whose joint pdf is

$$f(x_1, x_2) = \begin{cases} 6(1 - x_2) & \text{if } 0 < x_1 < x_2 < 1 \\ 0 & \text{elsewhere} \end{cases} \qquad (3.2.9)$$

Obviously, one has $\mathcal{X}_1 = \mathcal{X}_2 = (0, 1)$. We can write:

$$f_1(x_1) = \int_{x_1}^1 f(x_1, x_2) dx_2 = \int_{x_1}^1 6(1 - x_2) dx_2 = 3(1 - x_1)^2$$
$$f_2(x_2) = \int_0^{x_2} f(x_1, x_2) dx_1 = \int_0^{x_2} 6(1 - x_2) dx_1 = 6x_2(1 - x_2)$$

for $0 < x_1 < 1, 0 < x_2 < 1$. One uses the marginal pdf $f_2(x_2)$ to obtain:

$$E(X_2) = 6 \int_0^1 x_2^2(1 - x_2) dx_2 = \frac{1}{2}$$
$$E(X_2^2) = 6 \int_0^1 x_2^3(1 - x_2) dx_2 = \frac{3}{10}$$

so that
$$V(X_2) = E(X_2^2) - E^2(X_2) = \frac{3}{10} - \frac{1}{4} = \frac{1}{20}$$

One may similarly evaluate $E(X_1)$ and $V(X_1)$. ▲

Example 3.2.2 (Example 3.2.1 Continued) One easily verifies that

$$f_{1|2}(x_1) = 1/x_2 \text{ if } 0 < x_1 < x_2 \text{ and } 0 < x_2 < 1$$
$$f_{2|1}(x_2) = 2(1 - x_2)(1 - x_1)^{-2} \text{ if } x_1 < x_2 \leq 1 \text{ and } 0 < x_1 < 1$$

are respectively the two conditional pdfs. For fixed $0 < x_1 < 1$, the conditional mean of X_2 given that $X_1 = x_1$ is obtained as follows:

$$\mu_{2|1} = 2(1 - x_1)^{-2} \int_{x_1}^1 x_2(1 - x_2) dx_2 = \frac{1}{3}(1 - x_1)^{-1}(1 + x_1 - 2x_1^2)$$

One can evaluate the expression of $E[X_2^2 \mid X_1 = x_1]$ as:

$$2(1-x_1)^{-2} \int_{x_1}^{1} x_2^2(1-x_2)dx_2 = \tfrac{1}{6}(1-x_1)^{-1}(1+x_1+x_1^2-3x_1^3)$$

After some simplifications, we find:

$$\sigma_{2|1}^2 = V[X_2 \mid X_1 = x_1] = E[X_2^2 \mid X_1 = x_1] - \mu_{2|1}^2 = \tfrac{1}{18}(1-x_1)^2$$

Some details are left out. ▲

Anytime we wish to calculate the probability of an event $A(\subseteq \Re^2)$, we proceed as follows:

$$P(A) = \underset{A \cap (\mathcal{X}_1 \times \mathcal{X}_2)}{\int \int} f(x_1, x_2)dx_1 dx_2 \qquad (3.2.10)$$

Now, if we wish to evaluate the conditional probability of an event $B(\subseteq \Re)$ defined only through X_i given that $X_j = x_j$, then we would integrate the conditional pdf $f_{i|j}(x_i)$ over that part of the set B where $f_{i|j}(x_i)$ is positive. That is, one has

$$P(B \mid X_j = x_j) = \underset{B \cap \mathcal{X}_i}{\int} f_{i|j}(x_i)dx_i, i \neq j = 1, 2 \qquad (3.2.11)$$

Theorem 3.2.1 *Suppose that* $\mathbf{X} = (X_1, X_2)$ *has a bivariate pdf* $f(x_1, x_2)$ *with* $x_i \in \mathcal{X}_i$, *the support of* $X_i, i = 1, 2$. *Let* $E_1[.]$ *and* $V_1[.]$ *respectively denote the expectation and variance of* X_1. *Then, we have*

(i) $E[X_2] = E_1[E_{2|1}\{X_2 \mid X_1\}];$

(ii) $V(X_2) = V_1[E_{2|1}\{X_2 \mid X_1\}] + E_1[V_{2|1}\{X_2 \mid X_1\}].$

Example 3.2.3 Suppose that conditionally $X_1 = x_1$, the random variable X_2 is distributed as $N(\beta_0 + \beta_1 x_1, x_1^2)$ for any fixed $x_1 \in \Re$. Here, β_0, β_1 are two fixed real numbers. Suppose also that marginally, X_1 is distributed as $N(3, 10)$. How should one find $E[X_2]$ and $V[X_2]$? Theorem 3.2.1(i) will immediately imply:

$$E[X_2] = E_1[E_{2|1}\{X_2 \mid X_1\}] = E[\beta_0 + \beta_1 X_1] = \beta_0 + \beta_1 E[X_1]$$

which reduces to $\beta_0 + 3\beta_1$. Similarly, Theorem 3.2.1(ii) will imply:

$$V[X_2] = V[\beta_0 + \beta_1 X_1] + E[X_1^2] = \beta_1^2 V[X_1] + \{V[X_1] + E^2[X_1]\}$$

which reduces to $10\beta_1^2 + 19$. Here, the marginal distribution of X_2 is referred to as a *compound distribution*. One notes that we found the mean and variance of X_2 without obtaining its marginal distribution. ▲

3.2.2   Three and Higher Dimensions

The concepts from Section 3.2.1 extend easily in a *multivariate* situation. For example, let us suppose that $\mathbf{X} = (X_1, X_2, X_3, X_4)$ has a joint pdf $f(x_1, x_2, x_3, x_4)$ with $x_i \in \mathcal{X}_i$, the support of $X_i, i = 1, 2, 3, 4$. That is, $f(x_1, x_2, x_3, x_4) > 0$ whenever $\mathbf{x} \in \Pi_{i=1}^4 \mathcal{X}_i$, zero elsewhere, and

$$\int_{\mathcal{X}_4} \int_{\mathcal{X}_3} \int_{\mathcal{X}_2} \int_{\mathcal{X}_1} f(x_1, x_2, x_3, x_4) \Pi_{i=1}^4 dx_i = 1$$

The marginal pdf of (X_1, X_3), that is, the joint pdf of X_1 and X_3, for example, can be found as follows:

$$f_{1,3}(x_1, x_3) = \int_{\mathcal{X}_4} \int_{\mathcal{X}_2} f(x_1, x_2, x_3, x_4) dx_2 dx_4$$

The conditional pdf of (X_2, X_4) given that $X_1 = x_1, X_3 = x_3$ will be:

$$f_{2,4|1,3}(x_2, x_4) = \frac{f(x_1, x_2, x_3, x_4)}{f_{1,3}(x_1, x_3)} \quad \text{when } f_{1,3}(x_1, x_3) > 0$$

with $x_i \in \mathcal{X}_i, i = 1, ..., 4$. Notions of expectations and conditional expectations are incorporated in a natural way when $\mathbf{X}$ is multidimensional.

Theorem 3.2.2 *Suppose that* $\mathbf{X} = (X_1, ..., X_k)$ *is a k-dimensional discrete or continuous random variable. Also, suppose that we have real valued functions* $h_i(\mathbf{x})$ *and constants* $a_i, i = 0, 1, ..., p$. *Then, we have*

$$E\{a_0 + \Sigma_{i=1}^p a_i h_i(\mathbf{X})\} = a_0 + \Sigma_{i=1}^p a_i E\{h_i(\mathbf{X})\}$$

that is, expectation is a linear operation as long as all the expectations involved are finite.

Example 3.2.4 Let us denote $\mathcal{X}_1 = \mathcal{X}_3 = (0, 1), \mathcal{X}_2 = (0, 2)$ and define

$$a_1(x_1) = \begin{cases} 2x_1 & \text{if } 0 < x_1 < 1 \\ 0 & \text{otherwise} \end{cases} \qquad a_2(x_2) = \begin{cases} \frac{3}{8}x_2^2 & \text{if } 0 < x_2 < 2 \\ 0 & \text{otherwise} \end{cases}$$

$$a_3(x_3) = \begin{cases} x_3 + 2x_3^3 & \text{if } 0 < x_3 < 1 \\ 0 & \text{otherwise} \end{cases} \tag{3.2.12}$$

These are nonnegative functions and $\int_{\mathcal{X}_i} a_i(x_i) dx_i = 1, i = 1, 2, 3$. With $\mathbf{x} = (x_1, x_2, x_3)$ and $\mathcal{X} = \Pi_{i=1}^3 \mathcal{X}_i$, let us denote

$$f(\mathbf{x}) = \begin{cases} \frac{1}{3}a_1(x_1)a_2(x_2) + \frac{1}{3}a_2(x_2)a_3(x_3) \\ \quad + \frac{1}{6}a_3(x_3)a_1(x_1) & \text{if } \mathbf{x} \in \mathcal{X} \\ 0 & \text{otherwise} \end{cases} \tag{3.2.13}$$

One can easily verify:

$$\int_{x_3=0}^1 \int_{x_2=0}^2 \int_{x_1=0}^1 f(\mathbf{x}) \Pi_{i=1}^3 dx_i = 1$$

and observe that $f(\mathbf{x})$ is nonnegative for all $\mathbf{x} \in \Re^3$. In other words, $f(\mathbf{x})$ is a pdf of a three-dimensional random variable $\mathbf{X} = (X_1, X_2, X_3)$. The joint pdf of X_1, X_3 is then given by

$$
\begin{aligned}
g(x_1, x_3) \\
= \int_{x_2=0}^{2} f(\mathbf{x}) dx_2 \\
= \tfrac{1}{3}\{a_1(x_1) + a_3(x_3) + a_3(x_3)a_1(x_1)\}, x_i \in \mathcal{X}_i, i = 1,3
\end{aligned}
$$

One may directly check by double integration that $g(x_1, x_3)$ is a genuine pdf with its support being $\mathcal{X}_1 \times \mathcal{X}_3$. For any $0 < x_2 < 2$, the conditional pdf of X_2 given that $X_1 = x_1, X_3 = x_3$ is then given by:

$$
f_{2|1,3}(x_2) = f(\mathbf{x})/g(x_1, x_3) \text{ with fixed } 0 < x_1, x_3 < 1
$$

Also, for any $0 < x_3 < 1$, the marginal pdf of X_3 is given by:

$$
g_3(x_3) = \int_{x_1=0}^{1} g(x_1, x_3) dx_1 = \tfrac{1}{3}\{1 + 2a_3(x_3)\} \text{ for any } 0 < x_3 < 1
$$

Utilizing the expressions of $g(x_1, x_3), g_3(x_3)$, for any $0 < x_1 < 1$, we may write down the conditional pdf of X_1 given that $X_3 = x_3$ as follows:

$$
g_{1|3}(x_1) = g(x_1, x_3)/g_3(x_3) \text{ for fixed } 0 < x_3 < 1
$$

One may easily evaluate, for example, $E(X_1 X_3)$ or $E(X_1 X_3^2)$ respectively as double integrals:

$$
\int_{x_3=0}^{1} \int_{x_1=0}^{1} x_1 x_3 g(x_1, x_3) dx_1 dx_3 \text{ or } \int_{x_3=0}^{1} \int_{x_1=0}^{1} x_1 x_3^2 g(x_1, x_3) dx_1 dx_3 \blacktriangle
$$

3.2.3   Discrete Distributions

The previous concepts carry over easily in the case of bivariate or multivariate problems in *discrete* scenarios. However, a *joint probability mass function* (joint pmf) and a conditional pmf will now be interpreted as *joint probability* and *conditional* probability, respectively.

Example 3.2.5 We toss a fair coin twice and define $U_i = 1$ or 0 if the ith toss results in a head (H) or tail (T), $i = 1, 2$. We denote $X_1 = U_1 + U_2$ and $X_2 = U_1 - U_2$. One can verify that X_1 takes the values $0, 1,$ and 2 with the respective probabilities $\tfrac{1}{4}, \tfrac{1}{2},$ and $\tfrac{1}{4}$, whereas X_2 takes the values $-1, 0,$ and 1 with the respective probabilities $\tfrac{1}{4}, \tfrac{1}{2},$ and $\tfrac{1}{4}$. How about studying the random variables (X_1, X_2) together? Naturally, the pair (X_1, X_2) takes one of the nine $(= 3^2)$ possible pairs of vector values:

$(0, -1), (0, 0), (0, 1), (1, -1), (1, 0), (1, 1), (2, -1), (2, 0),$ and $(2, 1)$.

Table 3.2.1. Joint Distribution of X_1 and X_2

		X_1 values			Row Total
		0	1	2	
	-1	0	0.25	0	0.25
X_2 values	0	0.25	0	0.25	0.50
	1	0	0.25	0	0.25
Col. Total		0.25	0.50	0.25	1.00

Note that $X_1 = 0$ implies that we must have observed TT, in other words $U_1 = U_2 = 0$, so that $P(X_1 = 0 \cap X_2 = 0) = P(TT) = \frac{1}{4}$, whereas $P(X_1 = 0 \cap X_2 = -1) = P(X_1 = 0 \cap X_2 = 1) = 0$. The entries in Table 3.2.1 provide the joint probabilities, $P(X_1 = i \cap X_2 = j)$ for all $i = 0, 1, 2$ and $j = -1, 0, 1$.

We have $\mathcal{X}_1 = \{0, 1, 2\}, \mathcal{X}_2 = \{-1, 0, 1\}$ and the joint pmf is summarized as $f(x_1, x_2) = 0$ when $(x_1, x_2) = (0, -1), (0, 1), (1, 0), (2, -1), (2, 1)$, but $f(x_1, x_2) = 0.25$ when $(x_1, x_2) = (0, 0), (1, -1), (1, 1), (2, 0)$. The marginal pmf of X_1 is:

$$f_1(0) = \sum_{x_2 \in \mathcal{X}_2} P\{X_1 = 0 \cap X_2 = x_2\} = 0 + 0.25 + 0 = 0.25$$

$$f_1(1) = \sum_{x_2 \in \mathcal{X}_2} P\{X_1 = 1 \cap X_2 = x_2\} = 0.25 + 0 + 0.25 = 0.50$$

$$f_1(2) = \sum_{x_2 \in \mathcal{X}_2} P\{X_1 = 2 \cap X_2 = x_2\} = 0 + 0.25 + 0 = 0.25$$

which match with the column totals in the table. Similarly, row totals in Table 3.2.1 line up with the marginal pmf of X_2. ▲

Example 3.2.6 Consider two random variables X_1 and X_2 whose joint distribution has been provided in Table 3.2.2.

Table 3.2.2. Joint Distribution of X_1 and X_2

		X_1 values			Row Total
		-1	2	5	
	1	0.12	0.18	0.25	0.55
X_2 values					
	2	0.20	0.09	0.16	0.45
Col. Total		0.32	0.27	0.41	1.00

One has $\mathcal{X}_1 = \{-1, 2, 5\}, \mathcal{X}_2 = \{1, 2\}$. The marginal pmfs of X_1, X_2 are respectively given by $f_1(-1) = 0.32, f_1(2) = 0.27, f_1(5) = 0.41$, and $f_2(1) =$

0.55, $f_2(2) = 0.45$. The conditional pmf of X_1 given that $X_2 = 1$, for example, is given by $f_{1|2}(-1) = \frac{12}{55}, f_{1|2}(2) = \frac{18}{55}, f_{1|2}(5) = \frac{25}{55}$. We can apply the notion of the conditional expectation to evaluate $E[X_1 \mid X_2 = 1]$:

$$E[X_1 \mid X_2 = 1] = (-1)(\tfrac{12}{55}) + (2)(\tfrac{18}{55}) + (5)(\tfrac{25}{55}) = \tfrac{149}{55} \approx 2.7091$$

We also have

$$E[X_1] = (-1)(0.32) + (2)(0.27) + (5)(0.41) = 2.27$$

Suppose that $g(x_1, x_2) = x_1 x_2$ and we want to evaluate $E[g(X_1, X_2)]$:

$$E[g(X_1, X_2)] = (-1)(1)(0.12) + (-1)(2)(0.20) + (2)(1)(0.18)$$
$$+(2)(2)(0.09) + (5)(1)(0.25) + (5)(2)(0.16) = 3.05$$

Instead, if we had $h(x_1, x_2) = x_1^2 x_2$, for example, then one can similarly check that $E[h(X_1, X_2)] = 16.21$. ▲

Now, we introduce a multivariate discrete distribution which appears in statistical literature often. It is the **multinomial distribution** which plays a key role in categorical data analysis and other areas.

A discrete random variable $\mathbf{X} = (X_1, ..., X_k)$ is said to have a *multinomial distribution* involving n and $p_1, ..., p_k$, denoted by $\text{Mult}_k(n, p_1, ..., p_k)$, if the joint pmf $P(X_1 = x_1, ..., X_k = x_k)$ is given by:

$$f(\mathbf{x}) = \frac{n!}{\Pi_{i=1}^{k}(x_i!)} \Pi_{i=1}^{k} p_i^{x_i} \text{ for } x_i = 0, 1, ..., n, 0 < p_i < 1,$$

$$i = 1, ..., k, \ \Sigma_{i=1}^{k} x_i = n \text{ and } \Sigma_{i=1}^{k} p_i = 1$$

(3.2.14)

The fact that the function $f(\mathbf{x})$ given by Equation (3.2.14) is a genuine multivariate pmf follows immediately from the multinomial theorem.

> The *marginal distribution* of X_i is Binomial(n, p_i)
> $\Rightarrow E(X_i) = np_i$ and $V(X_i) = np_i(1 - p_i)$
> for each fixed but otherwise arbitrary $i = 1, ..., k$.

(3.2.15)

Example 3.2.7 Suppose that we roll a fair die 20 times. Let us define X_i = number of times the die lands up with the face with i dots on it, $i = 1, ..., 6$. It is not hard to see that *marginally*, each X_i has a Binomial $(20, \frac{1}{6})$ distribution, $i = 1, ..., 6$. It is clear, however, that $\Sigma_{i=1}^{6} X_i$ must be 20. That is, $X_1, ..., X_6$ are not all free-standing random variables! In fact, $\mathbf{X} = (X_1, ..., X_6)$ has a $\text{Mult}_6(20, p_1, ..., p_6)$ distribution with $p_1 = ... = p_6 = \frac{1}{6}$. ▲

Theorem 3.2.3 *Suppose* $\mathbf{X} = (X_1, ..., X_k)$ *has a* $\text{Mult}_k(n, p_1, ..., p_k)$ *distribution. Then, any subset of* $\mathbf{X}$ *variables of size* r, *namely* $(X_{i_1}, ..., X_{i_r})$, *has a multinomial distribution in the sense that* $(X_{i_1}, ..., X_{i_r}, X_{r+1}^*)$ *is*

$\text{Mult}_{r+1}(n, p_{i_1}, ..., p_{i_r}, 1 - \Sigma_{j=1}^r p_{i_j})$ where $1 \leq i_1 < i_2 < ... < i_r \leq k$ and $X_{r+1}^* = n - \Sigma_{j=1}^r X_{i_j}$.

Theorem 3.2.4 *Suppose* $\mathbf{X} = (X_1, ..., X_k)$ *has a* $\text{Mult}_k(n, p_1, ..., p_k)$ *distribution. Consider any subset of* X *variables of size* r, *namely* $(X_{i_1}, ..., X_{i_r})$. *Then, the conditional joint distribution of* $(X_{i_1}, ..., X_{i_r})$ *given all remaining* X *variables is also multinomial:*

$$P\{X_{i_1} = x_{i_1}, X_{i_2} = x_{i_2}, ..., X_{i_r} = x_{ir} \mid \sum_{l \neq i_1, i_2, ..., i_r}^{k} X_l = t\}$$

$$= \frac{(n-t)!}{\Pi_{j=1}^r x_{i_j}!} \Pi_{j=1}^r (p'_{i_j})^{x_{i_j}}, \quad x_{i_j} = 0, 1, ..., n - t, \Sigma_{j=1}^r x_{i_j} = n - t$$

with $0 \leq t \leq n$, $p'_{i_j} = p_{i_j}/\{\Sigma_{j=1}^r p_{i_j}\}$ *where* $1 \leq i_1 < i_2 < ... < i_r \leq k$.

Example 3.2.8 (Example 3.2.7 Continued) We are interested in counting how many times the faces with the numbers 1 and 5 land up. Our focus is on the three-dimensional random variable (X_1, X_5, X_3^*) where $X_3^* = n - (X_1 + X_5)$. By Theorem 3.2.3, (X_1, X_5, X_3^*) has a $\text{Mult}_3(20, \frac{1}{6}, \frac{1}{6}, \frac{2}{3})$ distribution. ▲

3.3   Covariances and Correlation Coefficient

Definition 3.3.1 *The covariance between random variables* X_1 *and* X_2, *denoted by* $Cov(X_1, X_2)$, *is defined as*

$$Cov(X_1, X_2) = E\left[(X_1 - \mu_1)(X_2 - \mu_2)\right] \tag{3.3.1}$$

where $\mu_i = E(X_i), i = 1, 2$.

It is clear that (i) $Cov(X_1, X_2) = Cov(X_2, X_1)$, (ii) $Cov(X_1, X_1) = V(X_1)$, and (iii) $Cov(X_1, X_2)$ can be alternatively expressed as:

$$Cov(X_1, X_2) = E(X_1 X_2) - E(X_1)E(X_2) \tag{3.3.2}$$

provided that the expectations $E(X_1 X_2), E(X_1)$, and $E(X_2)$ are finite.

Theorem 3.3.1 *Suppose that* $X_i, Y_i, i = 1, 2$ *are random variables. Then, we have:*

(i) $Cov(X_1, c) = 0$ *where* $c \in \Re$ *is fixed, if* $E(X_1)$ *is finite;*

(ii) $Cov(X_1 + X_2, Y_1 + Y_2) = Cov(X_1, Y_1) + Cov(X_1, Y_2)$
       $+ Cov(X_2, Y_1) + Cov(X_2, Y_2)$ *provided that* $Cov(X_i, Y_j)$
       is finite for $i, j = 1, 2$.

So, covariance is a bilinear operation, that is, it is linear in both coordinates.

Example 3.3.1 (Example 3.2.1 Continued) $E(X_1) = \int_0^1 x_1 f_1(x_1) dx_1 = 3\int_0^1 x_1(1-x_1)^2 dx_1 = \frac{1}{4}$ and we had $E(X_2) = \frac{1}{2}$. We obtain $E(X_1 X_2)$ as

$$\int_{x_2=0}^1 \int_{x_1=0}^{x_2} x_1 x_2 f(x_1, x_2) dx_1 dx_2 = 6 \left\{ \int_{x_1=0}^1 x_2(1-x_2) \int_{x_1=0}^{x_2} x_1 dx_1 \right\} dx_2$$

$$= 3 \int_{x_1=0}^1 x_2^3(1-x_2) dx_2 = \frac{3}{20}$$

So, $Cov(X_1, X_2) = E(X_1 X_2) - E(X_1)E(X_2) = \frac{3}{20} - (\frac{1}{4})(\frac{1}{2}) = \frac{1}{40}$. ▲

Example 3.3.2 (Example 3.2.6 Continued) We know that $E(X_1) = 2.27$ and also $E(X_1 X_2) = 3.05$. Similarly, we can find $E(X_2) = (1)(0.55) + 2(0.45) = 1.45$ and so, $Cov(X_1, X_2) = 3.05 - (2.27)(1.45) = -0.2415$. ▲

$Cov(X_1, X_2)$, if finite, may be any real number, positive, negative, or zero. Also, $Cov(X_1, X_2)$ has the same "unit" that may be associated with the random variable $X_1 X_2$. A standardized version of covariance, commonly known as *correlation coefficient*, was made popular by Karl Pearson. Refer to Stigler (1989) for a history of the invention of "correlation."

Definition 3.3.2 *The correlation coefficient between random variables X_1 and X_2, denoted by ρ_{X_1, X_2} or simply by ρ, is defined as follows:*

$$\rho \equiv \rho_{X_1, X_2} = \frac{Cov(X_1, X_2)}{\sigma_1 \sigma_2} \tag{3.3.3}$$

whenever one has $-\infty < Cov(X_1, X_2) < \infty, 0 < \sigma_1^2 (\equiv V[X_1]) < \infty$ and $0 < \sigma_2^2 (\equiv V[X_2]) < \infty$. The pair X_1, X_2 are called negatively correlated, uncorrelated, or positively correlated if ρ_{X_1, X_2} is negative, zero, or positive, respectively.

Theorem 3.3.2 *Consider random variables X_1 and X_2 for which we can assume that $-\infty < Cov(X_1, X_2) < \infty, 0 < V(X_1), V(X_2) < \infty$. Then, one has:*

(i)   *Let $Y_i = c_i + d_i X_i$ where $-\infty < c_i < \infty$ and $0 < d_i < \infty$ are fixed numbers with $i = 1, 2$. Then, $\rho_{Y_1, Y_2} = \rho_{X_1, X_2}$;*

(ii) $|\rho_{X_1, X_2}| \leq 1$;

(iii)   *In part (ii), the equality holds, that is, ρ_{X_1, X_2} is $+1$ or -1, if and only if X_1 and X_2 are linearly related. In other words, ρ_{X_1, X_2} is $+1$ (or -1) if and only if $X_1 = a + bX_2$ with probability (w.p.) one and some real numbers a and b.*

Example 3.3.3 (Example 3.3.2 Continued) One can check that $V(X_1) = 6.4971$ and $V(X_2) = 0.6825$. We also found earlier that $Cov(X_1, X_2) = -0.2415$. Using Equation (3.3.3), we get $\rho_{X_1, X_2} = Cov(X_1, X_2)/(\sigma_1 \sigma_2) = -0.2415/\sqrt{(6.4971)(0.6825)} \approx -0.11468$. ▲

Theorem 3.3.3 *Let $a_1, ..., a_k$ and $b_1, ..., b_k$ be arbitrary but fixed real numbers. Consider a k-dimensional random variable $\mathbf{X} = (X_1, ..., X_k)$. Then, we have:*

(i) $E\left\{\Sigma_{i=1}^{k}a_{i}X_{i}\right\} = \Sigma_{i=1}^{k}a_{i}E\{X_{i}\};$

(ii) $Cov(X_{i}, X_{j}) = E[X_{i}X_{j}] - E[X_{i}]E[X_{j}]$ for all $i, j;$

(iii) $V\left\{\Sigma_{i=1}^{k}a_{i}X_{i}\right\} = \Sigma_{i=1}^{k}a_{i}^{2}V\{X_{i}\} + 2\sum_{1\leq i<j\leq k}\sum a_{i}a_{j}Cov(X_{i}, X_{j});$

(iv) $Cov\left(\Sigma_{i=1}^{k}a_{i}X_{i}, \Sigma_{j=1}^{k}b_{j}X_{j}\right) = \Sigma_{i=1}^{k}\Sigma_{j=1}^{k}a_{i}b_{j}Cov(X_{i}, X_{j}).$

3.3.1 Multinomial Distribution

Suppose that $\mathbf{X} = (X_{1}, ..., X_{k})$ has a $Mult_{k}(n, p_{1}, ..., p_{k})$ distribution with the associated pmf given by Equation (3.2.14). Let us proceed to evaluate the covariance between X_{i} and X_{j} for all fixed $i \neq j = 1, ..., k.$

Consider n marbles which are tossed so that each marble lands in one and only one of k boxes. The probability that a marble lands in box #i is $p_{i}, i = 1, ..., k, \Sigma_{i=1}^{k}p_{i} = 1.$ Let X_{l} be the number of marbles landing in the box #$l, l = i, j,$ and define

$$Y_{l} = \begin{cases} 1 & \text{if } l^{th} \text{ marble lands in box } \#i \\ 0 & \text{otherwise} \end{cases}$$

$$Z_{l} = \begin{cases} 1 & \text{if } l^{th} \text{ marble lands in box } \#j \\ 0 & \text{otherwise} \end{cases}$$

We can express $X_{i} = \Sigma_{l=1}^{n}Y_{l}$ and $X_{j} = \Sigma_{l=1}^{n}Z_{l}$ and rewrite $Cov(X_{i}, X_{j})$ as

$$\Sigma_{l=1}^{n}\Sigma_{m=1}^{n}Cov(Y_{l}, Z_{m}) = nCov(Y_{1}, Z_{1}) = n\{E[Y_{1}Z_{1}] - E[Y_{1}]E[Z_{1}]\}.$$

It is easy to see that $E[Y_{1}Z_{1}] = 0, E[Y_{1}] = p_{i}, E[Z_{1}] = p_{j},$ and hence $Cov(X_{i}, X_{j}) = n\{0 - p_{i}p_{j}\} = -np_{i}p_{j}$ for all fixed $i \neq j = 1, ..., k.$

A negative covariance should intuitively make sense because more marbles in box #i would imply fewer marbles elsewhere!

3.4 Independence of Random Variables

Suppose that $\mathbf{X} = (X_{1}, ..., X_{k})$ has a joint pmf or pdf $f(x_{1}, ..., x_{k})$ with $x_{i} \in \mathcal{X}_{i}(\subseteq \Re), i = 1, ..., k.$

Definition 3.4.1 We say that $X_{1}, ..., X_{k}$ form a collection of independent random variables if and only if

$$f(x_{1}, ..., x_{k}) = \Pi_{i=1}^{k}f_{i}(x_{i}) \text{ for each } x_{i} \in \mathcal{X}_{i}, i = 1, ..., k \qquad (3.4.1)$$

Also, we say that $X_{1}, ..., X_{k}$ form a collection of dependent random variables if and only if these are not independent.

> In order to prove that $X_{1}, ..., X_{k}$ are dependent, one needs to show existence of a particular set of $x_{1}, ..., x_{k},$ with $x_{i} \in \mathcal{X}_{i},$ $i = 1, ..., k$ such that $f(x_{1}, ..., x_{k}) \neq \Pi_{i=1}^{k}f_{i}(x_{i}).$

Example 3.4.1 (Example 3.2.6 Continued) Recall that $P(X_1 = -1) = 0.32$, $P(X_2 = 1) = 0.55$ and $P(X_1 = -1 \cap X_2 = 1) = 0.12$. Hence, we have $P(X_1 = -1)P(X_2 = 1) \neq P(X_1 = -1 \cap X_2 = 1)$. That is, $f(-1, 1) \neq f_1(-1)f_2(1)$. Thus, X_1, X_2 are *dependent*. ▲

Example 3.4.2 (Example 3.2.1 Continued) With $x_1 = 0.4$, $x_2 = 0.5$, we have $f(x_1, x_2) = 3$, but $\Pi_{i=1}^2 f_i(x_i) = 1.62 \neq f(x_1, x_2)$. Thus, X_1, X_2 are *dependent*. ▲

Example 3.4.3 Consider X_1, X_2 with a joint pdf

$$f(x_1, x_2) = \begin{cases} 4x_1 x_2 & \text{if } 0 < x_1, x_2 < 1 \\ 0 & \text{elsewhere} \end{cases}$$

In this case, one can show that X_1, X_2 are *independent*. ▲

> It is possible that X_i, X_j are independent for all $i \neq j = 1, 2, 3$, but X_1, X_2, X_3 are dependent.

Example 3.4.4 Let $f_i(x_i)$ be nonnegative integrable functions, *not* identically equal to unity, for $0 < x_i < 1$ such that $\int_{x_i=0}^1 f_i(x_i)dx_i = 1$, $i = 1, 2, 3$. Define $g(x_1, x_2, x_3)$ as

$$\frac{1}{4}\{f_1(x_1)f_2(x_2) + f_2(x_2)f_3(x_3) + f_1(x_1)f_3(x_3) + 1\} \qquad (3.4.2)$$

for $0 < x_i < 1$, $i = 1, 2, 3$, and note that $g(x_1, x_2, x_3)$ is a bonafide pdf. Now, focus on $\mathbf{X} = (X_1, X_2, X_3)$ whose joint pdf is given by $g(x_1, x_2, x_3)$. One may verify that (i) X_i, X_j are *independent* for all $i \neq j = 1, 2, 3$, but (ii) X_1, X_2, X_3 are *dependent*. ▲

Theorem 3.4.1 *Suppose* $\mathbf{X}_1, ..., \mathbf{X}_p$ *are independent vector valued random variables, not necessarily all having the same dimension. Consider real valued functions* $g_i(\mathbf{x}_i)$, $i = 1, ..., p$. *Then, we have:*

$$E[\Pi_{i=1}^p g_i(\mathbf{X}_i)] = \Pi_{i=1}^p E[g_i(\mathbf{X}_i)]$$

as long as $E[g_i(\mathbf{X}_i)]$, $i = 1, ..., p$ *are finite.*

> A consequence of this theorem is: if two real valued random variables X_1, X_2 are *independent*, then $Cov(X_1, X_2)$, *when finite*, is necessarily zero. Additionally if the variances are finite, the correlation coefficient, ρ_{X_1, X_2}, is indeed zero.

Theorem 3.4.2 *Suppose that* $\mathbf{X}_1, \mathbf{X}_2$ *are independent vector valued random variables, not necessarily of the same dimension. Then,*

(i) $Cov(g_1(\mathbf{X}_1), g_2(\mathbf{X}_2)) = 0$ *where* $g_1(.), g_2(.)$ *are real valued functions, if* $E[g_1 g_2]$, $E[g_1]$ *and* $E[g_2]$ *are finite;*

(ii) $\mathbf{g}_1(\mathbf{X}_1), \mathbf{g}_2(\mathbf{X}_2)$ *are independent where* $\mathbf{g}_1(.), \mathbf{g}_2(.)$ *do not need to be restricted as real valued or one-to-one.*

> One may wonder whether the requirements of finite moments stated in Theorem 3.4.2(i) are crucial for the conclusion to hold. See the following example for the case in point.

Example 3.4.5 Suppose that X_1 is distributed as a standard normal variable with its pdf $f_1(x_1) = \{\sqrt{2\pi}\}^{-1} exp(-x_1^2/2)$ and X_2 is distributed as a Cauchy variable with its pdf $f_2(x_2) = \pi^{-1}(1+x_2^2)^{-1}$, $-\infty < x_1, x_2 < \infty$. Suppose also that these random variables are independent. But, we cannot claim that $Cov(X_1, X_2) = 0$ because $E[X_2]$ is *not* finite. ▲

Definition 3.4.2 *The supports $\mathcal{X}_1, ..., \mathcal{X}_k$ are called unrelated to each other if the following condition holds: consider any subset of the random variables, $X_{i_j}, j = 1, ..., p, p = 1, ..., k - 1$. Conditionally, given that $X_{i_j} = x_{i_j}$, the support of the remaining random variable X_l stays the same as $\mathcal{X}_l$ for any $l \neq i_j, j = 1, ..., p, p = 1, ..., k - 1$.*

Theorem 3.4.3 *Let $X_1, ..., X_k$ have a joint pmf or pdf $f(x_1, ..., x_k)$, $x_i \in \mathcal{X}_i, i = 1, ..., k$. Assume that the supports are unrelated to each other. Then, $X_1, ..., X_k$ are independent if and only if*

$$f(x_1, ..., x_k) = \Pi_{j=1}^{k} h_j(x_j) \text{ for all } x_i \in \mathcal{X}_i, i = 1, ..., k \qquad (3.4.3)$$

for some functions $h_1(x_1), ..., h_k(x_k)$.

Example 3.4.6 Suppose that the joint pdf of X_1, X_2, X_3 is:

$$cexp\{-2x_1 - \pi x_3^2\}/[\sqrt{x_1}(1+x_2^2)]I(x_1 \in \Re^+, (x_2, x_3) \in \Re^2)$$

where $c(> 0)$ is a constant. Let us denote

$$h_1(x_1) = \{\sqrt{x_1}\}^{-1} exp\{-2x_1\}, h_2(x_2) = (1 + x_2^2)^{-1},$$
$$h_3(x_3) = exp\{-\pi x_3^2\}$$

so that $f(x_1, x_2, x_3) = ch_1(x_1)h_2(x_2)h_3(x_3)$ for all $x_1 \in \Re^+, (x_2, x_3) \in \Re^2$. The supports, $\mathcal{X}_1 = \Re^+, \mathcal{X}_2 = \Re, \mathcal{X}_3 = \Re$, are also *unrelated*. Hence, X_1, X_2, and X_3 are independent. One notes that $h_1(x_1), h_2(x_2)$, and $h_3(x_3)$ are not probability densities! ▲

> Look at Example 3.4.7 to see what may happen when the supports are related!

Example 3.4.7 (Example 3.4.2 Continued) Recall that $\mathcal{X}_1 = \mathcal{X}_2 = (0, 1)$. Denote $h_1(x_1) = 6, h_2(x_2) = (1 - x_2)$ and we may feel tempted to claim that X_1, X_2 are independent. But, that will be wrong! On the whole space $\mathcal{X}_1 \times \mathcal{X}_2$, we cannot claim that $f(x_1, x_2) = h_1(x_1)h_2(x_2)$. One may check this by considering $x_1 = \frac{1}{2}, x_2 = \frac{1}{4}$ and then one has $f(x_1, x_2) = 0$ whereas $h_1(x_1)h_2(x_2) = \frac{9}{2}$. The relationship $f(x_1, x_2) = h_1(x_1)h_2(x_2)$ holds in the subspace where $0 < x_1 < x_2 < 1$. In Example 3.4.2, we verified that X_1, X_2 were *dependent*! ▲

3.5   Bivariate Normal Distribution

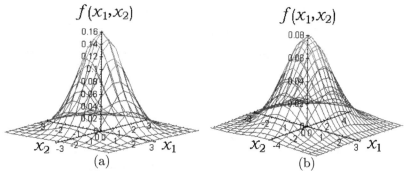

Figure 3.5.1. Bivariate normal density:
(a) $N_2(0,0,1,1,0)$; (b) $N_2(0,0,1,4,0)$.

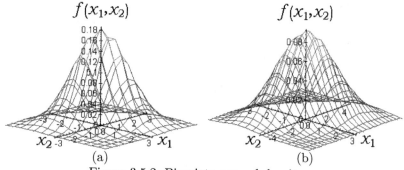

Figure 3.5.2. Bivariate normal density:
(a) $N_2(0,0,1,1,0.5)$; (b) $N_2(0,0,1,4,0.5)$.

Let (X_1, X_2) be a two-dimensional continuous random variable with a joint pdf:

$$f(x_1, x_2) = c \; exp\left[-\tfrac{1}{2}(1 - \rho^2)^{-1}\{u_1^2 - 2\rho u_1 u_2 + u_2^2\}\right] \qquad (3.5.1)$$

where

$$u_1 = (x_1 - \mu_1)/\sigma_1, u_2 = (x_2 - \mu_2)/\sigma_2$$
$$c = \{2\pi\sigma_1\sigma_2(1 - \rho^2)^{\frac{1}{2}}\}^{-1}, \quad -\infty < x_1, x_2 < \infty$$
$$-\infty < \mu_1, \mu_2 < \infty, \; 0 < \sigma_1, \sigma_2 < \infty, \; -1 < \rho < 1$$

The expression in Equation (3.5.1) is a *bivariate normal density* with parameters $\mu_1, \mu_2, \sigma_1^2, \sigma_2^2$ and ρ, denoted by $N_2(\mu_1, \mu_2, \sigma_1^2, \sigma_2^2, \rho)$. The pdf is centered at (μ_1, μ_2) in the x_1, x_2 plane which has been plotted in Figures 3.5.1 and 3.5.2. Tong's (1990) book may serve as a valuable resource, especially because it includes associated tables of percentage points.

Figure 3.5.2(a) appears more concentrated around its center than its counterpart in Figure 3.5.1(a). This is due to the fact that we used $\rho =$

0.5 in the former picture and $\rho = 0$ in the latter picture while the other parameters were held fixed.

Theorem 3.5.1 *Suppose that* (X_1, X_2) *has the bivariate normal distribution* $N_2(\mu_1, \mu_2, \sigma_1^2, \sigma_2^2, \rho)$. *Then, one has:*

(i)   *the marginal distribution of* X_i *is given by* $N(\mu_i, \sigma_i^2), i = 1, 2;$

(ii)   *the conditional distribution of* $X_i \mid X_j = x_j$ *is normal with*

mean $\mu_i + \dfrac{\rho\sigma_i}{\sigma_j}(x_j - \mu_j)$ *and variance* $\sigma_i^2(1 - \rho^2)$, *for all*

fixed $x_j \in \Re, i \neq j = 1, 2.$

The conditional mean of $X_i \mid X_j = x_j$, that is, $\mu_i + \dfrac{\rho\sigma_i}{\sigma_j}(x_j - \mu_j)$, is also known as the *regression function* of X_i on $X_j, i \neq j = 1, 2$. We note that the regression functions in the case of a bivariate normal distribution are straight lines, namely $(\mu_i - \dfrac{\rho\sigma_i}{\sigma_j}\mu_j) + \dfrac{\rho\sigma_i}{\sigma_j}x_j, i \neq j = 1, 2.$

> Marginal distributions of X_1, X_2 may be normal, but this does
> not necessarily imply that (X_1, X_2) is distributed as N_2.

Example 3.5.1 It is not difficult to construct dependent random variables X_1, X_2 such that marginally each is normally distributed, but jointly (X_1, X_2) is not distributed as N_2. We write $f(x_1, x_2; \mu_1, \mu_2, \sigma_1^2, \sigma_2^2, \rho)$ for the pdf in Equation (3.5.1). Next, we fix $0 < \alpha, \rho < 1$ and define $g(x_1, x_2; \rho)$ as

$$\alpha\, f(x_1, x_2; 0, 0, 1, 1, \rho) + (1 - \alpha)\, f(x_1, x_2; 0, 0, 1, 1, -\rho) \qquad (3.5.2)$$

for $x_1, x_2 \in \Re$. We can express

$$\int\int_{\Re^2} g(x_1, x_2; \rho)dx_1 dx_2 = \alpha(1) + (1 - \alpha)(1) = 1$$

Also, $g(x_1, x_2; \rho)$ is nonnegative for all $(x_1, x_2) \in \Re^2$. Thus, $g(x_1, x_2; \rho)$ is a genuine pdf on $\Re^2$.

Let X_1, X_2 be the random variables whose joint pdf is $g(x_1, x_2; \rho)$ from Equation (3.5.2). By direct integration, one can verify that marginally, both X_1 and X_2 are indeed distributed as a standard normal variable, but $g(x_1, x_2; \rho)$ does not correspond to a bivariate normal pdf! ▲

3.6   Correlation Coefficient and Independence

Theorem 3.6.1 *Suppose that* (X_1, X_2) *has the bivariate normal distribution* $N_2(\mu_1, \mu_2, \sigma_1^2, \sigma_2^2, \rho)$ *with* $-\infty < \mu_1, \mu_2 < \infty, 0 < \sigma_1, \sigma_2 < \infty$ *and* $-1 < \rho < 1$. *Then,* X_1 *and* X_2 *are distributed independently if and only if the correlation coefficient* $\rho = 0$.

> However, a zero correlation coefficient alone between two arbitrary random variables does not necessarily imply that these variables are independent.

Example 3.6.1 Suppose that X_1 is $N(0,1)$ and let $X_2 = X_1^2$. Then, we have $Cov(X_1, X_2) = E(X_1^3) - E(X_1)E(X_2) = 0$, since $E(X_1) = 0$ and $E(X_1^3) = 0$. That is, ρ_{X_1,X_2} is zero. But the fact that X_1, X_2 are dependent can be easily verified as follows: one can surely claim that $P\{X_2 > 4\} > 0$, however, the conditional probability, $P\{X_2 > 4 \mid -2 \le X_1 \le 2\}$, happens to be zero. Thus, we note that

$$P\{X_2 > 4 \mid -2 \le X_1 \le 2\} \ne P\{X_2 > 4\}$$

Hence, X_1, X_2 are dependent random variables. ▲

3.7   Exponential Family

The *exponential family of distributions* is very rich for statistical modeling. Interesting properties will be emphasized in Chapter 6 and beyond. Let the random variable X have its pmf or pdf given by $f(x; \boldsymbol{\theta})$, $x \in \mathcal{X} \subseteq \Re$, $\boldsymbol{\theta} = (\theta_1, ..., \theta_k) \in \Theta \subseteq \Re^k$. Here, $\boldsymbol{\theta}$ is a vector valued parameter with k components. Note that $\mathbf{X}, \mathbf{x}$ may be vector valued too.

Definition 3.7.1 *We say that $f(x; \boldsymbol{\theta})$ belongs to a k-parameter exponential family if and only if*

$$f(x; \boldsymbol{\theta}) = a(\boldsymbol{\theta})g(x)exp\{\Sigma_{i=1}^k b_i(\boldsymbol{\theta})R_i(x)\} \qquad (3.7.1)$$

with some appropriate forms for $g(x) \ge 0$, $a(\boldsymbol{\theta}) \ge 0$, $b_i(\boldsymbol{\theta})$ and $R_i(x)$, $i = 1, ..., k$, $x \in \mathcal{X}$ and $\boldsymbol{\theta} \in \Theta$. It is crucial to note that $a(\boldsymbol{\theta}), b_i(\boldsymbol{\theta}), i = 1, ..., k$, cannot involve x, and $g(x), R_1(x), ..., R_k(x)$ cannot involve $\boldsymbol{\theta}$.

In order to involve statistically meaningful parameterizations, one would **assume** the following conditions:

Condition 1: Neither b_i nor R_i satisfy linear constraints;

Condition 2: Θ contains a k-dimensional rectangle. $\qquad (3.7.2)$

Many situations call for a *one-parameter exponential family*. See Table 3.7.1. A pmf or pdf $f(x; \theta)$ from a *one-parameter exponential family* looks like:

$$a(\theta)g(x)exp\{b(\theta)R(x)\} \text{ for } x \in \mathcal{X} \text{ and } \theta \in \Theta \qquad (3.7.3)$$

with appropriate real valued functions $a(\theta) \ge 0, b(\theta)$, $g(x) \ge 0$ and $R(x)$. Here, we assume that $\mathcal{X} \subseteq \Re$ and Θ is a subinterval of $\Re$. Again, $a(\theta), b(\theta)$ cannot involve x and $g(x), R(x)$ cannot involve θ.

Table 3.7.1. Selected Members of a One-Parameter
Exponential Family Defined in Equation (3.7.3)

θ	$a(\theta)$	$b(\theta)$	$g(x)$	$R(x)$
		Bernoulli(p)		
p	$1-\theta$	$log\left(\frac{\theta}{1-\theta}\right)$	1	x
		Binomial(n,p)		
p	$(1-\theta)^n$	$log\left(\frac{\theta}{1-\theta}\right)$	$\binom{n}{x}$	x
		Poisson(λ)		
λ	$e^{-\theta}$	$log(\theta)$	$(x!)^{-1}$	x
		NB$(\mu,k), k$ known		
μ	$\left(\frac{k}{\theta+k}\right)^k$	$log\left(\frac{\theta}{\theta+k}\right)$	$\binom{k+x-1}{k-1}$	x
		Geometric(p)		
p	$\theta(1-\theta)^{-1}$	$log(1-\theta)$	1	x
		N$(\mu,\sigma^2), \mu$ known		
σ	$\frac{1}{\theta}exp\left\{-\frac{\mu^2}{2\theta^2}\right\}$	θ^{-2}	$\{2\pi\}^{-\frac{1}{2}}$	$-\frac{1}{2}x^2$ $+\mu x$
		N$(\mu,\sigma^2), \sigma$ known		
μ	$\frac{1}{\sqrt{2\pi}}exp\left\{-\frac{\theta^2}{2\sigma^2}\right\}$	$\frac{\theta}{\sigma^2}$	$\frac{1}{\sigma}exp\left\{-\frac{x^2}{2\sigma^2}\right\}$	x
		Gamma$(\alpha,\beta), \alpha$ known		
β	$\{\theta^\alpha\Gamma(\alpha)\}^{-1}$	$-\theta^{-1}$	$x^{\alpha-1}$	x
		Gamma$(\alpha,\beta), \beta$ known		
α	$\{\beta^\theta\Gamma(\theta)\}^{-1}$	θ	$\frac{1}{x}exp\{-\frac{x}{\beta}\}$	$log(x)$

Example 3.7.1 Let X be distributed as $N(\mu,\sigma^2)$ with $k=2, \boldsymbol{\theta}=(\mu,\sigma) \in \Re \times \Re^+$. Then, the pdf of X has the same form as in Equation (3.7.1) with $x \in \Re$, $\theta_1 = \mu$, $\theta_2 = \sigma$, $R_1(x) = x, R_2(x) = x^2, b_1(\boldsymbol{\theta}) = \theta_1\theta_2^{-2}, b_2(\boldsymbol{\theta}) = -\frac{1}{2}\theta_2^{-2}, a(\boldsymbol{\theta}) = \{2\pi\theta_2^2\}^{-\frac{1}{2}}exp\{-\theta_1^2(2\theta_2^2)^{-1}\}, g(x) = 1.$ ▲

Example 3.7.2 Let X be distributed as Poisson(λ), $\lambda > 0$. With $\mathcal{X} = \{0,1,2,...\}$, $\theta = \lambda$, and $\Theta = (0,\infty)$, the pmf of X resembles Equation (3.7.3) where

$$g(x) = (x!)^{-1}, a(\theta) = exp\{-\theta\}, b(\theta) = log(\theta), R(x) = x. ▲$$

Example 3.7.3 Suppose that X is distributed as $N(\mu,1)$, $\mu \in \Re$. With $\mathcal{X} = \Re$, $\theta = \mu$, and $\Theta = \Re$, the pdf of X resembles Equation (3.7.3) where

$$g(x) = exp\{-\frac{1}{2}x^2\}, a(\theta) = \{\sqrt{2\pi}\}^{-1}exp\{-\frac{1}{2}\theta^2\}, b(\theta) = \theta, R(x) = x. ▲$$

One should not expect that every probability distribution
would necessarily belong to an exponential family.

Example 3.7.4 Consider X having a negative exponential pdf, namely, $\theta^{-1}exp\{-(x-\theta)/\theta\}I(x>\theta)$ with $\theta>0$. The term $I(x>\theta)$ cannot be absorbed within $a(\theta), b(\theta), g(x)$ or $R(x)$, and so this distribution does not belong to a one-parameter exponential family. ▲

Example 3.7.5 Suppose that X has a $N(\theta, \theta^2)$ distribution with $\theta(>0)$. The pdf $f(x;\theta)$ can be expressed as

$$\{\theta\sqrt{2\pi}\}^{-1}exp\{-\tfrac{1}{2}(x-\theta)^2/\theta^2\} = \{\sqrt{2\pi e}\}^{-1}\theta^{-1}exp\left\{-\frac{x^2}{2\theta^2}+\frac{x}{\theta}\right\}$$

which does not have the same form as in Equation (3.7.3) and it does not belong to a one-parameter exponential family. ▲

The distribution in the Example 3.7.5 belongs to a *curved exponential family,* introduced by Efron (1975, 1978). We should mention that there are notions referred to as *natural parameterization,* a *natural parameter space,* and a *natural exponential family.* A serious discussion of these topics needs substantial mathematical understanding beyond the scope of this book. References include Lehmann (1983, 1986) and Lehmann and Casella (1998), as well as Barndorff-Nielson (1978).

3.8   Selected Probability Inequalities

These inequalities apply to both discrete and continuous random variables. One will see applications in later chapters.

Markov Inequality: *Suppose that W is a real valued random variable such that $P(W \geq 0) = 1$ and $E(W)$ is finite. Then, for any fixed $\delta\ (>0)$, one has:*

$$P(W \geq \delta) \leq \delta^{-1}E(W) \qquad (3.8.1)$$

Tchebysheff's Inequality: *Suppose that X is a real valued random variable with a finite second moment. We denote its mean μ and variance $\sigma^2(>0)$. Then, for any fixed real number $\varepsilon(>0)$, one has*

$$P\{|X-\mu| \geq \varepsilon\} \leq \sigma^2/\varepsilon^2 \qquad (3.8.2)$$

Thus, with $k > 0$, if we substitute $\varepsilon = k\sigma$ in Equation (3.8.2), we can immediately conclude:

$$P\{|X-\mu| < k\sigma\} \geq 1 - k^{-2} \qquad (3.8.3)$$

Sometimes Equation (3.8.3) is referred to as Tchebysheff's Inequality.

Cauchy-Schwartz Inequality: *Suppose that X_1, X_2 are real valued random variables such that $E[X_1^2], E[X_2^2]$ and $E[X_1X_2]$ are finite. Then, we have*

$$E^2[X_1X_2] \leq E[X_1^2]E[X_2^2] \qquad (3.8.4)$$

In Equation (3.8.4), the equality holds if and only if $X_1 = kX_2$ w.p.1 for some constant k.

This inequality can be used to claim that

$$Cov^2(X_1, X_2) \leq V[X_1]V[X_2] \qquad (3.8.5)$$

which is referred to as the Covariance Inequality. The equality in Equation (3.8.5) holds if and only if $X_1 = a + bX_2$ w.p.1 with constants a, b.

Jensen's Inequality: *Suppose that X is a real valued random variable and $g(x), x \in \Re$ is a convex function. We assume that $E[X]$ is finite. Then, one has*

$$E[g(X)] \geq g(E[X]) \qquad (3.8.6)$$

> In Jensen's Inequality, that is, in Equation (3.8.6), equality holds when $g(x)$ is linear in x or X is degenerate. The inequality in Equation (3.8.6) is reversed when $g(x)$ is concave.

Example 3.8.1 Suppose X is Poisson$(\lambda), \lambda > 0$. Then, $E[X^3] > \lambda^3$ because $g(x) = x^3, x \in \Re^+$ is convex and $E[X] = \lambda$. ▲

Example 3.8.2 Suppose X is $N(-1, \sigma^2), \sigma^2 > 0$. Then, $E[|X|] > 1$ because $g(x) = |x|, x \in \Re$ is convex. ▲

Bonferroni Inequality: *Suppose that $\mathcal{B}$ is a Borel sigma-field of subsets of $\mathbf{S}$. Let $A_1, ..., A_k$ be events from $\mathcal{B}$. Then,*

$$P\{\cap_{i=1}^{k} A_i\} \geq \Sigma_{i=1}^{k} P(A_i) - (k - 1) \qquad (3.8.7)$$

3.9   Exercises and Complements

3.2.1 Let X_1, X_2 have a joint pdf

$$f(x_1, x_2) = \begin{cases} 4x_1 x_2 & \text{if } 0 < x_1, x_2 < 1 \\ 0 & \text{elsewhere} \end{cases}$$

(i)   Evaluate $E(X_i)$ and $V(X_i), i = 1, 2$;
(ii)   Evaluate $E[X_1(1 - X_1)]$ and $E[(X_1 + X_2)^2]$;
(iii)   Show that $f_{i|j}(x_i) = f_i(x_i)$ for $i \neq j = 1, 2$.

3.2.2 Let X_1, X_2 have a joint pdf

$$f(x_1, x_2) = \begin{cases} 1 & \text{if } 0 < x_1, x_2 < 1 \\ 0 & \text{elsewhere} \end{cases}$$

Show that $P\{X_1 X_2 > a\} = 1 - a + a \log(a)$ for any $0 < a < 1$.

3.2.3 With $c(>0)$, let X_1, X_2 have a joint pdf

$$f(x_1, x_2) = \begin{cases} c exp\{-2x_1 - \frac{1}{2}x_2\} & \text{if } 0 < x_1, x_2 < \infty \\ 0 & \text{elsewhere} \end{cases}$$

Find c. Derive the expressions of $f_1(x_1)$, $f_2(x_2)$, $f_{1|2}(x_1)$, and $f_{2|1}(x_2)$. Does either X_1, X_2 have a standard distribution?

3.2.4 Let X_1, X_2 have a joint pdf

$$f(x_1, x_2) = \begin{cases} 3x_1 & \text{if } 0 < x_2 < x_1 < 1 \\ 0 & \text{elsewhere} \end{cases}$$

Find $f_1(x_1)$, $f_2(x_2)$, $f_{2|1}(x_2)$ and evaluate $P\{X_2 < 0.2 \mid X_1 = 0.5\}$.

3.2.5 Conditionally given $X_1 = x_1$, a random variable X_2 is distributed as Poisson(x_1) for any fixed $x_1 \in \Re^+$. Marginally, X_1 is distributed as Gamma$(\alpha = 3, \beta = 10)$. Obtain $E[X_2]$ and $V[X_2]$. {*Hint*: Use Theorem 3.2.1.}

3.2.6 With $c(>0)$, let X_1, X_2, X_3 have a joint pdf given by

$$f(x_1, x_2, x_3) = \begin{cases} c exp\{-2x_1 - 4x_2^2 - \frac{1}{2}x_3\} & \text{if } 0 < x_1, x_3 < \infty, \\ & \quad -\infty < x_2 < \infty \\ 0 & \text{elsewhere} \end{cases}$$

Find c and marginal pdfs of X_1, X_2, X_3. Does either X_1, X_2, X_3 have a standard distribution? Evaluate $E[(X_1 + X_2 + X_3)^2]$.

3.2.7 (Example 3.2.4 Continued) Denote $\mathcal{X}_1 = \mathcal{X}_3 = \mathcal{X}_4 = \mathcal{X}_5 = (0, 1)$, $\mathcal{X}_2 = (0, 2)$. Recall functions $a_1(x_1), a_2(x_2)$, and $a_3(x_3)$ from Equation (3.2.12). Additionally, denote

$$a_4(x_4) = \begin{cases} 1 & \text{if } 0 < x_4 < 1 \\ 0 & \text{otherwise} \end{cases} \qquad a_5(x_5) = \begin{cases} 4x_5^3 & \text{if } 0 < x_5 < 1 \\ 0 & \text{otherwise} \end{cases}$$

With $\mathbf{x} = (x_1, x_2, x_3, x_4, x_5)$ and $\mathcal{X} = \Pi_{i=1}^5 \mathcal{X}_i$, let us denote

$$f(\mathbf{x}) = \begin{cases} \frac{1}{5}a_1(x_1)a_2(x_2) + \frac{1}{5}a_2(x_2)a_3(x_3) \\ \quad + \frac{1}{10}\{a_3(x_3)a_1(x_1) + a_4(x_4) + a_4(x_4)a_5(x_5)\} & \text{if } \mathbf{x} \in \mathcal{X} \\ 0, & \text{otherwise} \end{cases}$$

Show that $f(\mathbf{x})$ is a genuine pdf. Write down (i) the joint pdf of (X_2, X_3, X_5), (ii) the conditional pdf of X_3, X_5 given X_1, X_4, and (iii) the conditional pdf of X_3 given X_1, X_2, X_4.

3.2.8 An urn contains sixteen marbles of the same size and weight of which two are red, five are yellow, three are green, and six are blue. One marble is randomly picked from the urn, its color is noted, and this marble is returned to the urn. Another marble is randomly picked, its color noted, and this marble is also returned to the urn. The experiment continues in this fashion. This process is called *sampling with replacement*. After the experiment is run n times, one looks at the number of selected red (X_1), yellow (X_2), green (X_3), and blue (X_4) marbles.

  (*i*)   What is the joint distribution of (X_1, X_2, X_3, X_4)?

  (*ii*)   What is the mean and variance of X_3?

  (*iii*) If $n = 10$, what is the joint distribution of (X_1, X_2, X_3, X_4)?

  (*iv*)   If $n = 15$, what is the conditional distribution of (X_1, X_2, X_3) given that $X_4 = 5$?

3.2.9 The mgf $M_{\mathbf{X}}(\mathbf{t})$ of $\mathbf{X} = (X_1, ..., X_k)$ with $\mathbf{t} = (t_1, ..., t_k)$ is defined as

$$E[exp\{t_1 X_1 + ... + t_k X_k\}]$$

Show that $M_{\mathbf{X}}(\mathbf{t}) = \{p_1 e^{t_1} + ... + p_k e^{t_k}\}^n$ if $\mathbf{X}$ is $\text{Mult}_k(n, p_1, ..., p_k)$.

3.2.10 Let X_1 and X_2 have a joint pmf

$$f(x_1, x_2) = \binom{4}{x_1}\binom{3}{x_2}\binom{2}{3 - x_1 - x_2} / \binom{9}{3}$$

with x_1, x_2 integers, $0 \le x_1 \le 3, 0 \le x_2 \le 3$, but $1 \le x_1 + x_2 \le 3$.

  (*i*)   Find marginal pmfs for X_1 and X_2;

  (*ii*)   Find conditional pmfs $f_{1|2}(x_1)$ and $f_{2|1}(x_2)$;

  (*iii*) Evaluate $E[X_1 \mid X_2 = x_2]$ and $E[X_2 \mid X_1 = x_1]$.

3.2.11 Let X_1 and X_2 have a joint pmf

$$f(x_1, x_2) = p^2 q^{x_2}, x_1 = 0, 1, 2, ..., x_2 \text{ and } x_2 = 0, 1, 2, ...$$

with $0 < p < 1, q = 1 - p$. Find marginal distributions of X_1 and X_2.

3.3.1 Prove Theorem 3.3.1. {*Hint*: (i) $Cov(X_1, c) = E(cX_1) - E(X_1)E(c) = cE(X_1) - cE(X_1) = 0$; (ii) $Cov(X_1 + X_2, Y_1) = E\{(X_1 + X_2)Y_1\} - E\{(X_1 + X_2)\}E(Y_1) = Cov(X_1, Y_1) + Cov(X_2, Y_1).$}

3.3.2 Prove Theorem 3.3.2. {*Hint*: Use Cauchy-Schwartz Inequality.}

3.3.3 With any two random variables X_1 and X_2, show that

$$Cov(X_1 + X_2, X_1 - X_2) = V(X_1) - V(X_2)$$

provided that $Cov(X_1, X_2), V(X_1)$, and $V(X_2)$ are finite.

3.3.4 (Exercise 3.3.3 Continued) Consider X_1, X_2 for which one has $V(X_1) = V(X_2)$. Use Exercise 3.3.3 to construct pairs of uncorrelated random variables.

3.3.5 Let X_1, X_2 have a joint pdf

$$f(x_1, x_2) = \begin{cases} 3(x_1^2 x_2 + x_1 x_2^2) & \text{if } 0 < x_1, x_2 < 1 \\ 0 & \text{elsewhere} \end{cases}$$

Show that $\rho_{X_1, X_2} = -\frac{5}{139}$.

3.4.1 With $c(>0)$, let X_1, X_2 have a joint pdf

$$f(x_1, x_2) = \begin{cases} cx_1 x_2^2 & \text{if } 0 < x_1, x_2 < 2 \\ 0 & \text{elsewhere} \end{cases}$$

Prove whether or not X_1, X_2 are independent.

3.4.2 With $c(>0)$, let X_1, X_2, X_3 have a joint pdf

$$f(x_1, x_2, x_3) = \begin{cases} c \exp\{-2x_1 - 4x_2^2 - \frac{1}{2}x_3\} & \text{if } 0 < x_1, x_3 < \infty \text{ and} \\ & -\infty < x_2 < \infty \\ 0 & \text{elsewhere} \end{cases}$$

Prove whether or not X_1, X_2, X_3 are independent.

3.4.3 (Example 3.4.4 Continued) Fix the functions:

$$f_1(x_1) = 2x_1, \; f_2(x_2) = 3x_2^2 \text{ and } f_3(x_3) = \frac{5}{2}x_3^{3/2}, 0 < x_1, x_2, x_3 < 1$$

Combine these to form a new function $g(x_1, x_2, x_3)$ as in Equation (3.4.2).

(i)   Verify that $g(x_1, x_2, x_3)$ is a genuine pdf;
(ii)   By integration, find the expressions of all pairwise marginal pdfs $g_{i,j}(x_i, x_j)$ and single marginal pdfs $g_i(x_i)$, for $i \neq j = 1, 2, 3$;
(iii)   Show that X_1, X_2, X_3 are pairwise independent, but they are not independent.

3.4.4 Start with nonnegative integrable functions $f_i(x_i)$ for $0 < x_i < 1$, *not* identically equal to unity, such that $\int_{x_i=0}^1 f_i(x_i)dx_i = 1, i = 1, 2, 3, 4$. With $\mathbf{x} = (x_1, x_2, x_3, x_4)$ we define:

$$g(\mathbf{x}) = \frac{1}{8}\{f_1(x_1)f_2(x_2)f_3(x_3)f_4(x_4) + f_1(x_1)f_2(x_2) + f_2(x_2)f_3(x_3)$$
$$+ f_1(x_1)f_3(x_3) + f_1(x_1) + f_2(x_2) + f_3(x_3) + f_4(x_4)\}$$

for $0 < x_i < 1, i = 1, 2, 3, 4$.

(i)   Verify that $g(x_1, x_2, x_3, x_4)$ is a genuine pdf;
(ii)   By integration, find the marginal pdf $g_{1,2,3}(x_1, x_2, x_3)$;
(iii)   Show that X_1, X_2, X_3 are independent;
(iv)   Consider X_1, X_2, X_3, X_4 with a joint pdf $g(x_1, x_2, x_3, x_4)$. Show that X_1, X_2, X_3, X_4 are not independent.

3.4.5 Prove Theorem 3.4.1. {*Hint*: Write $f(\mathbf{x}_1, ..., \mathbf{x}_p) = \Pi_{i=1}^p f_i(\mathbf{x}_i)$ for $x_i \in \mathcal{X}_i$. Then, $E[\Pi_{i=1}^p g_i(\mathbf{X}_i)] = \int_{\mathcal{X}_p} \cdots \int_{\mathcal{X}_1} \Pi_{i=1}^p g_i(\mathbf{x}_i) \Pi_{i=1}^p f_i(\mathbf{x}_i) \Pi_{i=1}^p d\mathbf{x}_i = \Pi_{i=1}^p \int_{\mathcal{X}_i} g_i(\mathbf{x}_i) f_i(\mathbf{x}_i) d\mathbf{x}_i = \Pi_{i=1}^p E[g_i(\mathbf{X}_i)].$}

3.4.6 Prove Theorem 3.4.2.

3.4.7 Prove Theorem 3.4.3.

3.4.8 Let X_1, X_2 have a joint pmf

$$f(x_1, x_2) = (2^{x_1-1}3^{x_2})^{-1}, \text{ for } x_1, x_2 = 1, 2, ...$$

Prove whether or not X_1, X_2 are independent.

3.4.9 Let X_1, X_2 have a joint pmf

$$f(x_1, x_2) = 2/\{n(n+1)\}, \text{ for } x_2 = 1, ..., x_1; x_1 = 1, 2, ..., n$$

Prove that X_1, X_2 are dependent.

3.4.10 Let X_1, X_2 have a joint pmf

$$f(x_1, x_2) = \frac{e^{-(a+b)}a^{x_1}b^{x_2-x_1}}{x_1!(x_2 - x_1)!}, \text{ for } x_1 = 0, 1, 2, ..., x_2; \text{ and}$$

$$x_2 = 0, 1, 2, ...; \text{ with positive numbers } a \text{ and } b$$

Prove whether or not X_1, X_2 are independent.

3.5.1 Prove Theorem 3.5.1.

3.5.2 Let (X_1, X_2) be distributed as $N_2(0, 0, 1, 1, \rho)$ with $\rho \in (-1, 1)$. Find the expression of the mgf of $X_1 X_2$, that is, $E\{exp[tX_1 X_2]\}$, for t belonging to a subinterval of $\Re$.

3.5.3 Suppose that the joint pdf of (X_1, X_2) is

$$f(x_1, x_2) = \{2\pi\sqrt{3}\}^{-1} exp\{-\tfrac{1}{6}[(2x_1 - x_2)^2 + 2x_1 x_2]\}$$

for $-\infty < x_1, x_2 < \infty$. Evaluate $E[X_i], V[X_i]$ for $i = 1, 2$, and ρ_{X_1,X_2}.

3.5.4 Suppose that the joint pdf $f(x_1, x_2)$ of (X_1, X_2) is

$$kexp\{-\tfrac{1}{216}[16(x_1 - 2)^2 - 12(x_1 - 2)(x_2 + 3) + 9(x_2 + 3)^2]\}$$

for $-\infty < x_1, x_2 < \infty$ where $k(> 0)$. Evaluate $E[X_i], V[X_i]$ for $i = 1, 2$, and ρ_{X_1,X_2}.

3.5.5 Let (X_1, X_2) be distributed as $N_2(3, 1, 16, 25, \tfrac{3}{5})$. Evaluate the following: $P\{3 < X_2 < 8 \mid X_1 = 7\}$ and $P\{-3 < X_1 < 3 \mid X_2 = -4\}$.

3.5.6 Let X_1 be distributed as $N(\mu, \sigma^2)$ and conditionally, the distribution of X_2 given $X_1 = x_1$ is $N(x_1, \sigma^2)$. Show that (X_1, X_2) is distributed as $N_2(\mu, \mu, \sigma^2, 2\sigma^2, 1/\sqrt{2})$.

3.5.7 Suppose that the joint pdf $f(x_1, x_2)$ of (X_1, X_2) is

$$(2\pi)^{-1}exp\{-\tfrac{1}{2}[x_1{}^2 + x_2{}^2]\}(1 + x_1 x_2 exp\{-\tfrac{1}{2}[x_1{}^2 + x_2^2 - 2]\})$$

for $-\infty < x_1, x_2 < \infty$.

(i)   Verify that $f(x_1, x_2)$ is a genuine pdf;

(ii)   Show that the marginal distributions of X_1, X_2 are both univariate normal;

(iii)   Does $f(x_1, x_2)$ match with the pdf given by Equation (3.5.1)?

3.6.1 Prove Theorem 3.6.1.

3.6.2 Let X_1 have a probability distribution

X_1 values:	-5	-2	0	2	5
Probabilities:	0.25	0.2	0.1	0.2	0.25

Define $X_2 = X_1^2$. Show that

(i) $E[X_1] = E[X_1^3] = 0$;

(ii) $Cov(X_1, X_2) = 0$, that is, the random variables X_1 and X_2 are uncorrelated;

(iii) X_1 and X_2 are dependent random variables.

3.6.3 Let X_1 have an arbitrary discrete distribution, symmetric about zero and having its third moment finite. Define $X_2 = X_1^2$. Show that

(i) $E[X_1] = E[X_1^3] = 0$;

(ii) $Cov(X_1, X_2) = 0$, that is, the random variables X_1 and X_2 are uncorrelated;

(iii) X_1 and X_2 are dependent random variables.

3.6.4 In the types of examples found in Exercises 3.6.2 and 3.6.3, what if someone had defined $X_2 = X_1^4$ or $X_2 = X_1^6$ instead? Would X_1 and X_2 be uncorrelated, but dependent?

3.6.5 Consider X_1, X_2 whose joint probability distribution is concentrated on four points $(0, 0), (0, 1), (1, 0)$, and $(1, 1)$ with positive probabilities. Show that $Cov(X_1, X_2) = 0$ will imply that X_1, X_2 are independent.

3.6.6 Let X_1, X_2 have the joint pdf

$$f(x_1, x_2) = (4\pi)^{-1}[x_1^2 + x_2^2] exp\{-\tfrac{1}{2}[x_1{}^2 + x_2{}^2]\}$$

for $-\infty < x_1, x_2 < \infty$. Show that X_1 and X_2 are uncorrelated, but these are dependent random variables.

3.6.7 Let (U_1, U_2) be distributed as $N_2(5, 15, 8, 8, \rho)$ for some $\rho \in (-1, 1)$. Let $X_1 = U_1 + U_2$ and $X_2 = U_1 - U_2$. Show that X_1 and X_2 are uncorrelated.

3.7.1 Verify the entries given in Table 3.7.1.

3.7.2 Let X have a uniform distribution on the interval $(0, \theta)$ with $\theta > 0$. Verify that this distribution does not belong to a one-parameter exponential family.

3.7.3 Consider Beta(α, β) distribution. Does its pdf belong to an exponential family when

(*i*) α is known, but β is unknown?

(*ii*) β is known, but α is unknown?

(*iii*) α and β are both unknown?

3.7.4 Let X have a uniform distribution on $(-\theta, \theta)$ with $\theta(> 0)$ unknown. Show that its pdf does not belong to a one-parameter exponential family.

3.7.5 The joint pdf of X_1, X_2 is

$$f(x_1, x_2) = exp\{-\theta x_1 - \theta^{-1} x_2\} I(x_1 > 0 \cap x_2 > 0)$$

with $\theta(> 0)$ unknown.

(*i*)   Are X_1, X_2 independent?

(*ii*)   Does this pdf belong to a one-parameter exponential family?

3.7.6 Express a bivariate normal pdf in the form of an appropriate member of a one-parameter or multiparameter exponential family in the following situations:

(*i*) $\mu_1, \mu_2, \sigma_1, \sigma_2, \rho$ are all unknown parameters;

(*ii*) $\mu_1 = \mu_2 = 0$ and σ_1, σ_2, ρ are unknown parameters;

(*iii*) $\sigma_1 = \sigma_2 = 1$ and μ_1, μ_2, ρ are unknown parameters;

(*iv*) $\sigma_1 = \sigma_2 = \sigma$ and $\mu_1, \mu_2, \sigma, \rho$ are unknown parameters;

(*v*) $\mu_1 = \mu_2 = 0, \sigma_1 = \sigma_2 = 1$, but ρ is an unknown parameter.

3.8.1. Prove Markov Inequality (3.8.1). {*Hint*: Suppose that A stands for the event $[W \geq \delta]$ and write $W \geq \delta I_A$, that is, $W - \delta I_A \geq 0$ w.p.1. Thus,

$$E(W - \delta I_A) = E(W) - \delta P(A) \geq 0$$

which implies that $P(A) \leq \delta^{-1} E(W).$}

3.8.2 Let X be a real valued random variable such that with some $r > 0$ and $\tau \in T(\subseteq \Re)$, one has $\psi_r = E\{|X - \tau|^r\}$ which is finite. Show that for any fixed real number $\varepsilon(> 0)$,

$$P\{|X - \tau| \geq \varepsilon\} \leq \psi_r / \varepsilon^r.$$

{*Hint*: Use Markov Inequality.}

3.8.3 Prove the Cauchy-Schwartz Inequality {*Hint*: If $E[X_2^2] = 0$, then $X_2 = 0$ w.p.1 so that both sides of Equation (3.8.4) hold. If $E[X_2^2] > 0$, consider $E[(X_1 + \lambda X_2)^2]$ where λ is any real number. But, whatever be λ, $E[(X_1 + \lambda X_2)^2]$ must be nonnegative!}

3.8.4 Prove Jensen's Inequality stated in Equation (3.8.6). {*Hint*: The curve $y = g(x)$ must lie above the tangents at all points $(u, g(u))$ for any $u \in \Re$. With $u = E[X]$, consider the point $(u, g(u))$ at which the equation of the tangent line would be given by

$$y = g(E[X]) + b(x - E[X])$$

for some appropriate b. But, then we can claim that

$$g(x) \geq g(E[X]) + b(x - E[X])$$

for all $x \in \Re$.}

4

Sampling Distribution

4.1 Introduction

Let $X_1, ..., X_n$ be $n (\geq 2)$ *independent and identically distributed* (iid) observations from a population. Now, we introduce some methods to find the probability distribution of a function of $X_1, ..., X_n$. For example, we handle a sample sum, $\Sigma_{i=1}^n X_i$, a sample mean, $\overline{X} = n^{-1} \Sigma_{i=1}^n X_i$, or a sample variance, $S^2 = (n-1)^{-1} \Sigma_{i=1}^n (X_i - \overline{X})^2$. We also explore the distributions of $X_{n:1}$, the smallest observation, and $X_{n:n}$, the largest observation. These probability distributions are called *sampling distributions*.

Section 4.2 demonstrates the usefulness of a moment generating function (mgf) to derive distributions in both discrete and continuous cases. Section 4.3 provides techniques for finding distributions of *order statistics*. Section 4.4 develops transformation techniques. A highlight consists of the *Helmert (orthogonal) transformation* in a normal distribution. Section 4.5 includes sampling distributions for both *one-sample* and *two-sample* problems. Section 4.6 briefly touches upon *multivariate normal distribution*. Selected reviews in matrices and vectors are included in Section 4.7.

4.2 Moment Generating Function Approach

A proof of the following important result is left out as an exercise.

Theorem 4.2.1 *A real valued random variable X_i has its mgf $M_{X_i}(t)$ for $i = 1, ..., n$. Let $X_1, ..., X_n$ be independent. Then, the mgf of $U = \Sigma_{i=1}^n a_i X_i$ is given by $\Pi_{i=1}^n M_{X_i}(t a_i)$ where $a_1, ..., a_n$ are arbitrary but fixed real numbers.*

First, $M_U(t)$ is found by invoking Theorem 4.2.1. Next, we match $M_U(t)$ with that of a standard distribution, and then with the help of Theorem 2.4.1 we identify the probability distribution of the random variable U. In many problems, this approach works extremely well.

Example 4.2.1 Let X_i be independent Binomial(n_i, p), $i = 1, ..., k$. Let $U = \Sigma_{i=1}^k X_i$, $q = 1 - p$, and $n = \Sigma_{i=1}^k n_i$. Then, $M_U(t) = \Pi_{i=1}^k (q + pe^t)^{n_i} = (q + pe^t)^n$, which matches with the mgf of a Binomial(n, p) random variable. So, the random variable U is distributed as Binomial(n, p). ▲

Example 4.2.2 We roll a fair die n times and let X_i be the number that lands in the ith toss, $i = 1, ..., n$. Let U be the total score, that is, $\Sigma_{i=1}^n X_i$. What is $P(U = 15)$ when $n = 4$? Now, the technique of full enumeration will be tedious. Instead, we start with the mgf $M_{X_i}(t) = \frac{1}{6}e^t(1+e^t+...+e^{5t})$ so that $M_U(t) = \frac{1}{6^n}e^{nt}(1 + e^t + ... + e^{5t})^n$. The coefficient of e^{kt} in an expansion of $M_U(t)$ must coincide with $P(X = k)$, $k = n, n + 1, ..., 6n$. So, if $n = 4$, then $P(X = 15) = \frac{1}{6^4}\{$coefficient of e^{11t} in the expansion of $(1+e^t+...+e^{5t})^4\} = \frac{140}{6^4}$. But, if $n = 10$, then $P(X = 15) = \frac{1}{6^{10}}\{$coefficient of e^{5t} in the expansion of $(1 + e^t + ... + e^{5t})^{10}\} = \frac{2002}{6^{10}}$. One may find the *probability mass function* (pmf) of the random variable U as an exercise. ▲

Example 4.2.3 Suppose $X_1, ..., X_k$ are independent random variables, X_i distributed as $N(\mu_i, \sigma_i^2)$, $i = 1, ..., k$. Write $U = \Sigma_{i=1}^k X_i$, $\mu = \Sigma_{i=1}^k \mu_i$, and $\sigma^2 = \Sigma_{i=1}^k \sigma_i^2$. One has $M_U(t) = \Pi_{i=1}^k\{e^{t\mu_i+\frac{1}{2}t^2\sigma_i^2}\} = e^{t\mu+\frac{1}{2}t^2\sigma^2}$, which matches with the $N(\mu, \sigma^2)$ mgf. So, the random variable U is distributed as $N(\mu, \sigma^2)$. ▲

Example 4.2.4 Let Z have a standard normal distribution and let $U = Z^2$. With $t < \frac{1}{2}$, we can express $M_U(t) = \int_{-\infty}^{\infty} e^{tz^2}\{\sqrt{2\pi}\}^{-1}e^{-\frac{1}{2}z^2}dz = \int_{-\infty}^{\infty}\{\sqrt{2\pi}\}^{-1}e^{-\frac{1}{2}(1-2t)z^2}dz = (1 - 2t)^{-1/2}$, which matches with the mgf of a χ_1^2 random variable. So, Z^2 has the χ_1^2 distribution. ▲

Theorem 4.2.2 (Reproductive Property of Independent Normal, Gamma, and Chi-Square Distributions) *Suppose that $X_1, ..., X_n$ are independent random variables. Write $U = \Sigma_{i=1}^n X_i$ and $\overline{X} = n^{-1}U$. Then, one concludes:*

(i)   *If $X_1, ..., X_n$ have a common $N(\mu, \sigma^2)$ distribution, then U is distributed as $N(n\mu, n\sigma^2)$ and $\overline{X}$ is distributed as $N(\mu, \frac{1}{n}\sigma^2)$;*

(ii)   *If $X_1, ..., X_n$ have a common $\mathrm{Gamma}(\alpha, \beta)$ distribution, then U is distributed as $\mathrm{Gamma}(n\alpha, \beta)$;*

(iii)   *If X_i has a $\mathrm{Gamma}(\frac{1}{2}\nu_i, 2)$ distribution, that is, a Chi-square distribution with ν_i degrees of freedom for $i = 1, ..., n$, then U is distributed as $\mathrm{Gamma}(\frac{1}{2}\nu, 2)$ with $\nu = \Sigma_{i=1}^n \nu_i$, which is a Chi-square distribution with ν degrees of freedom.*

These properties for the sums of independent normal, gamma, and Chi-square random variables will prove to be very useful in latter chapters. Their proofs are left out as exercises.

4.3   Order Statistics

A distribution function approach works well for some problems involving continuous random variables. Let X be a continuous real valued random variable and let Y be a real valued function of X. We express the df $G(y)$ of Y as the probability of an event defined via X. Once $G(y)$ is found, one obtains $dG(y)/dy$, if $G(y)$ is differentiable, to find the pdf of Y.

Let $X_1, ..., X_n$ be iid continuous random variables with the same pdf $f(x)$ and *distribution function* (df) $F(x)$. Let $X_{n:1} \leq X_{n:2} \leq ... \leq X_{n:n}$ stand for the ordered random variables. We call $X_{n:i}$ the ith-*order statistic* and denote $Y_i = X_{n:i}$ for $i = 1, ..., n$, $\mathbf{Y} = (Y_1, ..., Y_n)$. The joint pdf of $Y_1, ..., Y_n$ is:

$$f_{\mathbf{Y}}(y_1, ..., y_n) = \begin{cases} n! \Pi_{i=1}^{n} f(y_i) & \text{if } -\infty < y_1 \leq ... \leq y_n < \infty \\ 0 & \text{elsewhere} \end{cases} \quad (4.3.1)$$

In Equation (4.3.1), the multiplier $n!$ arises because $y_1, ..., y_n$ can be arranged in $n!$ ways and the pdf of any single arrangement is $\Pi_{i=1}^{n} f(y_i)$. Often we are interested in the *smallest* and *largest* order statistics. For the largest order statistic Y_n, we can find the distribution as follows: $P(Y_n \leq y)$ is:

$$P(X_i \leq y \text{ for each } i = 1, ..., n)$$
$$= \Pi_{i=1}^{n} P(X_i \leq y), \text{ since } X_1, ..., X_n \text{ are independent}$$
$$= \{F(y)\}^n$$

and hence the pdf of Y_n would be:

$$g(y) = \frac{d}{dy} P(Y_n \leq y) = \frac{d}{dy} [\{F(y)\}^n] = n\{F(y)\}^{n-1} f(y) \quad (4.3.2)$$

in the space of y values. Similarly, for the smallest order statistic Y_1, we can express $P(Y_1 > y)$ as:

$$P(X_i > y \text{ for each } i = 1, ..., n)$$
$$= \Pi_{i=1}^{n} P(X_i > y), \text{ since } X_1, ..., X_n \text{ are independent}$$
$$= \{1 - F(y)\}^n$$

Thus, the pdf of Y_1 would be:

$$h(y) = \frac{d}{dy} P(Y_1 \leq y) = \frac{d}{dy} [1 - \{1 - F(y)\}^n] = n\{1 - F(y)\}^{n-1} f(y) \quad (4.3.3)$$

in the space of y values.

> In Exercise 4.3.5, we have indicated how one may find the joint pdf of order statistics, $Y_i = X_{n:i}$ and $Y_j = X_{n:j}$.

Using Exercise 4.3.5, one may derive the joint pdf of Y_1 and Y_n. However, in order to write down the joint pdf of (Y_1, Y_n) quickly, we show a *heuristic* approach:

Since y_1, y_n are assumed fixed, the remaining $n - 2$ order statistics can be located anywhere between y_1 and y_n, and also these could be any $n - 2$ from the original n observations. Now, $P\{y_1 < X_i < y_n\} = F(y_n) - F(y_1)$ for $i = 1, ..., n$. Hence, with $k = n!/(n - 2)!$, the joint pdf of (Y_1, Y_n) can be expressed as:

$$f(y_1, y_n)$$
$$= \begin{cases} k\{F(y_n) - F(y_1)\}^{n-2} f(y_1) f(y_n) & \text{if } -\infty < y_1 < y_n < \infty \\ 0 & \text{elsewhere} \end{cases}$$
$$(4.3.4)$$

Example 4.3.1 Suppose $X_1, ..., X_n$ are iid Uniform$(0, \theta)$ with $\theta > 0$. Let us consider the largest and smallest order statistics Y_n and Y_1, respectively. Now note that

$$f(x) = \begin{cases} \theta^{-1} & \text{if } 0 < x < \theta \\ 0 & \text{elsewhere} \end{cases} \qquad F(x) = \begin{cases} 0 & \text{if } x \le 0 \\ \frac{x}{\theta} & \text{if } 0 < x < \theta \\ 1 & \text{if } x \ge \theta \end{cases}$$

From Equations (4.3.2) and (4.3.3), the marginal pdf of Y_n and Y_1 will be respectively $g(y) = ny^{n-1}\theta^{-n}I(0 < y < \theta)$ and $h(y) = n(\theta - y)^{n-1}\theta^{-n}I(0 < y < \theta)$. Using Equation (4.3.4), the joint pdf of (Y_1, Y_n) would be

$$f(y_1, y_n) = n(n-1)(y_n - y_1)^{n-2}\theta^{-n}I(0 < y_1 < y_n < \theta). \quad \blacktriangle$$

Example 4.3.2 The concept clearly extends for n independent, but *not* identically distributed random variables. Consider independent random variables X_1, X_2 where the respective pdfs are $f_1(x) = \frac{1}{3}x^2 I(-1 < x < 2)$ and $f_2(x) = \frac{5}{33}x^4 I(-1 < x < 2)$. One has the distribution functions $F_1(x) = \int_{-1}^{x} \frac{1}{3}t^2 dt = \frac{1}{9}(x^3 + 1)$, $F_2(x) = \int_{-1}^{x} \frac{5}{33}t^4 dt = \frac{1}{33}(x^5 + 1)$ for $-1 < x < 2$. Hence, for $-1 < y < 2$, the df $F(y)$ of $Y_2 = X_{2:2}$, the larger order statistic, can be found as follows:

$$P(Y_2 \le y) = P(X_1 \le y \cap X_2 \le y) = F_1(y)F_2(y) = \frac{1}{297}(y^3 + 1)(y^5 + 1) \quad (4.3.5)$$

since X_1, X_2 are independent. By differentiating $F(y)$, one can find the pdf of $X_{2:2}$. Similarly, one may find the pdf of $X_{2:1}$. $\quad \blacktriangle$

4.4   Transformation

Start with real valued *continuous* random variables $X_1, ..., X_n$ and have transformed real valued *continuous* random variables $Y_i = g_i(X_1, ..., X_n)$, $i = 1, ..., n$. See Figure 4.4.1. The transformation from $(x_1, ..., x_n) \in \mathcal{X}$ to $(y_1, ..., y_n) \in \mathcal{Y}$ is assumed *one-to-one*. For example, with $n = 3$, we may have on hand three transformed random variables $Y_1 = X_1 + X_2, Y_2 = X_1 - X_2$, and $Y_3 = X_1 + X_2 + X_3$ with $\mathcal{X} = \mathcal{Y} = \Re$.

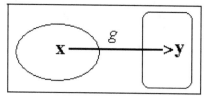 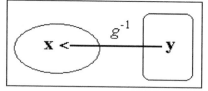

Figure 4.4.1. $g : \mathcal{X} \to \mathcal{Y}$ and $g^{-1} : \mathcal{Y} \to \mathcal{X}$.

Write down the joint pdf $f(x_1, ..., x_n)$ of $X_1, ..., X_n$ and replace all the variables $x_1, ..., x_n$ appearing within the expression of $f(x_1, ..., x_n)$ in terms of the transformed variables $g_i^{-1}(y_1, ..., y_n)$, $i = 1, ..., n$. Since the transformation $(x_1, ..., x_n)$ to $(y_1, ..., y_n)$ is one-to-one, we can *uniquely* express each x_i in terms of $(y_1, ..., y_n)$ and write $x_i \equiv b_i(y_1, ..., y_n)$, $i = 1, ..., n$. Thus, $f(x_1, ..., x_n)$ will be first replaced by $f(b_1, ..., b_n)$. Then, we multiply $f(b_1, ..., b_n)$ by the *absolute* value of the determinant of the Jacobian matrix of transformation, involving the transformed variables $y_1, ..., y_n$ *only*. Let us define a matrix

$$
J = \begin{pmatrix}
\frac{\partial x_1}{\partial y_1} & \frac{\partial x_1}{\partial y_2} & \cdot & \cdot & \frac{\partial x_1}{\partial y_n} \\
\frac{\partial x_2}{\partial y_1} & \frac{\partial x_2}{\partial y_2} & \cdot & \cdot & \frac{\partial x_2}{\partial y_n} \\
\cdot & \cdot & \cdot & \cdot & \cdot \\
\cdot & \cdot & \cdot & \cdot & \cdot \\
\frac{\partial x_n}{\partial y_1} & \frac{\partial x_n}{\partial y_2} & \cdot & \cdot & \frac{\partial x_n}{\partial y_n}
\end{pmatrix}
\tag{4.4.1}
$$

> The understanding is that the x_i variable in this matrix and elsewhere is replaced by $b_i(y_1, ..., y_n)$, $i = 1, ..., n$.

Let $det(J)$ stand for the *determinant* of the matrix J and $\mid det(J) \mid$ stand for the absolute value of $det(J)$. Then, a joint pdf of the transformed random variables $Y_1, ..., Y_n$ is given by:

$$
g(y_1, ..., y_n) = f(b_1(y_1, ..., y_n), ..., b_n(y_1, ..., y_n)) \mid det(J) \mid
\tag{4.4.2}
$$

for $y \in \mathcal{Y}$.

Suppose that the transformation from $(X_1, ..., X_n) \rightarrow (Y_1, ..., Y_n)$ is *not* one-to-one. Then, the result given in Equation (4.4.2) will need some minor adjustments. We begin by partitioning the $\mathcal{X}$ space into $A_1, ..., A_k$ in such a way that the associated transformation from $A_i \rightarrow \mathcal{Y}$ becomes one-to-one for each $i = 1, ..., k$. Given $(y_1, ..., y_n)$, let the associated x_j be expressed as $x_j = b_{ij}(y_1, ..., y_n)$, $j = 1, ..., n$, $i = 1, ..., k$. Suppose that the corresponding Jacobian matrices are denoted by J_i, $i = 1, ..., k$. Then, one essentially applies Equation (4.4.2) on each piece $A_1, ..., A_k$ and the pdf $g(y_1, ..., y_n)$ of Y is obtained as follows:

$$
g(y_1, ..., y_n) = \Sigma_{i=1}^{k} f(b_{i1}(y_1, ..., y_n), ..., b_{in}(y_1, ..., y_n)) \mid det(J_i) \mid
\tag{4.4.3}
$$

for $y \in \mathcal{Y}$.

Example 4.4.1 (One-Variable One-to-One Transformation) Let X be a single continuous random variable with its pdf and df respectively $f(x)$ and $F(x)$, $x \in (a, b)$. We write $Y = F(X)$ so that this transformation from x to y is one-to-one and $x = F^{-1}(y)$ for $y \in (0, 1)$. We obviously have $\mathcal{X} = (a, b)$, $\mathcal{Y} = (0, 1)$, and $\frac{dy}{dx} = f(x) = f(F^{-1}(y))$. Then, for $0 < y < 1$, the pdf of Y becomes:

$$
g(y) = f(F^{-1}(y)) \left| \frac{dx}{dy} \right| = f(F^{-1}(y)) \left| 1/\{f(F^{-1}(y))\} \right| = 1
$$

Thus, Y is distributed uniformly on $(0, 1)$. Next, let $U = q(Y)$ where $q(y) = -log(y)$ for $0 < y < 1$. Again, this transformation from y to u is one-to-one. Thus, the pdf of U is obtained as follows:

$$h(u) = \left| \frac{d}{du} q^{-1}(u) \right| = e^{-u} \text{ for } 0 < u < \infty$$

which shows that U is distributed as a standard exponential variable. Verify that $V = -2log(U)$ has the χ_2^2 distribution! ▲

Suppose that X has a continuous random variable with its df $F(x)$. Then, $F(X)$ is distributed as Uniform$(0, 1)$, $-log\{F(X)\}$ is distributed as standard exponential, and $-2log\{F(X)\}$ is distributed as χ_2^2.

Example 4.4.2 (One-Variable Not One-to-One Transformation)
Suppose that Z is a standard normal variable and $Y = Z^2$. Here, $\mathcal{Z} = \Re, \mathcal{Y} = \Re^+$. This transformation from z to y is *not* one-to-one, but the two pieces of transformations from $z \in [0, \infty)$ to $y \in \Re^+$ and from $z \in (-\infty, 0)$ to $y \in \Re^+$ are individually both one-to-one. Also recall that the pdf $\phi(z)$ of Z is symmetric about $z = 0$. Hence, for $0 < y < \infty$, we write down the pdf of Y as follows:

$$g(y) = \phi(y^{\frac{1}{2}}) \left| \frac{1}{2} y^{-\frac{1}{2}} \right| + \phi(-y^{\frac{1}{2}}) \left| -\frac{1}{2} y^{-\frac{1}{2}} \right| = \phi(y^{\frac{1}{2}}) y^{-\frac{1}{2}} = \frac{1}{\sqrt{2\pi}} e^{-\frac{1}{2}y} y^{-\frac{1}{2}}$$

which coincides with the pdf of a χ_1^2 variable. ▲

Example 4.4.3 (Two Independent Variables Transformation)
Let X_1, X_2 be independent random variables and X_i be distributed as Gamma$(\alpha_i, \beta), \alpha_i > 0, \beta > 0, i = 1, 2$. Define transformed variables $Y_1 = X_1 + X_2, Y_2 = X_1/(X_1 + X_2)$ so that the transformation from (x_1, x_2) to (y_1, y_2) is one-to-one. Here, $\mathcal{X} = \Re^+ \times \Re^+, \mathcal{Y} = \Re^+ \times (0, 1)$. One can *uniquely* express $x_1 = y_1 y_2$ and $x_2 = y_1(1 - y_2)$. It is easy to verify:

$$J = \begin{pmatrix} y_2 & y_1 \\ 1 - y_2 & -y_1 \end{pmatrix}$$

Thus, $det(J) = -y_1 y_2 - y_1(1 - y_2) = -y_1$. The joint pdf of X_1, X_2 is:

$$f(x_1, x_2) = c \, e^{-(x_1+x_2)/\beta} x_1^{\alpha_1-1} x_2^{\alpha_2-1}$$

for $0 < x_1, x_2 < \infty$ with $c = \{\beta^{\alpha_1+\alpha_2} \Gamma(\alpha_1)\Gamma(\alpha_2)\}^{-1}$. Hence, the joint pdf of Y_1, Y_2 is:

$$\begin{aligned} g(y_1, y_2) &= c \, e^{-y_1/\beta} (y_1 y_2)^{\alpha_1-1} \{y_1(1 - y_2)\}^{\alpha_2-1} | -y_1 | \\ &= c \, e^{-y_1/\beta} y_1^{\alpha_1+\alpha_2-1} y_2^{\alpha_1-1} (1 - y_2)^{\alpha_2-1} \end{aligned} \qquad (4.4.4)$$

for $0 < y_1 < \infty, 0 < y_2 < 1$. The terms involving y_1 and y_2 in Equation (4.4.4) factorize, and also neither variable's domain involves the other variable. It follows that Y_1, Y_2 are independent random variables, Y_1 is distributed as Gamma$(\alpha_1 + \alpha_2, \beta)$, and Y_2 is distributed as Beta(α_1, α_2) since we can rewrite c as $\{\beta^{\alpha_1+\alpha_2}\Gamma(\alpha_1 + \alpha_2)\}^{-1}\{b(\alpha_1, \alpha_2)\}^{-1}$ where $b(\alpha_1, \alpha_2)$ stands for the beta function. ▲

> In the next example one has X_1 and X_2 dependent.
> Also, the transformed variables Y_1 and Y_2 are dependent.

Example 4.4.4 (Two-Variable Not One-to-One Transformation)
Let X_1, X_2 be iid $N(0,1)$ and denote $Y_1 = X_1/X_2, Y_2 = X_2^2$. Obviously, the transformation $(x_1, x_2) \rightarrow (y_1, y_2)$ is *not* one-to-one. Fixing y_1, y_2, the inverse solution would be $(y_1\sqrt{y_2}, \sqrt{y_2})$ or $(-y_1\sqrt{y_2}, -\sqrt{y_2})$. In Equation (4.4.3), we take $A_1 = \{(x_1, x_2) \in \Re^2 : x_2 > 0\}, A_2 = \{(x_1, x_2) \in \Re^2 : x_2 < 0\}$. One can show that $g(y_1, y_2) = \frac{1}{2\pi}e^{-\{1+y_1^2\}y_2/2}I(-\infty < y_1 < \infty, 0 < y_2 < \infty)$. Also, $g_1(y_1) = \frac{1}{\pi\{1+y_1^2\}}I(-\infty < y_1 < \infty)$ and $g_2(y_2)\frac{1}{\sqrt{2\pi}}e^{-y_2/2}(y_2)^{-1/2}I(0 < y_2 < \infty)$. Some details are left out. ▲

Example 4.4.5 (Two Dependent Variables Transformation) Let X_1 and X_2 have their joint pdf:

$$f(x_1, x_2) = \begin{cases} 2e^{-(x_1+x_2)} & \text{if } 0 < x_1 < x_2 < \infty \\ 0 & \text{elsewhere} \end{cases}$$

Let $Y_1 = X_1$ and $Y_2 = X_1 + X_2$. We first wish to obtain the joint pdf of Y_1, Y_2, and then derive the marginal pdfs of Y_1, Y_2. See Figure 4.4.2.

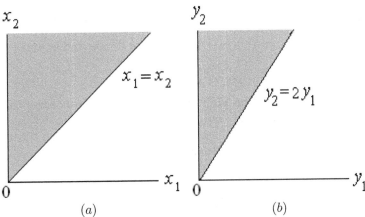

Figure 4.4.2. (a) The open-ended shaded space $\mathcal{X}$ where x_1, x_2 are defined; (b) The open-ended shaded space $\mathcal{Y}$ where y_1, y_2 are defined.

The one-to-one transformation $(x_1, x_2) \rightarrow (y_1, y_2)$ leads to the inverse: $x_1 = y_1, x_2 = y_2 - y_1$ so that $|det(J)| = 1$. Now, $x_2 > 0$ implies that

$0 < 2y_1 < y_2 < \infty$ since $y_1 < y_2 - y_1$. Thus, the joint pdf of Y_1, Y_2 is:

$$g(y_1, y_2) = 2e^{-y_2} I(0 < 2y_1 < y_2 < \infty)$$

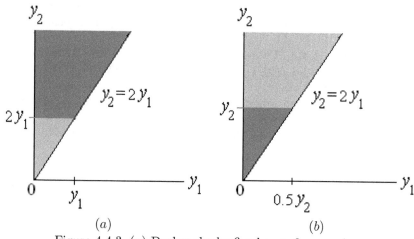

(a)   (b)

Figure 4.4.3. (a) Darker shade: fixed $y_1 > 0$: y_2 varies
from $2y_1$ to ∞; (b) Darker shade: fixed $y_2 > 0$: y_1
varies from 0 to $\frac{1}{2}y_2$.

The marginal pdfs of Y_1, Y_2 can be easily verified as follows:

$$g_1(y_1) = \int_{2y_1}^{\infty} h(y_1, y_2) dy_2 = 2e^{-2y_1} \text{ for } 0 < y_1 < \infty$$
$$g_2(y_2) = \int_0^{y_2/2} h(y_1, y_2) dy_1 = y_2 e^{-y_2} \text{ for } 0 < y_2 < \infty$$

See Figure 4.4.3. Some details are left out. ▲

Example 4.4.6 (The Helmert Orthogonal Transformation) This
consists of a very special kind of *orthogonal* transformation from a set of
n iid $N(\mu, \sigma^2)$ random variables $X_1, ..., X_n$, with $n \geq 2$, to a new set of n
random variables $Y_1, ..., Y_n$ defined as follows:

$$
\begin{aligned}
Y_1 &= \tfrac{1}{\sqrt{n}}(X_1 + ... + X_n) \\
Y_2 &= \tfrac{1}{\sqrt{2}}(X_1 - X_2) \\
Y_3 &= \tfrac{1}{\sqrt{6}}(X_1 + X_2 - 2X_3) \\
&\quad \cdot \\
&\quad \cdot \\
&\quad \cdot \\
Y_n &= \tfrac{1}{\sqrt{n(n-1)}}\{X_1 + ... + X_{n-1} - (n-1)X_n\}
\end{aligned}
$$

(4.4.5)

These $Y_1, ..., Y_n$ are called the *Helmert variables*. Denote the matrix:

$$
A = \begin{pmatrix}
\frac{1}{\sqrt{n}} & \frac{1}{\sqrt{n}} & \frac{1}{\sqrt{n}} & \cdot & \cdot & \frac{1}{\sqrt{n}} \\
\frac{1}{\sqrt{2}} & -\frac{1}{\sqrt{2}} & 0 & \cdot & \cdot & 0 \\
\frac{1}{\sqrt{6}} & \frac{1}{\sqrt{6}} & -\frac{2}{\sqrt{6}} & \cdot & \cdot & 0 \\
\cdot & \cdot & \cdot & & & \cdot \\
\cdot & \cdot & \cdot & \cdot & \cdot & \cdot \\
\frac{1}{\sqrt{n(n-1)}} & \frac{1}{\sqrt{n(n-1)}} & \frac{1}{\sqrt{n(n-1)}} & \cdot & \cdot & -\frac{n-1}{\sqrt{n(n-1)}}
\end{pmatrix} \tag{4.4.6}
$$

Then, one has $\mathbf{y} = A\mathbf{x}$ where $\mathbf{x}' = (x_1, ..., x_n)$ and $\mathbf{y}' = (y_1, ..., y_n)$.

> $A_{n \times n}$ is orthogonal. So, A' is the inverse of A. This implies that $\Sigma_{i=1}^n x_i^2 = \mathbf{x}'\mathbf{x} = \mathbf{x}' A' A\mathbf{x} = \mathbf{y}'\mathbf{y} = \Sigma_{i=1}^n y_i^2$. In $\Re^n$, a sphere in $\mathbf{x}$-coordinates continues to look like a sphere in $\mathbf{y}$-coordinates when the $\mathbf{x}$ axes are rotated orthogonally to match with the new $\mathbf{y}$ axes.

Observe that the matrix J defined in Equation (4.4.1) coincides with matrix A' and hence we can write $\mid det(J) \mid = \mid det(A') \mid = \{det(AA')\}^{\frac{1}{2}} \mid = 1$.
Now, the joint pdf of $X_1, ..., X_n$ is:

$$
f(x_1, ..., x_n) = \{\sigma\sqrt{2\pi}\}^{-n} exp\left\{-\tfrac{1}{2\sigma^2}\Sigma_{i=1}^n(x_i^2 - 2\mu x_i + \mu^2)\right\}
$$

for $-\infty < x_1, ..., x_n < \infty$, and thus the joint pdf of $Y_1, ..., Y_n$ is:

$$
\begin{aligned}
&\{\sigma\sqrt{2\pi}\}^{-n} \ exp\left\{-\tfrac{1}{2\sigma^2}(\Sigma_{i=1}^n y_i^2 - 2\mu\sqrt{n}y_1 + n\mu^2)\right\} \\
&= \{\sigma\sqrt{2\pi}\}^{-n} \ exp\left\{-\tfrac{1}{2\sigma^2}(\Sigma_{i=2}^n y_i^2 + (y_1 - \sqrt{n}\mu)^2)\right\}
\end{aligned} \tag{4.4.7}
$$

where $-\infty < y_1, ..., y_n < \infty$. In Equation (4.4.7), the terms involving $y_1, ..., y_n$ factorize and none of the variables' domains involve the other variables. It is clear (Theorem 3.4.3) that $Y_1, ..., Y_n$ are independently distributed. We also observe that Y_1 is distributed as $N(\sqrt{n}\mu, \sigma^2)$, whereas $Y_2, ..., Y_n$ are iid $N(0, \sigma^2)$. ▲

Theorem 4.4.1 (Distribution of $\overline{X}, S^2$ in a Normal Population)
Suppose that $X_1, ..., X_n$ are iid $N(\mu, \sigma^2)$ random variables, $n \geq 2$. Define $\overline{X} = n^{-1}\Sigma_{i=1}^n X_i$ and $S^2 = (n-1)^{-1}\Sigma_{i=1}^n(X_i - \overline{X})^2$. Then, we have:

(i)   *The sample mean $\overline{X}$ is distributed independently of the sample variance S^2;*

(ii)   *The sampling distribution of $\overline{X}$ is $N(\mu, \tfrac{1}{n}\sigma^2)$ and that of $(n-1)S^2/\sigma^2$ is χ_{n-1}^2.*

Proof (i) Using the Helmert variables from Equation (4.4.5), we can rewrite $\overline{X} = n^{-\frac{1}{2}}Y_1$ and express $(n-1)S^2$ as:

$$
\Sigma_{i=1}^n X_i^2 - n\overline{X}^2 = \Sigma_{i=1}^n Y_i^2 - Y_1^2 = \Sigma_{i=2}^n Y_i^2 \tag{4.4.8}
$$

It is clear that $\overline{X}$ is a function of Y_1 alone. From Equation (4.4.8), we note that S^2 depends functionally only on $(Y_2, ..., Y_n)$. But, since Y_1 and $(Y_2, ..., Y_n)$ are independent, we conclude that $\overline{X}$ and S^2 are independent.

(ii) Recall that $\overline{X} = n^{-\frac{1}{2}}Y_1$ where Y_1 is distributed as $N(\sqrt{n}\mu, \sigma^2)$ and so the sampling distribution of $\overline{X}$ follows immediately. From Equation (4.4.8), it is clear that $(n-1)S^2/\sigma^2 = \Sigma_{i=2}^n Y_i^2/\sigma^2$ which is a sum of $(n-1)$ independent χ_1^2 variables. Using the reproductive property of independent Chi-squares, it follows that $(n-1)S^2/\sigma^2$ has a χ_{n-1}^2 distribution. ■

Remark 4.4.1 It is important to note that the sample variance S^2 is the average of $Y_2^2, ..., Y_n^2$ and that each Helmert variable *independently* contributes one degree of freedom toward the total $(n-1)$ degrees of freedom associated with S^2. Having n observations $X_1, ..., X_n$, the decomposition in Equation (4.4.8) shows how exactly S^2 can be split up into $(n-1)$ *independent and identically distributed* components having one degree of freedom each. This is a key idea which eventually leads to *Analyses of Variance* techniques, used widely in statistics.

> Exercise 4.6.5 gives another transformation which proves that $\overline{X}, S^2$ are independent when $X_1, ..., X_n$ are iid $N(\mu, \sigma^2)$.

Remark 4.4.2 Another interesting feature may be highlighted among the Helmert variables $Y_2, ..., Y_n$. Only the last Helmert variable Y_n functionally involves the last observation, X_n. This particular feature has an important implication. Suppose that an additional observation X_{n+1} becomes available. Then, with $\overline{X}_{new} = (n+1)^{-1}\Sigma_{i=1}^{n+1}X_i$, the new sample variance S^2_{new} and its decomposition would be expressed as:

$$S^2_{new} = n^{-1}\Sigma_{i=1}^{n+1}(X_i - \overline{X}_{new})^2 \Rightarrow nS^2_{new} = \Sigma_{i=2}^n Y_i^2 + Y_{n+1}^2 \qquad (4.4.9)$$

where $Y_{n+1} = \{X_1 + ... + X_n - nX_{n+1}\}/\sqrt{(n+1)n}$. Note that $Y_2, ..., Y_n$ remain exactly the same as in Equation (4.4.5). So, the Helmert transformation shows exactly how the sample variance is recursively affected.

> Let $X_1, ..., X_n$ be iid random variables with $n \geq 2$. Then, $\overline{X}, S^2$ are independent $\Rightarrow X_1, ..., X_n$ are normally distributed.

Remark 4.4.3 Suppose that from iid random samples $X_1, ..., X_n$, one obtains the sample mean $\overline{X}$ and the sample variance S^2. If it is also assumed that $\overline{X}$ and S^2 are independently distributed, then effectively one is not assuming any less than normality of the original iid random variables. That is, independence of $\overline{X}$ and S^2 is a *characteristic property* of the normal distribution alone. This is a deep result in probability theory. For a formal proof of this characterization of the normal distribution and other historical notes, one may refer to Zinger (1958), Lukacs (1960), and Ramachandran (1967, Section 8.3). What can be concluded if $\overline{X}$ and S^2 are independent, but $X_1, ..., X_n$ are not necessarily iid? Refer to Mukhopadhyay (2005) who discussed some interesting possibilities.

4.5   Special Sampling Distributions

Definition 4.5.1 (*Student's t Distribution*) *Suppose that* X *is a standard normal variable,* Y *is a Chi-square variable with* ν *degrees of freedom, and* X, Y *are independent. Then,* $W = \dfrac{X}{\sqrt{\{Y\nu^{-1}\}}}$ *is said to have a Student's t distribution or simply a t distribution with* ν *degrees of freedom.*

A case in point: note that $-W$ is $(-X)/\sqrt{(Y\nu^{-1})}$ and we can immediately conclude that W and $-W$ have identical distributions, that is, a Student's t distribution is symmetric around zero.

How about finding moments of a Student's t variable? Its pdf is not essential for deriving moments of W. For any positive integer k, observe that

$$E[W^k] = \nu^{k/2} E[X^k Y^{-k/2}] = \nu^{k/2} E[X^k] E[Y^{-k/2}] \qquad (4.5.1)$$

as long as $E[Y^{-k/2}]$ is finite. We can split the expectation because X, Y are assumed independent. But, it is clear that Equation (4.5.1) will lead to finite entities provided that an appropriate negative moment of a Chi-square variable exists. We had discussed similar matters involving a gamma distribution in Chapter 2.

One can verify that $E(W)$ is finite if $\nu > 1$, whereas $E(W^2)$ is finite when $\nu > 2$. One may also check the following:

$$\begin{aligned} E(W) &= 0 \text{ if } \nu > 1 \text{ and} \\ V(W) &= \tfrac{1}{2}\nu\Gamma(\tfrac{1}{2}\nu - 1)\{\Gamma(\tfrac{1}{2}\nu)\}^{-1} = \nu/(\nu - 2) \text{ if } \nu > 2 \end{aligned} \qquad (4.5.2)$$

Example 4.5.1 (**One-Sample Problem**) Let $X_1, ..., X_n$ be iid $N(\mu, \sigma^2)$, $-\infty < \mu < \infty, 0 < \sigma < \infty, n \geq 2$. Let us recall that $\sqrt{n}(\overline{X} - \mu)/\sigma$ is distributed as $N(0, 1)$, $(n - 1)S^2/\sigma^2$ is χ^2_{n-1}, and these are also independent (Theorem 4.4.1). We may rewrite:

$$\sqrt{n}(\overline{X} - \mu)/S = \{\sqrt{n}(\overline{X} - \mu)/\sigma\} / \{(n - 1)S^2/[\sigma^2(n - 1)]\}^{\frac{1}{2}}$$

Note that $\sqrt{n}(\overline{X} - \mu)/S$ has the same representation as that of W with $\nu = n - 1$. That is, $\sqrt{n}(\overline{X} - \mu)/S$ has a Student's t distribution with $(n - 1)$ degrees of freedom. ▲

Example 4.5.2 (**Two-Sample Problem**) Let $X_{i1}, ..., X_{in_i}$ be iid $N(\mu_i, \sigma^2)$, $n_i \geq 2, i = 1, 2$, and X_1 be independent of X_2. Denote:

$$\begin{aligned} \overline{X}_i &= n_i^{-1}\Sigma_{j=1}^{n_i}X_{ij}, \ S_i^2 = (n_i - 1)^{-1}\Sigma_{j=1}^{n_i}(X_{ij} - \overline{X}_i)^2 \\ S_P^2 &= (n_1 + n_2 - 2)^{-1}\left\{(n_1 - 1)S_1^2 + (n_2 - 1)S_2^2\right\} \end{aligned} \qquad (4.5.3)$$

for $i = 1, 2$ where S_P^2 is called the *pooled sample variance*. Now,

$$(n_i - 1)S_i^2/\sigma^2 \text{ is } \chi^2_{n_i-1}, i = 1, 2$$

and these are also independent. We claim:

$$\{(n_1 - 1)S_1^2 + (n_2 - 1)S_2^2\}\,\sigma^{-2} \text{ has } \chi^2_{n_1+n_2-2} \text{ distribution}$$

using the reproductive property of independent Chi-squares. Also, $(\overline{X}_1, \overline{X}_2)$ and S_P^2 are independent. Now, along the lines of Example 4.5.1,

$$\{n_1^{-1} + n_2^{-1}\}^{-\frac{1}{2}}[(\overline{X}_1 - \overline{X}_2) - (\mu_1 - \mu_2)]S_P^{-1} \text{ has } t_{n_1+n_2-2} \text{ distribution. } \blacktriangle$$

Definition 4.5.2 (*F Distribution*) *Let X, Y be independent, distributed respectively as $\chi^2_{\nu_1}$ and $\chi^2_{\nu_2}$. Then, $U = (X/\nu_1) \div (Y/\nu_2)$ is said to have an F_{ν_1,ν_2} distribution.*

From Definition 4.5.1 of a Student's t random variable W, it is clear that $W^2 = X^2(Y/\nu)^{-1}$ has the same form as in Definition 4.5.2 for U. Thus, t_ν^2 and $F_{1,\nu}$ distributions are the same.

F distribution is *not symmetric* in the same sense as a t distribution is. But, from Definition 4.5.2, it follows immediately that $1/U$ is distributed as F_{ν_2,ν_1}. That is, $1/F$ has an F distribution too.

How about finding the moments of an F_{ν_1,ν_2} variable? For any positive integer k, observe that

$$E[U^k] = (\nu_2/\nu_1)^k E[X^k Y^{-k}] = (\nu_2/\nu_1)^k E[X^k]E[Y^{-k}] \qquad (4.5.4)$$

as long as $E[Y^{-k}]$ is finite. We split expectation in Equation (4.5.4) because X, Y are independent. It is clear that the expression in Equation (4.5.4) will lead to finite entities provided that an appropriate negative moment of a Chi-square variable exists. One can verify that $E(U)$ is finite if $\nu_2 > 2$ and $E(U^2)$ is finite when $\nu_2 > 4$. One may check the following:

$$E[U] = \nu_2/(\nu_2 - 2) \text{ if } \nu_2 > 2 \text{ and}$$
$$V[U] = [2\nu_2^2(\nu_1 + \nu_2 - 2)]/[\nu_1(\nu_2 - 2)^2(\nu_2 - 4)] \text{ if } \nu_2 > 4 \qquad (4.5.5)$$

Example 4.5.3 (**Two-Sample Problem**) Suppose $X_{i1}, ..., X_{in_i}$ are iid $N(\mu_i, \sigma_i^2)$, for $i = 1, 2$, and X_1 is independent of X_2. For $n_i \geq 2$, denote $\overline{X}_i = n_i^{-1}\Sigma_{j=1}^{n_i} X_{ij}$, $S_i^2 = (n_i - 1)^{-1}\Sigma_{j=1}^{n_i}(X_{ij} - \overline{X}_i)^2$ as in Example 4.5.2. Now, (i) $(n_i - 1)S_i^2/\sigma_i^2$ is $\chi^2_{n_i-1}, i = 1, 2$, (ii) they are also independent, and hence in view of Definition 4.5.2, the random variable $U = (S_1/\sigma_1)^2 \div (S_2/\sigma_2)^2$ has an F_{n_1-1,n_2-1} distribution. $\blacktriangle$

4.6   Multivariate Normal Distribution

Tong's (1990) book is devoted to multivariate normal distribution and includes valuable tables. It also briefly discusses multivariate t and F distributions. The references to tables and other features for a multivariate normal, t, or F distribution can also be found in Johnson and Kotz (1972).

Definition 4.6.1 *A* $p(\geq 1)$ *random vector* $\mathbf{X} = (X_1, ..., X_p)$ *has the* p-*dimensional normal* distribution, *denoted by* N_p, *if and only if each linear function* $\Sigma_{i=1}^p a_i X_i$ *has a univariate* normal *distribution for all fixed, but arbitrary real numbers* $a_1, ..., a_p$.

Example 4.6.1 Suppose that $X_1, ..., X_n$ are iid $N(\mu, \sigma^2)$, $-\infty < \mu < \infty$, and $0 < \sigma < \infty$. Consider $\overline{X}$ and look at the bivariate distribution of $(X_1, \overline{X})$. Obviously, $E(X_1) = E(\overline{X}) = \mu$, $V(X_1) = \sigma^2$, and $V(\overline{X}) = \frac{1}{n}\sigma^2$. Also, we can write $Cov(X_1, \overline{X}) = \Sigma_{i=1}^n Cov(X_1, \frac{1}{n}X_i)$. But, the covariance between X_1 and X_j is zero for $1 < j \leq n$. So, $Cov(X_1, \overline{X}) = \frac{1}{n}Cov(X_1, X_1) = \frac{1}{n}\sigma^2$ and hence $\rho_{X_1, \overline{X}} = 1/\sqrt{n}$. Any linear function L of X_1 and $\overline{X}$ is also a linear function of n original iid normal random variables $X_1, ..., X_n$, and hence L itself is distributed normally. This follows from the reproductive property of independent normal variables (Theorem 4.2.2, part [i]). By Definition 4.6.1, $(X_1, \overline{X})$ is jointly distributed as a bivariate normal variable. Also, the conditional distribution of X_1 given $\overline{X} = \overline{x}$ is $N\left(\overline{x}, \sigma^2(1 - \frac{1}{n})\right)$ which provides the following facts: $E(X_1 \mid \overline{X} = \overline{x}) = \overline{x}$ and $V(X_1 \mid \overline{X} = \overline{x}) = \sigma^2(1 - \frac{1}{n})$. ▲

Example 4.6.2 (Example 4.6.1 Continued) A result such as $E(X_1 \mid \overline{X} = \overline{x}) = \overline{x}$ can be alternately derived as follows: obviously, $E(\overline{X} \mid \overline{X} = \overline{x}) = \overline{x}$, and this can be rewritten as $\overline{x} = \frac{1}{n}E\{\Sigma_{i=1}^n X_i \mid \overline{X} = \overline{x}\} = \frac{1}{n}nE(X_1 \mid \overline{X} = \overline{x}) = E(X_1 \mid \overline{X} = \overline{x})$, which was the intended result in the first place. This argument works because $E\{X_i \mid \overline{X} = \overline{x}\} = E(X_1 \mid \overline{X} = \overline{x})$ for each $i = 2, ..., n$. Observe that this particular approach merely used the fact that $X_1, ..., X_n$ were iid, but it did not exploit the fact that $X_1, ..., X_n$ were iid normal to begin with. ▲

For completeness, we now give the pdf of a p-dimensional normal random vector. Denote a p-dimensional *column vector* $\mathbf{X}$ whose *transpose* is $\mathbf{X}' = (X_1, ..., X_p)$, consisting of the real valued random variable X_i as its ith component, $i = 1, ..., p$. We denote $E[X_i] = \mu_i$, $V[X_i] = \sigma_{ii}$, and $Cov(X_i, X_j) = \sigma_{ij}, 1 \leq i \neq j \leq p$. Denote the mean vector $\boldsymbol{\mu}$ where $\boldsymbol{\mu}' = (\mu_1, ..., \mu_p)$ and then write down σ_{ij} as the $(i, j)^{th}$ element of a $p \times p$ matrix $\boldsymbol{\Sigma}, 1 \leq i \neq j \leq p$. Then, $\boldsymbol{\Sigma} = (\sigma_{ij})_{p \times p}$ is called a *variance-covariance* or *dispersion matrix* of $\mathbf{X}$.

Assume that $\boldsymbol{\Sigma}$ has *full rank* which is equivalent to saying that $\boldsymbol{\Sigma}^{-1}$ exists. Then, the random vector $\mathbf{X}$ has a p-dimensional normal distribution with mean vector $\boldsymbol{\mu}$ and dispersion matrix $\boldsymbol{\Sigma}$, denoted by $N_p(\boldsymbol{\mu}, \boldsymbol{\Sigma})$, provided that the joint pdf of $X_1, ..., X_p$ is:

$$f(\mathbf{x}) = c \; exp\left\{-\tfrac{1}{2}(\mathbf{x} - \boldsymbol{\mu})'\boldsymbol{\Sigma}^{-1}(\mathbf{x} - \boldsymbol{\mu})\right\}, \text{ with } \mathbf{x} \in \Re^p$$
$$\text{and } c = \{(2\pi)^{p/2}\left[det\left(\boldsymbol{\Sigma}\right)\right]^{1/2}\}^{-1} \qquad (4.6.1)$$

One may check that a bivariate normal pdf from Equation (3.5.1) can also be written in this form.

> Suppose that $X_1, ..., X_p$ are jointly distributed as $N_p(\boldsymbol{\mu}, \boldsymbol{\Sigma})$. Then, $X_1, ..., X_p$ are independent if and only if $\boldsymbol{\Sigma}$ is diagonal.
(4.6.2)

> Let $(X_1, ..., X_p)$ be distributed as $N_p(\boldsymbol{\mu}, \boldsymbol{\Sigma})$. Then, any subvector $(X_{i_1}, X_{i_2}, ..., X_{i_k})$ is jointly distributed as $N_k(\boldsymbol{\theta}, \mathbf{A})$ where $\boldsymbol{\theta}' = (\mu_{i_1}, \mu_{i_2}, ..., \mu_{i_k})$ and the matrix $\mathbf{A}_{k \times k}$ is that part of $\boldsymbol{\Sigma}$ corresponding to the rows and columns numbered $i_1, i_2, ..., i_k, 1 \leq i_1 < i_2 < ... < i_k \leq p$.

4.6.1   Sampling Distributions: Bivariate Normal

Now, we focus on a bivariate normal distribution. Let $(X_1, Y_1), ..., (X_n, Y_n)$ be iid $N_2(\mu_1, \mu_2, \sigma_1^2, \sigma_2^2, \rho)$ where $-\infty < \mu_1, \mu_2 < \infty, 0 < \sigma_1^2, \sigma_2^2 < \infty$ and $-1 < \rho < 1, n \geq 2$. Denote:

$$\overline{X} = n^{-1}\Sigma_{i=1}^n X_i \qquad\qquad \overline{Y} = n^{-1}\Sigma_{i=1}^n Y_i$$

$$S_1^2 = (n-1)^{-1}\Sigma_{i=1}^n(X_i - \overline{X})^2 \qquad S_2^2 = (n-1)^{-1}\Sigma_{i=1}^n(Y_i - \overline{Y})^2$$

$$S_{12} = (n-1)^{-1}\Sigma_{i=1}^n(X_i - \overline{X})(Y_i - \overline{Y}) \qquad r = S_{12}/(S_1 S_2)$$

$$(4.6.3)$$

> Here, $r \equiv \dfrac{S_{12}}{S_1 S_2}$ is defined as the *Pearson correlation coefficient*
>
> or simply a *sample correlation coefficient*.

One can claim the following sampling distributions:

$$\overline{X} \text{ is } N(\mu_1, \tfrac{1}{n}\sigma_1^2) \qquad\qquad \overline{Y} \text{ is } N(\mu_2, \tfrac{1}{n}\sigma_2^2)$$

$$(n-1)S_1^2\sigma_1^{-2} \text{ is } \chi_{n-1}^2 \qquad (n-1)S_2^2\sigma_2^{-2} \text{ is } \chi_{n-1}^2$$

$$(4.6.4)$$

Note that any linear function of $\overline{X}, \overline{Y}$ is also a linear function of the original iid random vectors $(X_i, Y_i), i = 1, ..., n$. By appealing to Definition 4.6.1 of a multivariate normal distribution, we can immediately claim the following result:

$$(\overline{X}, \overline{Y}) \text{ is distributed as } N_2(\mu_1, \mu_2, n^{-1}\sigma_1^2, n^{-1}\sigma_2^2, \rho^*) \text{ with some } \rho^*$$

$$(4.6.5)$$

How does one find ρ^*? Invoking the bilinear property of covariance from Theorem 3.3.1, part (iii), one can express $Cov(\overline{X}, \overline{Y})$ as:

$$n^{-2}\Sigma_{i=1}^n\Sigma_{j=1}^n Cov\,(X_i, Y_j) = n^{-2}\Sigma_{i=1}^n Cov\,(X_i, Y_i) = n^{-1}\rho\sigma_1\sigma_2$$

so that the population correlation coefficient between $\overline{X}, \overline{Y}$ simplifies to $n^{-1}\rho\sigma_1\sigma_2/\left(\sqrt{n^{-2}\sigma_1^2\sigma_2^2}\right) = \rho = \rho^*$.

The distribution of the *Pearson correlation coefficient*,

$$r = S_{12}/(S_1 S_2) \tag{4.6.6}$$

is quite complicated, especially when $\rho \neq 0$. Without explicitly writing down the pdf of r, it is still simple enough to see that the distribution of r cannot depend on μ_1, μ_2, σ_1^2 and σ_2^2. To check this claim, let us denote $U_i = (X_i - \mu_1)/\sigma_1$, $V_i = (Y_i - \mu_2)/\sigma_2$, and then observe that (U_i, V_i), $i = 1, ..., n$ are iid $N_2(0, 0, 1, 1, \rho)$. But, r can be equivalently expressed in terms of (U_i, V_i), $i = 1, ..., n$ as follows:

$$r = \frac{\Sigma_{i=1}^n (U_i - \overline{U})(V_i - \overline{V})}{\sqrt{\Sigma_{i=1}^n (U_i - \overline{U})^2}\sqrt{\Sigma_{i=1}^n (V_i - \overline{V})^2}} \tag{4.6.7}$$

$\boxed{(4.6.7) \Rightarrow \text{The distribution of } r \text{ depends } \textit{only} \text{ on } \rho.}$

Francis Galton introduced a numerical measure, r, which he termed "reversion" in a lecture at the Royal Statistical Society on February 9, 1877 and later called "regression". The term "cor-relation" or "correlation" probably appeared first in Galton's paper to the Royal Statistical Society on December 5, 1888. At the time, "correlation" was defined using deviations from the median instead of the mean. Karl Pearson defined and calculated correlation as in Equation (4.6.3) in 1897. In 1898, Pearson and his collaborators discovered that the standard deviation of r happened to be $(1 - \rho^2)/\sqrt{n}$ when n was large. "Student" derived the "probable error of a correlation coefficient" in 1908. Soper (1913) gave large sample approximations for the mean and variance of r which were better than those proposed earlier by Pearson. Refer to DasGupta (1980) for some historical details.

The unsolved problem of finding the exact pdf of r for normal variates came to R. A. Fisher's attention via Soper's 1913 paper. The pdf of r was published in the year 1915 by Fisher for all values of $\rho \in (-1, 1)$. Fisher, at the age of 25, brilliantly exploited n-dimensional geometry to come up with the solution, reputedly within one week. Fisher's genius immediately came into the limelight. Following the publication of Fisher's results, however, Karl Pearson set up a major cooperative study of correlation. One will notice that in the team formed for this cooperative project (Soper et al. [1917]) studying the distribution of the sample correlation coefficient, the young Fisher was not included. This happened in spite of the fact that Fisher was right there and he already earned quite some fame. Fisher felt hurt, as he was left out of this project. One thing led to another, and R. A. Fisher and Karl Pearson continued to criticize each other on philosophical grounds.

The pdf of r when $\rho = 0$ is given by

$$f(r) = c \, (1 - r^2)^{\frac{1}{2}(n-4)} \textit{ for } -1 < r < 1 \tag{4.6.8}$$

where $c = \Gamma\left(\frac{1}{2}(n-1)\right) \left\{\sqrt{\pi}\, \Gamma\left(\frac{1}{2}(n-2)\right)\right\}^{-1}$ for $n \geq 3$. Using Equation (4.6.8) and some simple transformation techniques, one obtains the following result:

$$r(n-2)^{1/2}(1-r^2)^{-1/2} \text{ has } t_{n-2} \text{ distribution when } \rho = 0 \qquad (4.6.9)$$

The verification of Equation (4.6.9) is left as an exercise. Fisher (1915) also included an exact pdf of r as an infinite power series when $\rho \neq 0$.

4.6.2 Sampling Distributions: Multivariate Normal

Now, we briefly discuss some sampling distributions in the context of a multivariate normal population. Suppose that $\mathbf{X}_1, ..., \mathbf{X}_n$ are iid $N_p(\boldsymbol{\mu}, \boldsymbol{\Sigma})$ where $\boldsymbol{\mu} \in \Re^p$ and $\boldsymbol{\Sigma}$ is a $p \times p$ positive definite (p.d.) matrix, $n \geq 2$. Denote

$$\overline{\mathbf{X}} = n^{-1}\Sigma_{i=1}^n \mathbf{X}_i \text{ and } \mathbf{W} = \Sigma_{i=1}^n (\mathbf{X}_i - \overline{\mathbf{X}})(\mathbf{X}_i - \overline{\mathbf{X}})^{'} \qquad (4.6.10)$$

Observe that $\overline{\mathbf{X}}$ is a p-dimensional column vector whereas $\mathbf{W}$ is a $p \times p$ matrix, both functionally depending on $\mathbf{X}_1, ..., \mathbf{X}_n$.

Theorem 4.6.1 Let $\mathbf{X}_1, ..., \mathbf{X}_n$ be iid $N_p(\boldsymbol{\mu}, \boldsymbol{\Sigma})$ where $\boldsymbol{\mu} \in \Re^p$ and $\boldsymbol{\Sigma}$ is a p.d. matrix, $n \geq 2$. Then, the following sampling distributions hold:

(i) $\overline{\mathbf{X}}$ is distributed as $N_p(\boldsymbol{\mu}, n^{-1}\boldsymbol{\Sigma})$;

(ii) $n(\overline{\mathbf{X}} - \boldsymbol{\mu})^{'}\boldsymbol{\Sigma}^{-1}(\overline{\mathbf{X}} - \boldsymbol{\mu})$ is distributed as χ_p^2;

(iii)   For any fixed $\mathbf{a} \in \Re^p - \{\mathbf{0}\}$, $\mathbf{a}^{'}\mathbf{W}\mathbf{a}/\mathbf{a}^{'}\boldsymbol{\Sigma}\mathbf{a}$ is distributed as χ_{n-1}^2;

(iv) $\overline{\mathbf{X}}, \mathbf{W}$ are distributed independently.

4.7 Selected Reviews in Matrices

We briefly summarize some useful notions involving matrices made up of *real numbers*. Suppose that $A_{m \times n}$ is such a matrix having m rows and n columns. Sometimes, we rewrite A as $(\mathbf{a}_1, ..., \mathbf{a}_n)$ where $\mathbf{a}_i$ stands for the ith column vector, $i = 1, ..., n$.

The *transpose* of A, denoted by $A^{'}$, stands for a matrix whose rows consist of $\mathbf{a}_1^{'}, ..., \mathbf{a}_n^{'}$. When $m = n$, we say that A is a *square* matrix.

The *determinant* of a square matrix $A_{n \times n}$ is denoted by $det(A)$. If we have two square matrices $A_{n \times n}$ and $B_{n \times n}$, then one has

$$det(AB) = det(BA) = det(A)det(B) \qquad (4.7.1)$$

The *rank* of a matrix $A_{m \times n} = (\mathbf{a}_1, ..., \mathbf{a}_n)$, denoted by $R(A)$, stands for the maximum number of linearly independent vectors among $\mathbf{a}_1, ..., \mathbf{a}_n$. It can be shown that

$$R(A) = R(A^{'}) = R(AA^{'}) = R(A^{'}A) \qquad (4.7.2)$$

A square matrix $A_{n \times n} = (\mathbf{a}_1, ..., \mathbf{a}_n)$ is called *nonsingular* or *full rank* if and only if $R(A) = n$, that is, all its column vectors $\mathbf{a}_1, ..., \mathbf{a}_n$ are linearly independent. In other words, $A_{n \times n}$ is nonsingular or full rank if and only if the column vectors $\mathbf{a}_1, ..., \mathbf{a}_n$ form a *minimal generator* of $\Re^n$.

A square matrix $B_{n \times n}$ is called an *inverse* of $A_{n \times n}$ if and only if $AB = BA = I_{n \times n}$, the identity matrix. The inverse matrix of $A_{n \times n}$, if it exists, is unique and it is customarily denoted by A^{-1}. It may be worthwhile to recall the following result.

$$\boxed{\text{For a matrix } A_{n \times n} \colon A^{-1} \text{ exists} \Leftrightarrow R(A) = n \Leftrightarrow det(A) \neq 0.}$$

For a 2×2 matrix $A = \begin{pmatrix} a_{11} & a_{12} \\ a_{21} & a_{22} \end{pmatrix}$, one has $det(A) = a_{11}a_{22} - a_{12}a_{21}$. Let us suppose that $a_{11}a_{22} \neq a_{12}a_{21}$, that is, $det(A) \neq 0$. Its inverse matrix is easily found:

$$\boxed{A^{-1} = \frac{1}{det(A)} \begin{pmatrix} a_{22} & -a_{12} \\ -a_{21} & a_{11} \end{pmatrix} \text{ whenever } a_{11}a_{22} \neq a_{12}a_{21}.} \quad (4.7.3)$$

A square matrix $A_{n \times n}$ is called *orthogonal* if and only if $A^{'}$ is the inverse of A. If $A_{n \times n}$ is orthogonal then $det(A) = \pm 1$.

A square matrix $A = (a_{ij})_{n \times n}$ is called *symmetric* if and only if $A^{'} = A$, that is, $a_{ij} = a_{ji}$, $1 \leq i \neq j \leq n$.

Let $A = (a_{ij}), i, j = 1, ..., n$. Denote the $l \times l$ submatrix $B_l = (a_{pq}), p, q = 1, ..., l, l = 1, ..., n$. The lth *principal minor* is defined as the $det(B_l), l = 1, ..., n$.

For a symmetric matrix $A_{n \times n}$, $\mathbf{x}^{'} A \mathbf{x}$ with $\mathbf{x} \in \Re^n$ is customarily called a *quadratic form*. A symmetric matrix $A_{n \times n}$ is called *positive semidefinite* (p.s.d.) if (a) $\mathbf{x}^{'} A \mathbf{x} \geq 0$ for all $\mathbf{x} \in \Re^n$, and (b) $\mathbf{x}^{'} A \mathbf{x} = 0$ for some nonzero $\mathbf{x} \in \Re^n$. A symmetric matrix $A_{n \times n}$ is called p.d. if (a) $\mathbf{x}^{'} A \mathbf{x} \geq 0$ for all $\mathbf{x} \in \Re^n$, and (b) $\mathbf{x}^{'} A \mathbf{x} = 0$ if and only if $\mathbf{x} = \mathbf{0}$.

$$\boxed{\begin{array}{c} \text{A symmetric matrix } A \text{ is } p.d. \text{ if and only if } \textit{all} \text{ principal} \\ \text{minors of } A \text{ are positive.} \end{array}} \quad (4.7.4)$$

A symmetric matrix $A_{n \times n}$ is called *negative definite* (n.d.) if (a) $\mathbf{x}^{'} A \mathbf{x} \leq 0$ for all $\mathbf{x} \in \Re^n$, and (b) $\mathbf{x}^{'} A \mathbf{x} = 0$ if and only if $\mathbf{x} = \mathbf{0}$. In other words, a symmetric matrix $A_{n \times n}$ is n.d. if and only if $-A$ is p.d.

$$\boxed{\begin{array}{c} \text{A symmetric matrix } A \text{ is } n.d. \text{ if and only if all odd-order} \\ \text{principal minors are negative and all even-order principal} \\ \text{minors are positive.} \end{array}} \quad (4.7.5)$$

Next, we summarize an important tool which helps one to find a **maximum or minimum of a real valued function of two variables.** The result follows:

Suppose that $f(\mathbf{x})$ is a real valued function of a two-dimensional variable $\mathbf{x} = (x_1, x_2) \in (a_1, b_1) \times (a_2, b_2) \subseteq \Re^2$, having continuous second-order partial derivatives $\frac{\partial^2}{\partial x_1^2} f(\mathbf{x})$, $\frac{\partial^2}{\partial x_2^2} f(\mathbf{x})$ and $\frac{\partial^2}{\partial x_1 \partial x_2} f(\mathbf{x}) \equiv \frac{\partial^2}{\partial x_2 \partial x_1} f(\mathbf{x})$, everywhere in an open rectangle $(a_1, b_1) \times (a_2, b_2)$. Suppose also that for some point $\boldsymbol{\xi} = (\xi_1, \xi_2) \in (a_1, b_1) \times (a_2, b_2)$, one has

$$\tfrac{\partial}{\partial x_1} f(\mathbf{x}) \mid_{\mathbf{x}=\boldsymbol{\xi}} = \tfrac{\partial}{\partial x_2} f(\mathbf{x}) \mid_{\mathbf{x}=\boldsymbol{\xi}} = 0$$

Now, denote the matrix of the second-order partial derivatives:

$$H \equiv H(\mathbf{x}) = \begin{pmatrix} \dfrac{\partial^2}{\partial x_1^2} f(\mathbf{x}) & \dfrac{\partial^2}{\partial x_1 \partial x_2} f(\mathbf{x}) \\ \dfrac{\partial^2}{\partial x_1 \partial x_2} f(\mathbf{x}) & \dfrac{\partial^2}{\partial x_2^2} f(\mathbf{x}) \end{pmatrix} \qquad (4.7.6)$$

Then,

(i) $f(\mathbf{x})$ has a minimum at $\mathbf{x} = \boldsymbol{\xi}$ if H at $\mathbf{x} = \boldsymbol{\xi}$ is p.d.

(ii) $f(\mathbf{x})$ has a maximum at $\mathbf{x} = \boldsymbol{\xi}$ if H at $\mathbf{x} = \boldsymbol{\xi}$ is n.d. $\qquad (4.7.7)$

4.8   Exercises and Complements

4.2.1 Prove Theorem 4.2.1.

4.2.2 Let $X_1, ..., X_k$ be iid Poisson(λ). Show that $\Sigma_{i=1}^k X_i$ has a Poisson distribution with the parameter $k\lambda$.

4.2.3 Let Z_1, Z_2 be iid $N(0,1)$. Show that $Z_1^2 + Z_2^2$ is distributed as χ_2^2.

4.2.4 Consider two related questions:

(i) Let U have its pdf $\frac{1}{2\sigma} e^{-|u|/\sigma} I(-\infty < u < \infty)$. Show that the mgf of U is $(1 - \sigma^2 t^2)^{-1}$ for $|t| < \sigma^{-1}$, with $0 < \sigma < \infty$;

(ii) Let X_1, X_2 be iid with a common pdf $\frac{1}{\beta} exp(-\frac{x}{\beta}) I(x > 0)$ where $\beta > 0$. What is the pdf of $V = X_1 - X_2$?

4.2.5 Let Z be a standard normal variable. Derive $E[exp(tZ^2)]$ by integration. Hence, obtain the distribution of Z^2.

4.3.1 Let X_1, X_2 be independent random variables. X_1 is Binomial(n_1, p) and X_2 is Binomial(n_2, p) with $0 < p < 1$. Find the pmf of $Y = X_1 + X_2$ by direct calculation and identify the distribution of Y. {Hint: With $0 \le y \le n_1 + n_2$, express $P(X_1 + X_2 = y)$ as $\Sigma_{a=0}^{n_1} P(X_1 = a \cap X_2 = y - a)$ and this simplifies to $p^y(1-p)^{n_1+n_2-y}\sum_{a=0}^{n_1}\binom{n_1}{a}\binom{n_2}{y-a}$. It reduces to $\binom{n_1+n_2}{y}p^y(1-p)^{n_1+n_2-y}$. So, Y has a Binomial(n_1+n_2, p) distribution.}

4.3.2 Let Z be a standard normal variable. Denote $X = \max\{|Z|, 1/|Z|\}$.

(i)   Find the expression of $P\{X \le x\}$ for $x \ge 1$. Observe that $P\{X \le x\} = 0$ for $x < 1$;

(ii)   Use part (i) to show that the pdf of X is given by $f(x) = 2\{\phi(x) + x^{-2}\phi(x^{-1})\}I(x \ge 1)$.

{*Hint*: For $x \ge 1$, one writes $P\{X \le x\} = P\{x^{-1} \le |Z| \le x\} = 2P\{x^{-1} \le Z \le x\}$ which is $2\{\Phi(x) - \Phi(x^{-1})\}$. This is part (i). Differentiate this expression for part (ii).}

4.3.3 Suppose that X has its pdf:

$$f(x) = \begin{cases} \frac{3}{7}x^2 & \text{if } 1 < x < 2 \\ 0 & \text{elsewhere} \end{cases}$$

Obtain $G(y) = P(Y \le y)$ where $Y = X^2$ for all y in the real line. Hence, find the pdf of Y.

4.3.4 Let X_1, X_2 have their joint pdf:

$$f(x_1, x_2) = \begin{cases} 3x_1 & \text{if } 0 < x_2 < x_1 < 1 \\ 0 & \text{elsewhere} \end{cases}$$

What is the pdf of $U = X_1 - X_2$? {*Hint*: Obtain the df $G(u)$ of U. Obviously, $G(u) = 0$ (or 1) for $u \le 0$ (or for $u \ge 1$). With $0 < u < 1$, verify that $G(u) = \frac{1}{2}(3u - u^3)$.}

4.3.5 Let $X_1, ..., X_n$ be iid continuous random variables with a common pdf and df respectively given by $f(x)$ and $F(x)$. Derive the joint pdf of the ith- and j^{th}-order statistics $Y_i = X_{n:i}, Y_j = X_{n:j}, 1 \le i < j \le n$ by solving the following parts. Define $U_1 = \sum_{i=1}^{n}I(X_i \le u_1), U_2 = \sum_{i=1}^{n}I(u_1 < X_i \le u_2), U_3 = n - U_1 - U_2$. Show that:

(i) (U_1, U_2, U_3) is distributed as multinomial with $k = 3, p_1 = F(u_1), p_2 = F(u_2) - F(u_1),$ and $p_3 = 1 - p_1 - p_2 = 1 - F(u_2)$;

(ii) $P\{Y_i \le u_1 \cap Y_j \le u_2\} = P\{U_1 \ge i \cap (U_1 + U_2) \ge j\}$;

(iii) $P\{U_1 \ge i \cap (U_1 + U_2) \ge j\} = \sum_{k=i}^{j-1}\sum_{l=j-k}^{n-k}P\{U_1 = k \cap U_2 = l\}$ $+ P\{U_1 \ge j\}$;

(iv)   The expression in part (iii) can be rewritten as
$\sum_{k=i}^{j-1}\sum_{l=j-k}^{n-k}\frac{n!}{k!l!(n-k-l)!}\{F(u_1)\}^k\{F(u_2) - F(u_1)\}^l \times$ $\{1 - F(u_2)\}^{n-k-l}$;

(v)   The joint pdf of Y_i, Y_j can be obtained by evaluating
$\frac{\partial^2}{\partial y_i \partial y_j}\{P\{Y_i \le y_i \cap Y_j \le y_j\}\}$ first, and then by simplifying it.

4.3.6 Let $X_1, ..., X_n$ be iid Uniform$(0, \theta), \theta \in \Re$. Suppose Y_i is the ith-order statistic and define $U = Y_n - Y_1, V = \frac{1}{2}(Y_n + Y_1)$ where U and V are respectively referred to as the *range* and *midrange* for the data $X_1, ..., X_n$. This exercise shows how to find the pdf of U.

(i)   Show that the joint pdf of Y_1 and Y_n is given by $h(y_1, y_n) = n(n-1)\theta^{-n}(y_n - y_1)^{n-2}I(0 < y_1 < y_n < \theta)$;

(ii) Transform (Y_1, Y_n) to (U, V) where $U = Y_n - Y_1, V = \frac{1}{2}(Y_n + Y_1)$. Show that the joint pdf of (U, V) is given by $g(u, v) = n(n-1)\theta^{-n}u^{n-2}$ when $0 < u < \theta, \frac{1}{2}u < v < \theta - \frac{1}{2}u$, and zero otherwise;

(iii) Show that the pdf of U is given by $g_1(u) = n(n-1)\theta^{-n} \times (\theta - u)u^{n-2}I(0 < u < \theta)$. $\{$*Hint:* $g_1(u) = \int_{\frac{1}{2}u}^{\theta - \frac{1}{2}u} g(u, v)dv$, for $0 < u < \theta.\}$;

(iv) When $\theta = 1$, does U follow a standard distribution?

4.3.7 Let $X_1, ..., X_n$ be iid uniform random variables on $(\theta, \theta + 1), \theta \in \Re$. Let Y_i be the ith-order statistic and define $U = Y_n - Y_1, V = \frac{1}{2}(Y_n + Y_1)$. Recall that U, V are respectively referred to as the *range* and *midrange* for the data $X_1, ..., X_n$. This exercise shows how to find the pdf of U.

(i)   Show that the joint pdf of Y_1 and Y_n is given by $h(y_1, y_n) = n(n-1)(y_n - y_1)^{n-2}I(\theta < y_1 < y_n < \theta + 1)$;

(ii) Transform (Y_1, Y_n) to (U, V) where $U = Y_n - Y_1, V = \frac{1}{2}(Y_n + Y_1)$. Show that the joint pdf of (U, V) is given by $g(u, v) = n(n-1)u^{n-2}$ when $0 < u < 1, \theta + \frac{1}{2}u < v < \theta + 1 - \frac{1}{2}u$, and zero otherwise;

(iii) Show that the pdf of U is given by $g_1(u) = n(n-1) \times (1-u)u^{n-2}I(0 < u < 1)$. $\{$*Hint:* $g_1(u) = \int_{\theta + \frac{1}{2}u}^{\theta + 1 - \frac{1}{2}u} g(u, v)dv$, for $0 < u < 1.\}$;

(iv) Does U follow a standard distribution?

4.3.8 (Triangular Distribution) Let X_1, X_2 be iid uniform on $(0, 1)$. Show that the pdf of $U = X_1 + X_2$ is

$$g(u) = \begin{cases} u & \text{if } 0 < u < 1 \\ 2 - u & \text{if } 1 < u < 2 \\ 0 & \text{elsewhere} \end{cases}$$

The random variable U is said to follow a *triangular* distribution.

4.3.9 Suppose X_1, X_2, X_3 are iid Uniform$(0, 1)$.

(i)   Find the joint pdf of $X_{3:1}, X_{3:2}, X_{3:3}$;

(ii)   Derive the pdf of the median;

(iii)   Derive the pdf of the range $U = X_{3:3} - X_{3:1}$.

4.3.10 Let $X_1, ..., X_n$ be iid $N(0,1)$. Denote $Y_i = 4\Phi(X_i)$ where $\Phi(u) = \{\sqrt{2\pi}\}^{-1} \int_{-\infty}^{u} e^{-x^2/2} dx$, $u \in \Re$, $i = 1, ..., n$. Let $U = \Pi_{i=1}^{n} Y_i$.

(i)   Find the pdf of U;

(ii)   Evaluate $E[U]$ and $V[U]$;

(iii)   Find the marginal pdfs of $Y_{n:1}, Y_{n:n}$, and $Y_{n:n} - Y_{n:1}$.

4.3.11 Let X_1 be uniform on $(0,2)$, X_2 have its pdf $\frac{1}{2}xI(0 < x < 2)$, and X_3 have its pdf $\frac{3}{8}x^2 I(0 < x < 2)$. Let X_1, X_2, X_3 be independent. Derive the marginal pdfs of $X_{3:1}, X_{3:3}$. {*Hint*: Follow along Example 4.3.2.}

4.3.12 Let $X_1, ..., X_n$ be iid Uniform$(0,1)$. Define $Y_i = X_{n:i}/X_{n:i-1}$, $i = 1, ..., n$ with $X_{n:0} = 1$. Find the joint distribution of $Y_2, ..., Y_n$.

4.4.1 Let X have a pdf $\frac{1}{\beta}exp(-\frac{x}{\beta})I(x > 0)$, $\beta > 0$. Find the pdf of X^2.

4.4.2 Suppose that X has a pdf:

$$f(x) = \begin{cases} 2(1-x) & \text{if } 0 < x < 1 \\ 0 & \text{elsewhere} \end{cases}$$

Find the pdf of $W = \sqrt{X}$.

4.4.3 Suppose X has a Rayleigh density:

$$f(x) = \begin{cases} \frac{2x}{\theta} \, exp(-x^2/\theta) & \text{if } x > 0 \\ 0 & \text{elsewhere} \end{cases}$$

Derive the pdf of $U = X^2$.

4.4.4 Suppose X has its pdf $\frac{1}{2}e^{-|x|}$ for $x \in \Re$. Denote $Y = |X|$. Derive the pdf of Y. {*Hint*: $g(x) = |x| : \Re \to \Re^+$ is *not* one-to-one. For $0 < y < \infty$, write down the pdf of Y as follows: $\frac{1}{2}e^{-y}|+1| + \frac{1}{2}e^{-y}|-1| = e^{-y}$.}

4.4.5 (Example 4.4.6 Continued) Let $X_1, ..., X_n$ be iid $N(\mu, \sigma^2)$ with $n \geq 2$. Show that $(\overline{X} - 1)^2$ and $S + S^{2/3}$ are independently distributed.

4.4.6 Let X_1, X_2 be iid with a Gamma(α, β) distribution, $\alpha > 0, \beta > 0$. Denote $U = X_1 + X_2, V = X_2/X_1$. Find the marginal distributions of U, V.

4.4.7 Let X_1, X_2 be iid $N(\mu, \sigma^2)$. Define $U = X_1 + X_2, V = X_1 - X_2$. Find the joint pdf of (U, V). Evaluate the correlation coefficient between U, V. Evaluate $E\{(X_1 - X_2)^2 \mid X_1 + X_2\}$ and $E\{(X_1 + X_2)^2 \mid X_1 = X_2\}$.

4.4.8 Let X_1, X_2, X_3 be iid Gamma(α, β), $\alpha > 0, \beta > 0$. Define $U_1 = X_1 + X_2 + X_3, U_2 = X_2/(X_1 + X_2)$, and $U_3 = X_3/(X_1 + X_2 + X_3)$.

(i)   Show that one can express $x_1 = u_1(1 - u_2)(1 - u_3)$, $x_2 = u_1 u_2(1 - u_3)$, $x_3 = u_1 u_3$. Is this transformation one-to-one?

(ii)   Determine the matrix J from Equation (4.4.1) and show that $\mid det(J) \mid = u_1^2(1 - u_3)$;

(iii) Start with the joint pdf of X_1, X_2, X_3. Use transformation $(x_1, x_2, x_3) \rightarrow (u_1, u_2, u_3)$ to obtain the joint pdf of U_1, U_2, U_3;

(iv)   Are U_1, U_2, U_3 independent? Show that (a) U_1 is Gamma $(3\alpha, \beta)$, (b) U_2 is Beta(α, α), and (c) U_3 is Beta$(\alpha, 2\alpha)$.

4.4.9 Suppose that X_1, X_2 are iid standard exponential random variables. Denote $Y_1 = X_1 + X_2, Y_2 = X_2$. Derive the joint pdf of Y_1, Y_2. Hence, obtain the pdf of Y_1.

4.4.10 Suppose (X_1, X_2) has $N_2(0, 0, \sigma^2, \sigma^2, \rho)$ distribution with $0 < \sigma < \infty, -1 < \rho < 1$. Derive the joint pdf of $Y_1 = X_1 + X_2$ and $Y_2 = X_1 - X_2$.

4.4.11 Let X_1, X_2 be iid random variables. We are told that $X_1 + X_2$ and $X_1 - X_2$ are independently distributed. Show that the common pdf of X_1, X_2 must be a normal density. {*Hint*: Can Remark 4.4.3 be used?}

4.4.12 (Exponential Spacing) Let $X_1, ..., X_n$ be iid with a common pdf $\beta^{-1} exp(-x/\beta)I(x > 0)$, $\beta > 0$. Consider the order statistics $Y_i = X_{n:i}$, $i = 1, ..., n$. The observations $X_1, ..., X_n$ and $X_{n:1}, ..., X_{n:n}$ may be interpreted as *failure times* and *ordered failure times*, respectively. Let

$$U_1 = Y_1, U_2 = Y_2 - Y_1, U_3 = Y_3 - Y_2, ..., U_n = Y_n - Y_{n-1}$$

which are the *spacings* between successive failures. We have a one-to-one transformation from $(y_1, ..., y_n)$ to $(u_1, ..., u_n)$.

(i)   Show that the joint pdf of the order statistics $Y_1, ..., Y_n$ is given by $n!\beta^{-n} exp\left(-\Sigma_{i=1}^{n} y_i/\beta\right) I(0 < y_1 \leq y_2 \leq ... \leq y_n < \infty)$;

(ii)   Note that $y_i = \Sigma_{j=1}^{i} u_j$, $\Sigma_{i=1}^{n}(n - i + 1)u_i = \Sigma_{i=1}^{n} y_i$, $i = 1, ..., n$. Also, verify that $\mid det(J) \mid = 1$;

(iii) Show that the joint pdf $h(u_1, ..., u_n)$ of $U_1, .., U_n$ can be written as $n!\beta^{-n} exp\left\{-[nu_1 + (n - 1)u_2 + ... + 2u_{n-1} + u_n]/\beta\right\}$ for $0 < u_1, ..., u_n < \infty$;

(iv)   Hence, claim that $U_1, ..., U_n$ are *independently* distributed, and that U_i has an exponential distribution with mean parameter $(n - i + 1)^{-1}\beta, i = 1, ..., n$.

4.4.13 (Exercise 4.4.12 Continued) Evaluate $E[X_{n:k}]$, $k = 1, ..., n - 1$. {*Hint*: $X_{n:k+1} - X_{n:k} = Y_{k+1} - Y_k = U_{k+1}$, and hence $E[X_{n:k+1} - X_{n:k}] = E[U_{k+1}] = (n - k)^{-1}\beta$. Also, $X_{n:k} = \Sigma_{j=1}^{k} U_j$.}

4.4.14 (Exercise 4.4.12 Continued) Let $X_1, ..., X_n$ be iid with a common pdf:

$$\sigma^{-1} exp\{-(x - \mu)/\sigma\}I(x > \mu), -\infty < \mu < \infty, 0 < \sigma < \infty$$

In reliability applications, μ is often referred to as the minimum guarantee time or a minimum threshold. Consider the random variables $X_{n:1}$ and $T = \Sigma_{i=1}^{n}(X_i - X_{n:1})$. Find their joint distribution.

4.4.15 Suppose that X_1, X_2, X_3 are iid random variables. We are told that $X_1 + X_2 + X_3$ and $(X_1 - X_2, X_1 - X_3)$ are independently distributed. Show that the common pdf of X_1, X_2, X_3 must be normal. {*Hint*: Can the Remark 4.4.3 be used to solve this problem?}

4.4.16 (Polar Transformation) Consider a popular game of "hitting the bull's eye". One aims a dart at the center (or origin) and it lands at some point (Z_1, Z_2) whose distance from the center is $(Z_1^2 + Z_2^2)^{1/2}$. Suppose Z_1, Z_2 are independent $N(0, 1)$. Obtain $P\{Z_1^2 + Z_2^2 \le a\}$ by direct integration where $a > 0$. {*Hint*: Note that the joint pdf of (Z_1, Z_2) is $f(z_1, z_2) = (2\pi)^{-1} exp\{-\frac{1}{2}(z_1^2 + z_2^2)\}$ with $-\infty < z_1, z_2 < \infty$. The required probability is:

$$\underset{z_1^2+z_2^2 \le a}{\int \int} (2\pi)^{-1} exp\left\{-\tfrac{1}{2}(z_1^2 + z_2^2)\right\} dz_1 dz_2$$

Now, substitute $z_1 = \sqrt{r}\sin(\theta), z_2 = \sqrt{r}\cos(\theta)$ with $r > 0$ and $0 < \theta < 2\pi$, and let

$$A = \begin{pmatrix} \frac{\partial z_1}{\partial \theta} & \frac{\partial z_1}{\partial r} \\ \frac{\partial z_2}{\partial \theta} & \frac{\partial z_2}{\partial r} \end{pmatrix} = \begin{pmatrix} \sqrt{r}\cos(\theta) & \frac{1}{2\sqrt{r}}\sin(\theta) \\ -\sqrt{r}\sin(\theta) & \frac{1}{2\sqrt{r}}\cos(\theta) \end{pmatrix}$$

Then, rewrite the double integral as $(2\pi)^{-1} \int_{r=0}^{a} \int_{\theta=0}^{2\pi} exp\{-r/2\} det(A)d\theta dr$ which simplifies to

$$\tfrac{1}{2}(2\pi)^{-1} \int_{\theta=0}^{2\pi} d\theta \int_{r=0}^{a} exp\{-r/2\} dr$$
$$= \tfrac{1}{2}(2\pi)^{-1}(2\pi)2\{1 - e^{-\frac{1}{2}a}\} = 1 - e^{-\frac{1}{2}a}.\}$$

4.5.1 Suppose that $X_1, ..., X_n$ are iid $N(\mu, \sigma^2)$. Recall the Helmert variables $Y_1, ..., Y_n$ from Example 4.4.6.

(i)   Find the marginal pdf of $V = Y_i/Y_j$, $i \ne j = 2, ..., n$;

(ii)   Find the marginal pdf of $V = Y_i/|Y_j|$, $i \ne j = 2, ..., n$;

(iii)   Find the marginal pdf of $V = Y_i^2/Y_j^2$, $i \ne j = 2, ..., n$;

(iv)   Find the pdf of Y_i^2/T where $T = \underset{2 \le k \le n, k \ne i}{\Sigma} Y_k^2$, $i = 2, ..., n$;

(v)   Find the pdf of $(Y_i^2 + Y_j^2)/U$ where $U = \underset{2 \le k \le n, k \ne i, k \ne j}{\Sigma} Y_k^2$,

$i \ne j = 2, ..., n$.

4.5.2 With $n(\ge 4)$ iid random variables from $N(0, \sigma^2)$, consider the Helmert variables $Y_1, ..., Y_4$ from Example 4.4.6. Obtain the distribution of

$$W = Y_1/\sqrt{Y_2^2 + Y_3^2 + Y_4^2}$$

and that of W^2.

4.5.3 Consider the pdf $h(w)$ of t_ν, the Student's t with ν degrees of freedom, $w \in \Re$. If $\nu = 1$, show that

$$h(w) = \frac{1}{\pi(1 + w^2)}$$

which is the Cauchy density.

4.6.1 Show that $(3X_1, \overline{X} + 1)$ is distributed as

$$N_2(3\mu, \mu + 1, 9\sigma^2, \tfrac{1}{n}\sigma^2, \tfrac{1}{\sqrt{n}})$$

when $X_1, ..., X_n$ are iid $N(\mu, \sigma^2)$.

4.6.2 Find the distribution of $(aX_1 + bX_2, \overline{X})$ when $X_1, ..., X_n$ are iid $N(\mu, \sigma^2)$ if a, b are fixed nonzero real numbers.

4.6.3 Show that the bivariate normal density from Equation (3.5.1) can be expressed in the form given by Equation (4.6.1).

4.6.4 Let $X_1, ..., X_n$ be iid $N(\mu, \sigma^2)$. Find the joint distributions of
(i) $Y_2, ..., Y_n$ where $Y_i = X_i - X_1, i = 2, ..., n$;
(ii) $U_2, ..., U_n$ where $U_i = X_i - X_{i-1}, i = 2, ..., n$.

4.6.5 Suppose that $X_1, ..., X_k$ are iid $N(\mu, \sigma^2), k \geq 2$. Denote:

$$U_1 = \Sigma_{i=1}^k X_i, U_j = X_1 - X_j \text{ for } j = 2, ..., k$$

(i)   Show that $\mathbf{U} = (U_1, ..., U_k)$ has a k-dimensional normal distribution;
(ii)   Show that U_1 and $(U_2, ..., U_k)$ are independent;
(iii)   Use parts (i) and (ii) to derive the distribution of $\overline{X}$;
(iv)   Express S^2 as a function of $U_2, ..., U_k$ alone. Hence, show that $\overline{X}, S^2$ are independently distributed.

{*Hint*: Observe that $\binom{k}{2}S^2$ may be rewritten as $\sum_{1 \leq i < j \leq k} \tfrac{1}{2}(X_i - X_j)^2$.}

4.6.6 Let (X_1, X_2) have a joint pdf:

$$\alpha f(x_1, x_2; 0, 0, 1, 1, \rho) + (1 - \alpha)f(x_1, x_2; 0, 0, 1, 1, -\rho)$$

with $0 < \alpha < 1$, where $f(x_1, x_2; \mu_1, \mu_2, \sigma_1^2, \sigma_2^2, \rho)$ stands for a bivariate normal pdf with means μ_1, μ_2, variances σ_1^2, σ_2^2, and the correlation coefficient ρ with $0 < \rho < 1$. Show that $X_1 + X_2$ is *not* normally distributed even though both X_1, X_2 are standard normal variables.

5

Notions of Convergence

5.1 Introduction

This chapter introduces two fundamental concepts of convergence for a sequence of random variables $\{U_n; n \geq 1\}$. In Section 5.2, we discuss *convergence in probability* (denoted by $\overset{P}{\to}$) and the *weak laws of large numbers* (WLLN). Section 5.3 introduces the notion of *convergence in distribution* or *law* (denoted by $\overset{\mathcal{L}}{\to}$). Slutsky's Theorem sets some ground rules for manipulations involving both modes of convergence. Section 5.3 also includes the *central limit theorem* (CLT) for both sample mean (Theorem 5.3.3) and sample variance (Theorem 5.3.5). Some large-sample properties of Chi-square, t, and F distributions are included in Section 5.4.

5.2 Convergence in Probability

Definition 5.2.1 *A sequence of real valued random variables* $\{U_n; n \geq 1\}$ *is said to converge in probability to a real number u as $n \to \infty$, denoted by $U_n \overset{P}{\to} u$, if and only if*

$$P\{|U_n - u| > \varepsilon\} \to 0 \text{ as } n \to \infty, \text{ for every fixed } \varepsilon(>0) \qquad (5.2.1)$$

$U_n \overset{P}{\to} u$ means this: the probability that U_n will stay away from u, even by a small margin ε, can be made arbitrarily small for all $n \geq n_0 \equiv n_0(\varepsilon)$. For a fixed ε (> 0), denote $p_n(\varepsilon) = P\{|U_n - u| > \varepsilon\}$. Then, for U_n to converge to u in probability, we ask that the sequence $\{p_n(\varepsilon); n \geq 1\}$ of nonnegative real numbers converges to zero, for *all* fixed $\varepsilon(>0)$.

Theorem 5.2.1 (Weak WLLN) *Suppose $X_1, ..., X_n$ are independent and identically distributed (iid) real valued random variables with $E(X_1) = \mu$ and $V(X_1) = \sigma^2$, $-\infty < \mu < \infty$, $0 < \sigma < \infty$. Write $\overline{X}_n(= n^{-1}\Sigma_{i=1}^n X_i)$ for the sample mean, $n \geq 1$. Then, $\overline{X}_n \overset{P}{\to} \mu$ as $n \to \infty$.*

Proof Fix some arbitrary $\varepsilon(>0)$ and use Markov Inequality (3.8.1) to write:

$$P\{|\overline{X}_n - \mu| \geq \varepsilon\} = P\{(\overline{X}_n - \mu)^2 \geq \varepsilon^2\}$$
$$\leq \varepsilon^{-2} E\{(\overline{X}_n - \mu)^2\} = \sigma^2/(n\varepsilon^2) \qquad (5.2.2)$$

Then, we have

$$0 \leq P\{|\overline{X}_n - \mu| \geq \varepsilon\} \leq \sigma^2/(n\varepsilon^2) \to 0$$

as $n \to \infty$. So, $P\{|\overline{X}_n - \mu| \geq \varepsilon\} \to 0$ as $n \to \infty$ and hence $\overline{X}_n \overset{P}{\to} \mu$ as $n \to \infty$. ∎

Intuitively, the Weak WLLN helps us to conclude that $\overline{X}_n$ obtained from iid random variables can be expected to hang around the population mean μ with high probability, if n is large enough and $0 < \sigma < \infty$.

Example 5.2.1 Suppose $X_1, ..., X_n$ are iid $N(\mu, \sigma^2)$, $-\infty < \mu < \infty$, $0 < \sigma < \infty$. Write $S_n^2 = (n-1)^{-1}\sum_{i=1}^n (X_i - \overline{X}_n)^2, n \geq 2$. Using Equation (4.4.8), we can express $S_n^2 = (n-1)^{-1}\sum_{i=2}^n Y_i^2$ where $Y_2, ..., Y_n$ are the Helmert variables which are iid $N(0, \sigma^2)$. Observe that

$$V(Y_2^2) = E(Y_2^4) - E^2(Y_2^2) = 3\sigma^4 - (\sigma^2)^2 = 2\sigma^4$$

Now, since S_n^2 is a sample mean of iid random variables with finite variance, the Weak WLLN applies. So, $S_n^2 \overset{P}{\to} E(Y_2^2) = \sigma^2$. ▲

> Under mild additional conditions, one can conclude that $S_n^2 \overset{P}{\to} \sigma^2$, without the assumption of normality of $X_1, ..., X_n$. Refer to Example 5.2.7.

Example 5.2.2 Suppose $X_1, ..., X_n$ are iid Bernoulli$(p), 0 < p < 1$. The Weak WLLN implies that $\overline{X}_n \overset{P}{\to} E(X_1) = p$. ▲

Example 5.2.3 Let $X_1, ..., X_n$ be iid Uniform$(0, \theta), \theta > 0$. $T_n = X_{n:n}$, the largest order statistic, has its *probability density function* (pdf)

$$g(t) = nt^{n-1}\theta^{-n}I(0 < t < \theta) \tag{5.2.3}$$

For fixed $\varepsilon > 0$, one has:

$$P\{|T_n - \theta| \geq \varepsilon\} = P\{\theta - T_n \geq \varepsilon\} = \begin{cases} 0 & \text{if } \varepsilon \geq \theta \\ (1 - \frac{\varepsilon}{\theta})^n & \text{if } \varepsilon < \theta \end{cases} \tag{5.2.4}$$

If $0 < \varepsilon < \theta$, surely $\lim_{n\to\infty} (1 - \frac{\varepsilon}{\theta})^n = 0$. Now, Definition 5.2.1 implies: $X_{n:n} \overset{P}{\to} \theta$. ▲

Weak WLLN (Theorem 5.2.1) can be strengthened. See the next result.

Theorem 5.2.2 (Khintchine's WLLN) *Suppose $X_1, ..., X_n$ are iid real valued random variables with a finite mean μ. Then, $\overline{X}_n \overset{P}{\to} \mu$ as $n \to \infty$.*

Example 5.2.4 In order to appreciate Khintchine's WLLN, consider iid random variables $\{X_n; n \geq 1\}$ with a common *probability mass function* (pmf):

X values:	1	2	3	. . .	i	. . .	
Probabilities:	$\frac{c}{1^3}$	$\frac{c}{2^3}$	$\frac{c}{3^3}$	. . .	$\frac{c}{i^3}$	. . .	(5.2.5)

where $c^{-1} = \sum_{i=1}^\infty \frac{1}{i^3}, 0 < c < \infty$. $E(X_1) = c\sum_{i=1}^\infty \frac{1}{i^2}$ which is finite. But, $E(X_1^2) = c\sum_{i=1}^\infty \frac{1}{i}$, which is infinite. We have a situation where μ is finite but σ^2 is not finite. Khintchine's WLLN implies: $\overline{X}_n \xrightarrow{P} E(X_1)$ as $n \to \infty$, but Weak WLLN does not apply! ▲

Here are two issues to keep in mind: (i) U_n may converge to u in probability, but one may not take it for granted that $E(U_n) \to u$; (ii) $E(U_n)$ may converge to u, but one may not jump to conclude that $U_n \xrightarrow{P} u$.

Example 5.2.5 Consider two sequences of random variables:

$$
\begin{array}{llll}
U_n \text{ values:} & 1 & n & \quad V_n \text{ values:} & 1 & n^2 \\
\text{Probabilities:} & 1 - \frac{1}{n} & \frac{1}{n} & \quad \text{Probabilities:} & 1 - \frac{1}{n} & \frac{1}{n}
\end{array}
\qquad (5.2.6)
$$

Note that $U_n \xrightarrow{P} 1$ but $E(U_n) = 2 - \frac{1}{n} \to 2$ as $n \to \infty$. Next, $V_n \xrightarrow{P} 1$ but $E(V_n) = 1 - \frac{1}{n} + n \to \infty$ as $n \to \infty$. ▲

The notions "convergence in probability" and the "convergence in mean" are intricately related. Refer to Sen and Singer (1993, Chapter 2) and Serfling (1980). The proof of our next result involves many details. Refer to Mukhopadhyay (2000, Chapter 5).

Theorem 5.2.3 *Consider two sequences of real valued random variables* $\{U_n, V_n; n \geq 1\}$ *such that* $U_n \xrightarrow{P} u$, $V_n \xrightarrow{P} v$ *as* $n \to \infty$. *Then, we have:*

(i) $U_n \pm V_n \xrightarrow{P} u \pm v$;

(ii) $U_n V_n \xrightarrow{P} uv$;

(iii) $U_n/V_n \xrightarrow{P} u/v$ *if* $P(V_n = 0) = 0$ *for all* $n \geq 1, v \neq 0$.

Example 5.2.6 Suppose that $X_1, ..., X_n$ are iid $N(\mu, \sigma^2), -\infty < \mu < \infty, 0 < \sigma^2 < \infty, n \geq 2$. Consider $\overline{X}_n$ and S_n^2. We know that $\overline{X}_n \xrightarrow{P} \mu$, $S_n^2 \xrightarrow{P} \sigma^2$. Thus, $\overline{X}_n \pm S_n^2 \xrightarrow{P} \mu \pm \sigma^2$, $\overline{X}_n / S_n^2 \xrightarrow{P} \mu/\sigma^2$ as $n \to \infty$. ▲

Example 5.2.7 (Distribution-Free) Let $X_1, ..., X_n$ be iid with mean μ and finite variance $\sigma^2 (> 0), n \geq 2$. Write $S_n^2 = n(n-1)^{-1}\{n^{-1}\Sigma_{i=1}^n X_i^2 - \overline{X}_n^2\}$. By Khintchine's WLLN, $n^{-1}\Sigma_{i=1}^n X_i^2 \xrightarrow{P} E(X_1^2) = \mu^2 + \sigma^2$, and $\overline{X}_n^2 \xrightarrow{P} \mu^2$ from Theorem 5.2.3 (ii) with $U_n = V_n = \overline{X}_n$. Theorem 5.2.3 also implies:

$$
n^{-1}\Sigma_{i=1}^n X_i^2 - \overline{X}_n^2 \xrightarrow{P} \mu^2 + \sigma^2 - \mu^2 = \sigma^2
$$

So, $S_n^2 \xrightarrow{P} \sigma^2$ since $n(n-1)^{-1} \to 1$. ▲

Theorem 5.2.4 *Suppose* $\{U_n; n \geq 1\}$ *is a sequence of real valued random variables,* $U_n \xrightarrow{P} u$ *as* $n \to \infty$, *and* $g(.)$ *is a real valued continuous function. Then,* $g(U_n) \xrightarrow{P} g(u)$ *as* $n \to \infty$.

Proof A function $g(x)$ is continuous at $x = a$ if the following holds: given arbitrary but fixed $\varepsilon(> 0)$, there exists some positive number $\delta \equiv \delta(\varepsilon)$ such that

$$
|g(x) - g(u)| \geq \varepsilon \Rightarrow |x - u| \geq \delta
$$

Hence, for large enough $n (\geq n_0 \equiv n_0(\varepsilon))$, we can write

$$0 \leq P\{|g(U_n) - g(u)| \geq \varepsilon\} \leq P\{|U_n - u| \geq \delta\} \qquad (5.2.7)$$

The upper bound in Equation (5.2.7) $\to 0$ since $U_n \overset{P}{\to} u$. The proof is complete. ∎

Example 5.2.8 (Example 5.2.3 Continued) We have a Uniform$(0, \theta)$ population with $\theta > 0$ and $T_n = X_{n:n} \overset{P}{\to} \theta$. Hence, $X_{n:n}^2 \overset{P}{\to} \theta^2$ since $g(x) = x^2, x > 0$ is a continuous function. Also, $X_{n:n}^{\frac{1}{3}} \overset{P}{\to} \theta^{\frac{1}{3}}$ since $g(x) = x^{\frac{1}{3}}, x > 0$ is a continuous function. We can also claim that $\sin(X_{n:n}) \overset{P}{\to} \sin(\theta)$. ▲

Example 5.2.9 Let $X_1, ..., X_n$ be iid Poisson(λ) with $\lambda > 0$ and consider $\overline{X}_n$. By Weak WLLN, $\overline{X}_n \overset{P}{\to} \lambda$. Combining results from Theorems 5.2.3 and 5.2.4, we can write, for example,

$$\overline{X}_n^3 \left\{3\sqrt{\overline{X}_n} - \overline{X}_n + 5\right\} \overset{P}{\to} \lambda^3 \left\{3\sqrt{\lambda} - \lambda + 5\right\} \text{ and}$$

$$\tanh^{-1}(3\overline{X}_n + n^{-1}) \overset{P}{\to} \tanh^{-1}(3\lambda) \text{ as } n \to \infty$$

But, can one claim that $\overline{X}_n^{-3} \overset{P}{\to} \lambda^{-3}$? One should note that $P\{\overline{X}_n = 0\} = e^{-n\lambda}$, which is positive for every fixed $n \geq 1$ whatever be $\lambda(> 0)$, and Theorem 5.2.3, part (iii) does not apply! One can conclude, however, that $(\overline{X}_n + n^{-\gamma})^{-3} \overset{P}{\to} \lambda^{-3}$ with arbitrary but fixed $\gamma(> 0)$. ▲

5.3 Convergence in Distribution

Definition 5.3.1 *Consider* $\{U_n; n \geq 1\}$ *and* U, *all real valued random variables, with respective distribution functions (dfs)* $F_n(u) = P(U_n \leq u)$, $F(u) = P(U \leq u), u \in \Re$. U_n *is said to converge in distribution to* U *as* $n \to \infty$, *denoted by* $U_n \overset{\pounds}{\to} U$, *if and only if* $F_n(u) \to F(u)$ *pointwise at all continuity points of* $F(.)$, *and* $F(.)$ *is called the limiting (or asymptotic) distribution of* U_n.

> It is known that the totality of all discontinuity points of any df F can be at most countably infinite. Review Theorem 1.6.1.

Example 5.3.1 Suppose that $X_1, ..., X_n$ are iid Uniform$(0, \theta), \theta > 0$. The largest order statistic $T_n (= X_{n:n})$ has its pdf:

$$g(t) = nt^{n-1}\theta^{-n}I(0 < t < \theta)$$

The df of T_n is given by

$$G_n(t) = \begin{cases} 0 & \text{if } t \leq 0 \\ (t/\theta)^n & \text{if } 0 < t < \theta \\ 1 & \text{if } t \geq \theta \end{cases} \qquad (5.3.1)$$

Denote $U_n = n(\theta - T_n)/\theta$. Obviously, $P(U_n > 0) = 1$ and from Equation (5.3.1) we can write:

$$F_n(u) = P(U_n \le u) = \begin{cases} 0 & \text{if } u \le 0 \\ 1 - \left(1 - \frac{u}{n}\right)^n & \text{if } u > 0 \end{cases} \qquad (5.3.2)$$

So, $\lim\limits_{n\to\infty} F_n(u) = 0$ if $u \le 0$ and $1 - e^{-u}$ if $u > 0$ since $\lim\limits_{n\to\infty} \left(1 - \frac{u}{n}\right)^n = e^{-u}$.
Now, let U be a random variable with its pdf $f(u) = e^{-u}I(u > 0)$ and df

$$F(u) = \begin{cases} 0 & \text{if } u \le 0 \\ 1 - e^{-u} & \text{if } u > 0 \end{cases} \qquad (5.3.3)$$

All points $u \in \Re$ are continuity points of $F(u)$ and for all such u, we note that $\lim\limits_{n\to\infty} F_n(u) = F(u)$. Hence, $U_n \overset{\mathcal{L}}{\to} U$, that is, asymptotically U_n has a standard exponential distribution. ▲

In general, it may be hard to proceed along the lines of Example 5.3.1. There may be two hurdles. First, we must have an explicit expression of df of U_n, and second, we ought to be able to examine this df's asymptotic behavior as $n \to \infty$. Between these two concerns, the former is likely to create more headache. So, we are forced to pursue an indirect approach involving a *moment generating function* (mgf).

Theorem 5.3.1 *Suppose that the mgfs, $M_n(t) \equiv M_{U_n}(t)$, $M(t) \equiv M_U(t)$ are both finite for $|t| < h$ with some $h(> 0)$. Suppose that $M_n(t) \to M(t)$ for $|t| < h$ as $n \to \infty$. Then,*

$$U_n \overset{\mathcal{L}}{\to} U \text{ as } n \to \infty$$

A proof of this result is beyond the scope of this book. One may refer to Serfling (1980) or Sen and Singer (1993) for details.

Example 5.3.2 Suppose that $X_1, ..., X_n$ are iid Bernoulli(p) with $p = \frac{1}{2}$. Denote $U_n = 2\sqrt{n}(\overline{X}_n - 0.5)$ so that $M_n(t)$ is:

$$E\{exp(tU_n)\} = exp\left(-t\sqrt{n}\right) E\left\{exp\left(\tfrac{2t}{\sqrt{n}}\Sigma_{i=1}^n X_i\right)\right\}$$
$$= exp\left(-t\sqrt{n}\right) \tfrac{1}{2^n}\left\{1 + exp(2t/\sqrt{n})\right\}^n$$

since $E\left(exp(tX_1)\right) = \frac{1}{2}(1 + e^t)$. Thus,

$$M_n(t) = \tfrac{1}{2^n}\left\{exp(-t/\sqrt{n}) + exp(t/\sqrt{n})\right\}^n = \left\{1 + \tfrac{1}{2}n^{-1}t^2 + R_n\right\}^n \qquad (5.3.4)$$

with a remainder term R_n of order $O(n^{-2})$, that is, $\lim\limits_{n\to\infty} n^2 R_n$ is finite. In Equation (5.3.4), the expression in the last step converges to $exp(\frac{1}{2}t^2)$ as $n \to \infty$, because $\left(1 + \frac{a}{n}\right)^n \to e^a$ as $n \to \infty$. But, $M(t) = e^{\frac{1}{2}t^2}$ is the mgf of a standard normal variable. Thus, $2\sqrt{n}(\overline{X}_n - 0.5) \overset{\mathcal{L}}{\to} N(0,1)$. ▲

Example 5.3.3 Suppose that X_n is distributed as χ_n^2. Denote

$$U_n = \tfrac{1}{\sqrt{2n}}(X_n - n), n \geq 1.$$

The question is whether $U_n \overset{\pounds}{\to} U$, some appropriate random variable. For $t < 1/\sqrt{2}$, we start with the mgf of U_n:

$$M_n(t) = E\{exp(tU_n)\} = exp\left\{-t\sqrt{\tfrac{n}{2}}\right\} \left(1 - \tfrac{2t}{\sqrt{2n}}\right)^{-n/2}$$

With $s = \tfrac{t}{\sqrt{2n}}$, we rewrite:

$$M_n(t) = \left\{ \left(1 - t\sqrt{\tfrac{2}{n}}\right)^{-\sqrt{n/2}} e^{-t}\right\}^{\sqrt{n/2}}$$

For large n, we have:

$$log\,(M_n(t)) = \sqrt{\tfrac{n}{2}}\left\{-\sqrt{\tfrac{n}{2}}log\left[1 - t\sqrt{\tfrac{2}{n}}\right] - t\right\}$$

$$= \sqrt{\tfrac{n}{2}}\left\{-\sqrt{\tfrac{n}{2}}\left[-t\sqrt{\tfrac{2}{n}} - t^2\tfrac{1}{n} + O(n^{-3/2})\right] - t\right\} = \tfrac{1}{2}t^2 + O(\tfrac{1}{n^{1/2}})$$

using series expansion of $log(1-x)$ from Equation (1.6.15). Clearly, $M_n(t) \to exp\left(\tfrac{1}{2}t^2\right)$, the mgf of a standard normal variable. So,

$$\tfrac{1}{\sqrt{2n}}(X_n - n) \overset{\pounds}{\to} N(0,1)$$

which is the desired result. ▲

5.3.1   Combination of Two Modes of Convergence

Theorem 5.3.2 (Slutsky's Theorem) *Consider two sequences of random variables* $\{U_n, V_n; n \geq 1\}$, *another random variable* U, *all real valued, and a fixed real number* v. *Suppose that* $U_n \overset{\pounds}{\to} U$, $V_n \overset{P}{\to} v$ *as* $n \to \infty$. *Then, we have as* $n \to \infty$:

(i) $U_n \pm V_n \overset{\pounds}{\to} U \pm v$;

(ii) $U_n V_n \overset{\pounds}{\to} vU$;

(iii) $U_n V_n^{-1} \overset{\pounds}{\to} Uv^{-1}$ *if* $P(V_n = 0) = 0$ *for all* $n \geq 1, v \neq 0$.

Example 5.3.4 (Example 5.3.1 Continued) Suppose $X_1, ..., X_n$ are iid Uniform$(0, \theta)$. Let $T_n = X_{n:n}$ and $V_n = T_n^2\{T_n + \tfrac{1}{n}\}^{-1}$. We know that $T_n \overset{P}{\to} \theta$. Theorem 5.2.3 implies that $V_n \overset{P}{\to} \theta$. We proved that $U_n = n(\theta - T_n)/\theta \overset{\pounds}{\to} U$, a standard exponential variable. So, Slutsky's Theorem implies that $G_n = n(\theta - T_n)/T_n \overset{\pounds}{\to} U$ and $H_n = n(\theta - T_n)/V_n \overset{\pounds}{\to} U$. It may not be easy to apply Definition 5.3.1 directly. ▲

5.3.2   Central Limit Theorem

First, we discuss the central limit theorem for a *standardized* sample mean (Theorem 5.3.3) and then for a sample variance (Theorem 5.3.5).

> Let $X_1, ..., X_n$ be iid random samples from a population with mean μ and variance $\sigma^2, n \geq 2$. Consider $\overline{X}_n$ and S_n^2. $\sqrt{n}(\overline{X}_n - \mu)/\sigma$ is called a *standardized* version of *sample mean* when σ is known, and $\sqrt{n}(\overline{X}_n - \mu)/S_n$ is called a *Studentized* version when σ is unknown.

The *central limit theorem* (CLT) provides the asymptotic distribution for the standardized version of the sample mean $\overline{X}_n$. A proof of CLT under such generality requires knowledge of characteristic functions. Refer to Sen and Singer (1993, pp. 107-108). If one, however, assumes that the mgf of X_1 exists, then CLT can be proved fairly easily along the lines of Examples 5.3.2 and 5.3.3.

Theorem 5.3.3 (Central Limit Theorem) *Suppose that* $X_1, ..., X_n$ *are iid real valued random variables with the common mean* μ *and variance* $\sigma^2, 0 < \sigma < \infty$. *Then, as* $n \to \infty$, *we have:*

$$\frac{\sqrt{n}(\overline{X}_n - \mu)}{\sigma} \xrightarrow{\pounds} N(0, 1)$$

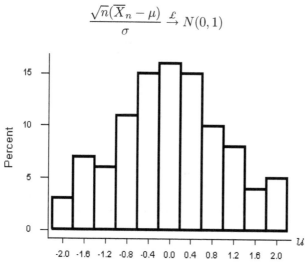

Figure 5.3.1. Histogram of 100 values of $u = \sqrt{n}(\overline{x}_n - 10)/2$ from a $N(10, 4)$ population when $n = 10$.

If the common pdf was normal with mean μ and variance σ^2, we showed in Chapter 4 that $\sqrt{n}(\overline{X}_n - \mu)/\sigma$ would be *exactly* a standard normal variable *whatever* be n. Using MINITAB Release 12.1, we drew random samples from $N(\mu = 10, \sigma^2 = 4)$ population with $n = 10$ and replicated the process 100 times. Having fixed n, in the ith replication we drew n random samples $x_{1i}, ..., x_{ni}$ which led to the value of the sample mean, $\overline{x}_{ni} = n^{-1}\Sigma_{j=1}^{n} x_{ji}$ for $i = 1, ..., 100$. We obtained a histogram of those

100 randomly observed values of the standardized sample mean, namely, $u_i = \sqrt{n}(\overline{x}_{ni} - 10)/2, i = 1, ..., 100$. Figure 5.3.1 shows the histogram which gives an *impression* of a standard normal pdf for u.

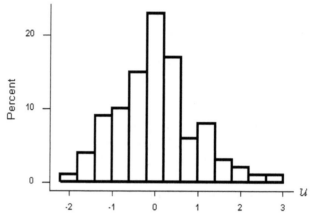

Figure 5.3.2. Histogram of 100 values of $u = \sqrt{n}(\overline{x}_n - 10)/2$ from a Gamma$(25, 0.4)$ population when $n = 10$.

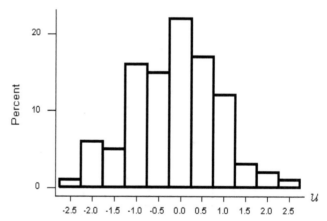

Figure 5.3.3. Histogram of 100 random $u = \sqrt{n}(\overline{x}_n - 10)/2$ values from a Uniform$(10 - a, 10 + a)$ population when $n = 10, a = \sqrt{12}$.

Then, we ran simulations with observations from a Gamma$(\alpha = 25, \beta = 0.4)$ population with its mean $\mu(= \alpha\beta) = 10$ and variance $\sigma^2(= \alpha\beta^2) = 4$. The histogram is shown in Figure 5.3.2. We also considered a Uniform$(10 - a, 10 + a)$ population with $a = \sqrt{12}$ with its mean $\mu = 10$ and variance $\sigma^2 = 4$. The histogram is shown in Figure 5.3.3. CLT suggests that the histograms from Figures 5.3.2 and 5.3.3 should look like a standard normal pdf for large n. We note that these histograms, though a little skewed to the right, create an *impression* of a standard normal pdf even though the sample size is only 10.

Example 5.3.5 Suppose $X_1, ..., X_n$ are iid $N(\mu, \sigma^2), -\infty < \mu < \infty, 0 < \sigma < \infty$. For $n \geq 2$, let S_n^2 be the sample variance. From (4.4.8), write $S_n^2 = (n-1)^{-1}\Sigma_{i=2}^n Y_i^2$ with $Y_2, ..., Y_n$ iid $N(0, \sigma^2)$. This shows that S_n^2 is a sample mean of $(n-1)$ iid random variables. CLT implies:

$$U_n = \sqrt{n-1}\left(S_n^2 - E(Y_2^2)\right) \xrightarrow{\pounds} N\left(0, V(Y_2^2)\right)$$

where $E(Y_2^2) = \sigma^2$ and $V(Y_2^2) = 2\sigma^4$. Let us now denote $V_n = \{n/(n-1)\}^{1/2}$ so that $V_n \xrightarrow{P} 1$. Then, Slutsky's Theorem gives:

$$U_n V_n \equiv \sqrt{n}\left(S_n^2 - \sigma^2\right) \xrightarrow{\pounds} N(0, 2\sigma^4). \;\blacktriangle$$

Theorem 5.3.4 (Mann-Wald Theorem) *Suppose that $\{T_n; n \geq 1\}$ is a sequence of real valued random variables such that*

$$\sqrt{n}(T_n - \theta) \xrightarrow{\pounds} N(0, \sigma^2) \text{ as } n \to \infty$$

where $0 < \sigma^2 < \infty$ may involve θ. Then,

$$\sqrt{n}\{g(T_n) - g(\theta)\} \xrightarrow{\pounds} N\left(0, \{\sigma g'(\theta)\}^2\right) \text{ as } n \to \infty$$

if $g(.)$ is a continuous real valued function such that $\frac{d}{d\theta}g(\theta)$, denoted by $g'(\theta)$, is finite and nonzero.

Proof We sketch a proof along the lines of Sen and Singer (1993, pp. 231-232). Observe that

$$\sqrt{n}\{g(T_n) - g(\theta)\} \equiv U_n V_n \tag{5.3.5}$$

where we denote:

$$U_n = \sqrt{n}(T_n - \theta), \; V_n = \{g(T_n) - g(\theta)\}/(T_n - \theta)$$

Next, note that $(T_n - \theta) = \frac{1}{\sqrt{n}}U_n$. Since $U_n \xrightarrow{\pounds} N(0, \sigma^2)$, by Slutsky's Theorem, we conclude that $T_n - \theta \xrightarrow{P} 0$. Thus, $V_n \xrightarrow{P} g'(\theta)$, by the definition of $g'(\theta)$ itself. Another application of Slutsky's Theorem and Equation (5.3.5) completes the proof. ∎

Example 5.3.6 Suppose that $X_1, ..., X_n$ are iid Poisson$(\lambda), \lambda > 0$, so that $\mu = \sigma^2 = \lambda$. The CLT implies: $\sqrt{n}(\overline{X}_n - \lambda)/\sqrt{\lambda} \xrightarrow{\pounds} N(0, 1)$. Then, the Mann-Wald Theorem implies: $\sqrt{n}(\overline{X}_n^3 - \lambda^3) \xrightarrow{\pounds} N(0, 9\lambda^5). \;\blacktriangle$

Theorem 5.3.5 (CLT for Sample Variance) *Suppose that $X_1, ..., X_n$ are iid random variables with the common mean μ, variance σ^2, $\mu_4 = E\{(X_1 - \mu)^4\}$, and we assume that $0 < \sigma^4 < \mu_4 < \infty$. Denote $\overline{X}_n =$*

$n^{-1}\Sigma_{i=1}^n X_i$ and $S_n^2 = (n-1)^{-1}\Sigma_{i=1}^n (X_i - \overline{X}_n)^2$ for $n \geq 2$. Then, as $n \to \infty$, we have:

$$\sqrt{n}\left(S_n^2 - \sigma^2\right) \xrightarrow{\pounds} N(0, \mu_4 - \sigma^4)$$

Proof Let $W_n = (n-1)n^{-1}S_n^2$. Denote $Y_i = (X_i - \mu)^2$, $i = 1, ..., n$, $\overline{Y}_n = n^{-1}\Sigma_{i=1}^n Y_i$, and write:

$$W_n = \tfrac{1}{n}\Sigma_{i=1}^n(X_i - \overline{X}_n)^2 = \tfrac{1}{n}\Sigma_{i=1}^n Y_i - \left\{\overline{X}_n - \mu\right\}^2$$
$$\Rightarrow \sqrt{n}(W_n - \sigma^2) = \sqrt{n}\left(\overline{Y}_n - \sigma^2\right) - \sqrt{n}\left\{\overline{X}_n - \mu\right\}^2 = U_n + V_n, \text{ say.}$$
$$(5.3.6)$$

Note that $Y_1, ..., Y_n$ are iid with mean σ^2 and variance $\mu_4 - \sigma^4$ which is assumed finite and positive. The CLT implies: $U_n \xrightarrow{\pounds} N(0, \mu_4 - \sigma^4)$.

Next, for any fixed $\varepsilon(> 0)$, we can write

$$P\{|V_n| > \varepsilon\} = P\{\sqrt{n}(\overline{X}_n - \mu)^2 > \varepsilon\} \leq E[\sqrt{n}(\overline{X}_n - \mu)^2]/\varepsilon \qquad (5.3.7)$$

using Markov Inequality from Equation (3.8.1). The upper bound in Equation (5.3.7) is $\sigma^2/(\sqrt{n}\varepsilon)$ so that $V_n \xrightarrow{P} 0$. Slutsky's Theorem and Equation (5.3.5) imply: $\sqrt{n}(W_n - \sigma^2) \xrightarrow{\pounds} N(0, \mu_4 - \sigma^4)$. We can write:

$$\sqrt{n}\left(S_n^2 - \sigma^2\right) = \sqrt{n}\left(\tfrac{n}{n-1}W_n - \sigma^2\right) = \sqrt{n}\left(W_n - \sigma^2\right) - \tfrac{\sqrt{n}}{n-1}W_n$$

and another application of Slutsky's Theorem completes the proof. ∎

Example 5.3.7 Under the setup of Theorem 5.3.5, we can conclude:

$$\sqrt{n}\left(S_n^2 - \sigma^2\right) S_n^{-2} \xrightarrow{\pounds} N(0, \mu_4\sigma^{-4} - 1). \blacktriangle$$

5.4 Convergence of Chi-Square, t, and F Distributions

5.4.1 Chi-Square Distribution

Consider U_ν which is distributed as χ_ν^2 so that $E(\nu^{-1}U_\nu) = 1$, $V(\nu^{-1}U_\nu) = 2\nu^{-1} \to 0$ as $\nu \to \infty$. The Weak WLLN implies that

$$\nu^{-1}\chi_\nu^2 \xrightarrow{P} 1 \text{ as } \nu \to \infty \qquad (5.4.1)$$

Let $X_1, ..., X_\nu, ...$ be iid χ_1^2 so that we may view χ_ν^2 as $\Sigma_{i=1}^\nu X_i$, and hence we may write $\nu^{-1}\chi_\nu^2 = \nu^{-1}\Sigma_{i=1}^\nu X_i = \overline{X}_\nu$. Now, CLT implies:

$$\sqrt{\nu}(\overline{X}_\nu - \mu)/\sigma \xrightarrow{\pounds} N(0, 1)$$

with $\mu = E(X_1) = 1$, $\sigma^2 = V(X_1) = 2$. Hence, $(2\nu)^{-1/2}(\chi_\nu^2 - \nu) \xrightarrow{\mathcal{L}} N(0,1)$. In Example 5.3.3, we came to the same conclusion with a different approach. For practical purposes, we would say:

> Distribution of $(\sqrt{2\nu})^{-1}(\chi_\nu^2 - \nu)$ is approximated by $N(0,1)$ distribution when ν is large.

5.4.2 Student's t Distribution

A Student's t random variable was described in Definition 4.5.1. Let $W_\nu = X \div (Y_\nu \nu^{-1})^{1/2}$ where X is standard normal, Y_ν is χ_ν^2, and X and Y_ν are independent. In other words, W_ν has a Student's t distribution with ν degrees of freedom! Since $Y_\nu \nu^{-1} \xrightarrow{P} 1$, so does $\sqrt{Y_\nu \nu^{-1}}$. Slutsky's Theorem implies: $W_\nu \xrightarrow{\mathcal{L}} N(0,1)$. For practical purposes, we would say:

> The t_ν distribution is approximated by the $N(0,1)$ distribution when ν is large.

(5.4.3)

5.4.3 F Distribution

An F_{ν_1,ν_2} random variable was introduced by Definition 4.5.2 as $\nu_1^{-1}X_{\nu_1} \div \nu_2^{-1}Y_{\nu_2}$ where X_{ν_1}, Y_{ν_2} are independent, X_{ν_1} is $\chi_{\nu_1}^2$ and Y_{ν_2} is $\chi_{\nu_2}^2$. Now, ν_1 is held fixed, but let $\nu_2 \to \infty$ so that $\nu_2^{-1}Y_{\nu_2} \xrightarrow{P} 1$. Slutsky's Theorem implies: as $\nu_2 \to \infty$, we conclude that $F_{\nu_1,\nu_2} \xrightarrow{\mathcal{L}} \nu_1^{-1}X_{\nu_1}$. That is, for practical purposes, we would say:

> $F_{\nu_1,\nu_2} \xrightarrow{\mathcal{L}} \chi_{\nu_1}^2 \div \nu_1$ as $\nu_2 \to \infty$, if ν_1 is held fixed, and
> $F_{\nu_1,\nu_2} \xrightarrow{\mathcal{L}} \nu_2 \div \chi_{\nu_2}^2$ as $\nu_1 \to \infty$, if ν_2 is held fixed.

(5.4.4)

5.5 Exercises and Complements

5.2.1 Suppose that $X_1, ..., X_n$ are iid with a common pmf:

X values:	-2	0	1	3	4
Probabilities:	0.20	0.05	0.10	0.15	0.50

Are there some numbers a, b such that $\overline{X}_n \xrightarrow{P} a$, $S_n^2 \xrightarrow{P} b$?

5.2.2 Let U_n be a real valued random variable such that $E(U_n) = a$, but $V(U_n) = b_n \to 0$ as $n \to \infty$. Show that $U_n \xrightarrow{P} a$.

5.2.3 Prove Khintchine's WLLN (Theorem 5.2.2). {*Hint*: See Feller (1968, pp. 246-248) or Rao (1973, p. 113).}

5.2.4 Suppose that $X_1, ..., X_n$ are iid $N(\mu, 1), \mu \in \Re - \{0\}$. Let

(i) $T_n = exp(\overline{X}_n)$ (ii) $T_n = log(|\overline{X}_n|)$

Is there some number a, in each case, such that $T_n \overset{P}{\to} a$?

5.2.5 Suppose that $X_1, ..., X_n$ are iid with a common pdf:

$$f(x) = \begin{cases} 2/x^3 & \text{if } 1 < x < \infty \\ 0 & \text{elsewhere} \end{cases}$$

Find some number b such that $\overline{X}_n \overset{P}{\to} b$.

5.2.6 (Example 5.2.3 Continued) Let $X_1, ..., X_n$ be iid Uniform$(0, \theta), \theta > 0$. Identify all $\gamma(> 0)$ so that $n^\gamma(X_{n:n} - \theta) \overset{P}{\to} 0$. Does $X_{n:n} - X_{n-1:n-1} \overset{P}{\to} 0$?

5.2.7 Suppose $X_1, ..., X_n$ are iid with a common pdf:

$$f(x) = \sigma^{-1}exp\{-(x - \mu)/\sigma\}I(x > \mu)$$

with $-\infty < \mu < \infty, 0 < \sigma < \infty$. Let $T_n = \Sigma_{i=1}^n(X_i - X_{n:1})$. Show that $nX_{n:1}/T_n \overset{P}{\to} \mu/\sigma$.

5.2.8 Suppose $X_1, ..., X_n$ are iid $N(0, 1)$ and denote:

$$T_n = n^{-1}\Sigma_{i=1}^n |X_i|, U_n = \sqrt{T_n}$$

Find numbers a, b such that $T_n \overset{P}{\to} a, U_n \overset{P}{\to} b$. {*Hint*: Use integration to check that $E[|X_1|] = \sqrt{2/\pi}$.}

5.2.9 Suppose that $X_1, ..., X_n$ are iid Poisson$(\lambda), \lambda > 0$. Does

$$I(\overline{X}_n > \tfrac{1}{2}\lambda) \overset{P}{\to} 0 \text{ or } 1?$$

5.2.10 Suppose that $X_1, ..., X_n$ are iid random variables with finite variance. Denote:

$$T_n = \frac{6}{n(n+1)(2n+1)}\Sigma_{i=1}^n i^2 X_i, n = 1, 2, ...$$

Show that $T_n \overset{P}{\to} E[X_1]$.

5.2.11 Suppose $X_1, ..., X_n$ are iid with a common pmf:

$$P\{X_1 = 2^{r-2log(r)}\} = 1/2^r \text{ for } r = 1, 2, ...$$

Does $\overline{X}_n \overset{P}{\to} c$ for some appropriate c?

5.3.1 (Exercise 5.2.1 Continued) Find numbers a and $b(> 0)$ such that

$$\sqrt{n}(\overline{X}_n - a)/b \overset{\pounds}{\to} N(0, 1)$$

5.3.2 (Exercise 5.2.4 Continued) Suppose $X_1, ..., X_n$ are iid $N(\mu, 1)$ with $\mu \in \Re - \{0\}$ and denote:

(i) $T_n = exp(\overline{X}_n)$;   (ii) $T_n = log(|\overline{X}_n|)$;

(iii) $T_n = exp(\overline{X}_n^2 + 2\overline{X}_n)$;   (iv) $T_n = |\overline{X}_n|^3 \, exp(\overline{X}_n)$.

Find expressions for $a_n, b_n(> 0)$ associated with each T_n such that

$$(T_n - a_n)/b_n \xrightarrow{\mathcal{L}} N(0, 1)$$

{*Hint*: Is the Mann-Wald Theorem applicable?}

5.3.3 (Example 5.3.7 Continued) Find expressions for $\eta(> 0)$ such that

$$\sqrt{n}(S_n - \sigma) \xrightarrow{\mathcal{L}} N(0, \eta^2)$$

{*Hint*: Is the Mann-Wald Theorem applicable?}

5.3.4 Suppose $X_1, ..., X_n$ are iid with a common pdf:

$$f(x) = \begin{cases} \{\sigma\sqrt{2\pi}\}^{-1}x^{-1}exp\left\{-\frac{1}{2\sigma^2}(log(x) - \mu)^2\right\} & \text{if } x > 0 \\ 0 & \text{elsewhere} \end{cases}$$

with $-\infty < \mu < \infty, 0 < \sigma < \infty$. Denote $T_n = (\Pi_{i=1}^n X_i)^{1/n}$.

(i)   Find $c(> 0)$ such that $T_n \xrightarrow{P} c$;

(ii)   Find $a_n, b_n(> 0)$ such that $(T_n - a_n)/b_n \xrightarrow{\mathcal{L}} N(0, 1)$.

5.3.5 Suppose $X_1, ..., X_n$ are iid Uniform$(0, \theta), \theta > 0$. Denote $T_n = (\Pi_{i=1}^n X_i)^{1/n}$.

(i)   Find $c(> 0)$ such that $T_n \xrightarrow{P} c$;

(ii)   Find $a_n, b_n(> 0)$ such that $(T_n - a_n)/b_n \xrightarrow{\mathcal{L}} N(0, 1)$.

5.3.6 Suppose that $X_1, ..., X_n$ are iid Uniform$(0, 1)$. Next, we denote $U_n = \sqrt{n}(\overline{X}_n - \frac{1}{2}), n \geq 1$. Now, show that

(i) $M_{X_1}(t) = \frac{1}{t}(e^t - 1), t \neq 0$;

(ii) $M_{U_n}(t) = \{\frac{1}{2}(e^a - e^{-a})/a\}^n$ where $a = \frac{1}{2}\frac{t}{\sqrt{n}}$;

(iii) $\lim_{n \to \infty} M_{U_n}(t) = \lim_{n \to \infty} \left[1 + \frac{t^2}{24n} + O(n^{-2})\right]^n = exp\{t^2/24\}$.

Use part (iii) to prove that $U_n \xrightarrow{\mathcal{L}} N(0, \frac{1}{12})$.

5.3.7 Suppose $X_1, ..., X_k$ are iid $N(0, 1)$ where $k = 2^n, n \geq 1$. Denote:

$$U_n = \frac{X_1}{X_2} + \frac{X_3}{X_4} + ... + \frac{X_{k-1}}{X_k}, \text{ and}$$

$$V_n = X_1^2 + X_2^2 + ... + X_n^2$$

Find the limiting distribution of $W_n = U_n/V_n$.

5.3.8 Suppose that $X_1, ..., X_n$ are iid $N(0, 1), n \geq 1$. Denote:

$$U_n = \frac{\sqrt{n}\{X_1 + X_2 + ... + X_n\}}{X_1^2 + X_2^2 + ... + X_n^2}$$

Show that $U_n \xrightarrow{\pounds} N(0, 1)$. {*Hint*: Use CLT and Slutsky's Theorem.}

5.3.9 Suppose that $X_1, ..., X_n$ are iid Bernoulli(p) with $0 < p < 1, n \geq 1$. Denote $V_n = n^{-1}\Sigma_{i=1}^n X_i$. Show that

$$\frac{\sqrt{n}\left(V_n^{-1} - p^{-1}\right)}{\sqrt{(1 - p)p^{-3}}} \xrightarrow{\pounds} N(0, 1)$$

{*Hint*: Would CLT and the Mann-Wald Theorem help?}

5.4.1 Let $X_1, ..., X_n$ be iid $N(\mu_1, \sigma^2)$, $Y_1, ..., Y_n$ be iid $N(\mu_2, 3\sigma^2)$ with $-\infty < \mu_1, \mu_2 < \infty, 0 < \sigma < \infty$, and X, Y be independent. Denote:

$$\overline{X}_n = n^{-1}\Sigma_{i=1}^n X_i, \overline{Y}_n = n^{-1}\Sigma_{i=1}^n Y_i, S_{1n}^2 = (n-1)^{-1}\Sigma_{i=1}^n(X_i - \overline{X}_n)^2,$$
$$S_{2n}^2 = (n-1)^{-1}\Sigma_{i=1}^n(Y_i - \overline{Y}_n)^2, \text{ and } T_n = S_{1n}^2 + \tfrac{1}{3}S_{2n}^2$$

for $n \geq 2$.

(i)   Show that $V_n = \frac{1}{2\sigma}\sqrt{n}(\overline{X}_n - \overline{Y}_n - \mu_1 + \mu_2)$ is distributed as $N(0, 1)$;

(ii)   Show that $(n-1)T_n/\sigma^2$ is distributed as $\chi^2_{2(n-1)}$;

(iii)   Are V_n, T_n independent?

(iv)   Show that $U_n = \sqrt{n}(\overline{X}_n - \overline{Y}_n - \mu_1 + \mu_2)/\sqrt{2T_n}$ is distributed as Student's $t_{2(n-1)}$;

(v)   Show that $T_n \xrightarrow{P} 2\sigma^2$;

(vi)   Show that $U_n \xrightarrow{\pounds} N(0, 1)$.

5.4.2 Suppose that $X_1, ..., X_m$ are iid exponential with mean $\beta(> 0)$, $Y_1, ..., Y_n$ are iid exponential with mean $\eta(> 0)$, and X is independent of Y. Denote $T_{m,n} = \overline{X}_m/\overline{Y}_n$.

(i)   Show that $T_{m,n}$ is distributed as $\frac{\beta}{\eta}F_{2m,2n}$;

(ii)   Determine the asymptotic distribution of $T_{m,n}$ as $n \to \infty$, when m is kept fixed;

(iii)   Determine the asymptotic distribution of $T_{m,n}$ as $m \to \infty$, when n is kept fixed.

6

Sufficiency, Completeness, and Ancillarity

6.1 Introduction

Many fundamental principles of statistical inference originated from the path-breaking contributions of Sir Ronald Aylmer Fisher in the 1920s. Those concepts are alive, well, and indispensable. The deepest of all statistical concepts is *sufficiency* that originated from Fisher (1920), and it blossomed further, again in the hands of Fisher (1922).

Section 6.2 introduces *Neyman factorization* of a *likelihood function*. In Section 6.3, a notion of *minimal sufficiency* and the fundamental results of Lehmann and Scheffé (1950) are discussed. These help in locating the *best* sufficient statistic, if it exists. Section 6.4 provides a quantification of *information* in one-parameter problems. Examples show that the information contained in the whole data is indeed preserved by a sufficient statistic. *Ancillarity* is discussed in Section 6.5, highlighting that ancillary statistics can be useful in statistical inference. *Location, scale,* and *location-scale families* are mentioned in Section 6.5.1. Section 6.6 introduces the concept of *completeness* and Section 6.6.2 highlights the celebrated Basu's Theorem.

6.2 Sufficiency

We begin with observable *independent and identically distributed* (iid) random variables $X_1, ..., X_n$ with a common *probability mass function* (pmf) or *probability density function* (pdf) $f(x), x \in \mathcal{X}$. The sample size n is assumed *known*. Practically speaking, we observe $X_1, ..., X_n$ from a *population* whose distribution $f(x)$ is indexed by a *parameter* (or *parameter vector*) θ (or $\boldsymbol{\theta}$) which captures important features of the population. A practical aspect of indexing $f(x)$ with θ (or $\boldsymbol{\theta}$) is that the population distribution would be completely specified once we know θ (or $\boldsymbol{\theta}$).

We let a pmf or pdf be $f(x; \theta)$ or $f(x; \boldsymbol{\theta})$ where the parameter θ (or $\boldsymbol{\theta}$) is *fixed* but unknown. In the case of a single parameter, we write $\theta \in \Theta$, the parameter space, $\Theta \subseteq \Re$. For example, one may have X distributed as $N(\mu, \sigma^2)$ with μ unknown, $-\infty < \mu < \infty$, but $\sigma(> 0)$ is known. Then, $f(x; \theta)$ will be the same as a $N(\mu, \sigma^2)$ pdf with $\theta = \mu$ and the parameter space $\Theta = \Re$. But, if both μ, σ^2 are unknown, then $f(x; \boldsymbol{\theta})$ will be the $N(\mu, \sigma^2)$ pdf with the parameter vector $\boldsymbol{\theta} = (\mu, \sigma^2)$ and $\Theta = \Re \times \Re^+$.

Our quest for gaining information about the unknown parameter θ (or $\boldsymbol{\theta}$) may be considered a core of statistical *inference*. The data $X_1, ..., X_n$, of course, have all *information* about θ even though we have not yet specified how to quantify "information." In Section 6.4, we partially address this. A dataset may be large or small, and data may be nice or cumbersome, but it is ultimately incumbent upon an experimenter to summarize the data so that all interesting features are captured by the summary. That is, ideally a summary should have the exact same "information" about θ as do the original data. Such a summary would be as good as the whole data and it will be called *sufficient* for θ.

Definition 6.2.1 *An observable real (or vector) valued function $T \equiv T(X_1, ..., X_n)$ is called a statistic.*

Some examples of statistics are $\overline{X}, X_1(X_2 - X_{n:n}), \Sigma_{i=1}^n X_i, S^2$, and so on. As long as numerical evaluation of T, having observed data $X_1 = x_1, ..., X_n = x_n$, does not involve any unknown entities, T will be called a statistic. Supposing that $X_1, ..., X_n$ are iid $N(\mu, \sigma^2)$ where μ is unknown, but σ is known, $T \equiv \overline{X}$ is a statistic, but its standardized form $\sqrt{n}(\overline{X} - \mu)/\sigma$ is *not*.

Definition 6.2.2 *A statistic T is called sufficient for an unknown parameter θ if and only if the conditional distribution of the random sample $\mathbf{X} \equiv (X_1, ..., X_n)$ given $T = t$ does not involve θ, for all $t \in \mathcal{T} \subseteq \Re$.*

In other words, *given* the value t of a *sufficient statistic T, conditionally* there is no more "information" or "juice" left in the original data regarding the unknown parameter θ. Put another way, one may think of $\mathbf{X}$ trying to tell us a story about θ, but once a sufficient summary T is available, the original story becomes redundant. Observe that $\mathbf{X}$ is sufficient for θ in this sense. But, we are aiming at a "shorter" summary statistic which has the same information available in $\mathbf{X}$.

Definition 6.2.3 *A vector valued statistic $\mathbf{T} \equiv (T_1, ..., T_k)$ with $T_i \equiv T_i(X_1, ..., X_n), i = 1, ..., k$, is called jointly sufficient for the unknown parameter θ (or $\boldsymbol{\theta}$) if and only if the conditional distribution of $\mathbf{X} \equiv (X_1, ..., X_n)$ given $\mathbf{T} = \mathbf{t}$ does not involve θ (or $\boldsymbol{\theta}$), for all $\mathbf{t} \in \mathcal{T} \subseteq \Re^k$.*

6.2.1 Neyman Factorization

A dataset consists of $X_1, ..., X_n$ from a population with a common pmf or pdf $f(x; \theta)$ where θ is an unknown parameter. The *Neyman Factorization Theorem* is widely used to find sufficient statistics.

Definition 6.2.4 *Having observed $X_i = x_i$, $i = 1, ..., n$, a likelihood function is defined as*

$$L(\theta) = \Pi_{i=1}^n f(x_i; \theta), \theta \in \Theta \qquad (6.2.1)$$

In a discrete case, $L(\theta)$ stands for $P_\theta\{X_1 = x_1 \cap ... \cap X_n = x_n\}$. In a continuous case, $L(\theta)$ stands for the joint pdf at the observed data $(x_1, ..., x_n)$

when θ obtains. One may note that once $\{x_i; i = 1, ..., n\}$ is observed, there are no random entities in Equation (6.2.1). A likelihood function $L(.)$ is simply a function of θ only.

It is not really essential that X be real valued or iid. In many examples, it will be so. But, if X happens to be vector valued or if it is not iid, then the corresponding joint pmf or pdf of $X_i = x_i$, $i = 1, ..., n$, would be the likelihood function, $L(\theta)$. We will give examples shortly.

> Sample size n is assumed known and fixed before data collection begins.

One may note that θ may be real or vector valued, however, we first pretend that θ is real valued. Fisher (1922) discovered the fundamental idea of factorization. Neyman (1935a) discovered a refined approach to factorize a likelihood function. Halmos and Savage (1949) and Bahadur (1954) developed more involved measure-theoretic treatments.

Theorem 6.2.1 (Neyman Factorization Theorem) *A real valued statistic $T = T(X_1, ..., X_n)$ is sufficient for an unknown parameter θ if and only if the following factorization holds*:

$$L(\theta) = g\left(T(x_1, ..., x_n); \theta\right) h(x_1, ..., x_n) \text{ for all } x_1, ..., x_n \in \mathcal{X} \quad (6.2.2)$$

where functions $g(.; \theta)$ and $h(.)$ are both nonnegative, $h(.)$ is free from θ, and $g\left(T(.); \theta\right)$ involves $x_1, ..., x_n$ only through $T(x_1, ..., x_n)$.

Proof For simplicity, we provide a proof only in a *discrete* case. Let us write $\mathbf{X} = (X_1, ..., X_n)$ and $\mathbf{x} = (x_1, ..., x_n)$. Let A and B respectively denote the events $\mathbf{X} = \mathbf{x}$ and $T(\mathbf{X}) = T(\mathbf{x})$, and observe that $A \subseteq B$.

Only if part: Suppose that T is sufficient for θ. Now, we write

$$
\begin{aligned}
L(\theta) \ &= P_\theta\{\mathbf{X} = \mathbf{x}\} \\
&= P_\theta\{\mathbf{X} = \mathbf{x} \cap T(\mathbf{X}) = T(\mathbf{x})\}, \text{ since } A \subseteq B \quad (6.2.3) \\
&= P_\theta\{T(\mathbf{X}) = T(\mathbf{x})\} P_\theta\{\mathbf{X} = \mathbf{x} \,|\, T(\mathbf{X}) = T(\mathbf{x})\}
\end{aligned}
$$

Denote $g\left(T(x_1, ..., x_n); \theta\right) = P_\theta\{T(\mathbf{X}) = T(\mathbf{x})\}$ and $h(x_1, ..., x_n) = P_\theta\{\mathbf{X} = \mathbf{x} \,|\, T(\mathbf{X}) = T(\mathbf{x})\}$. Since T is sufficient for θ, by Definition 6.2.2, the conditional probability $P_\theta\{\mathbf{X} = \mathbf{x} \,|\, T(\mathbf{X}) = T(\mathbf{x})\}$ *cannot* involve θ. Thus, $h(x_1, ..., x_n)$ so defined may involve only $x_1, ..., x_n$. So, the factorization given in Equation (6.2.2) holds. ◆

If part: Suppose that the factorization holds. Let $p(t; \theta)$ be the pmf of T. Observe that $p(t; \theta) = P_\theta\{T(\mathbf{X}) = t\} = \sum_{\mathbf{y}:T(\mathbf{y})=t} \{\Pi_{i=1}^n f(y_i; \theta)\} = \sum_{\mathbf{y}:T(\mathbf{y})=t} L(\theta)$. It is easy to see:

$$P_\theta\{\mathbf{X} = \mathbf{x} \,|\, T(\mathbf{X}) = t\} = 0 \text{ if } T(\mathbf{x}) \neq t \quad (6.2.4)$$

For all $\mathbf{x} \in \mathcal{X}$ such that $T(\mathbf{x}) = t$ and $p(t; \theta) \neq 0$, we can express $P_\theta\{\mathbf{X} = \mathbf{x} \mid T(\mathbf{X}) = t\}$ as:

$$
\begin{aligned}
L(\theta)/p(t; \theta) &= g\left(T(\mathbf{x}); \theta\right) h(\mathbf{x})/p(t; \theta) \\
&= g\left(t; \theta\right) h(\mathbf{x})/ \sum_{\mathbf{y}:T(\mathbf{y})=t} L(\theta) \\
&= g\left(t; \theta\right) h(\mathbf{x})/ \sum_{\mathbf{y}:T(\mathbf{y})=t} g(T(\mathbf{y}); \theta)h(\mathbf{y})
\end{aligned}
$$

because of factorization (6.2.2). Since $g\left(t; \theta\right) \neq 0$, one has:

$$
P_\theta\{\mathbf{X} = \mathbf{x} \mid T(\mathbf{X}) = t\} = g\left(t; \theta\right) h(\mathbf{x})/ \{g(t;\theta)\} \left\{ \sum_{\mathbf{y}:T(\mathbf{y})=t} h(\mathbf{y}) \right\}
$$

$$
= h(\mathbf{x})/ \left\{ \sum_{\mathbf{y}:T(\mathbf{y})=t} h(\mathbf{y}) \right\} \equiv q(\mathbf{x})
$$

(6.2.5)

with $q(\mathbf{x})$ free from θ. Equations (6.2.4) and (6.2.5) complete the proof. ∎

> In Theorem 6.2.1, we *do not* demand that $g\left(T(x_1, ..., x_n); \theta\right)$ is a pmf or pdf of $T(X_1, ..., X_n)$. It is essential, however, that $h(x_1, ..., x_n)$ must be free from θ.

It should be noted that the splitting of $L(\theta)$ may not be unique. Also, there may be different versions of sufficient statistics.

Remark 6.2.1 In Theorem 6.2.1, it was not essential that $X_1, ..., X_n$ or θ be real valued. If $\mathbf{X}_1, ..., \mathbf{X}_n$ are iid p-dimensional observations with a common pmf or pdf $f(\mathbf{x}; \boldsymbol{\theta})$, $\boldsymbol{\theta} \in \Theta \subseteq \mathcal{R}^q$, the Neyman Factorization Theorem will hold:

Denote the likelihood function, $L(\boldsymbol{\theta}) = \Pi_{i=1}^n f(\mathbf{x}_i; \boldsymbol{\theta})$.

> A vector valued statistic $\mathbf{T} = (T_1, ..., T_k)$ is *jointly sufficient* for $\boldsymbol{\theta} = (\theta_1, ..., \theta_q)$ if and only if $L(\boldsymbol{\theta}) = g(\mathbf{T}; \boldsymbol{\theta})h(\mathbf{x}_1, ..., \mathbf{x}_n)$ for all $\mathbf{x}_1, ..., \mathbf{x}_n \in \mathcal{X} \subseteq \mathcal{R}^p$, with both $g(.; \boldsymbol{\theta}), h(.)$ nonnegative, $g(.; \boldsymbol{\theta})$ depending upon $\mathbf{x}$ only through $\mathbf{T}$, and $h(.)$ is free from $\boldsymbol{\theta}$.

(6.2.6)

Example 6.2.1 Suppose $X_1, ..., X_n$ are iid Bernoulli(p) with p unknown, $0 < p < 1$. Here, $\mathcal{X} = \{0, 1\}$, $\theta = p$, and $\Theta = (0, 1)$. Then,

$$
L(p) = \Pi_{i=1}^n p^{x_i}(1-p)^{1-x_i} = p^{\sum_{i=1}^n x_i}(1-p)^{n-\sum_{i=1}^n x_i}
$$

(6.2.7)

It matches with factorization (6.2.2) where $g\left(\sum_{i=1}^n x_i; p\right) = p^{\sum_{i=1}^n x_i}(1-p)^{n-\sum_{i=1}^n x_i}$ and $h(x_1, ..., x_n) = 1$ for all $x_1, ..., x_n \in \{0, 1\}$. So, the statistic $T = \sum_{i=1}^n X_i$ is sufficient for p. We could instead express $L(\theta) = g(x_1, ..., x_n; p)h(x_1, ..., x_n)$ with $g(x_1, ..., x_n; p) = \Pi_{i=1}^n p^{x_i}(1-p)^{n-x_i}$ and

$h(x_1, ..., x_n) = 1$. So, one could also claim that $\mathbf{X} = (X_1, ..., X_n)$ was sufficient for p. But, $\Sigma_{i=1}^n X_i$ provides a significantly reduced summary compared with $\mathbf{X}$, the whole data. We will have more to say on this in Section 6.3. ▲

Example 6.2.2 Suppose $X_1, ..., X_n$ are iid Poisson(λ) with λ unknown, $0 < \lambda < \infty$. Here, $\mathcal{X} = \{0, 1, 2, ...\}$, $\theta = \lambda$, and $\Theta = (0, \infty)$. Then,

$$L(\lambda) = \Pi_{i=1}^n \left\{ e^{-\lambda} \lambda^{x_i}/x_i! \right\} = e^{-n\lambda} \lambda^{\Sigma_{i=1}^n x_i} \left(\Pi_{i=1}^n x_i! \right)^{-1} \qquad (6.2.8)$$

It matches with factorization (6.2.2) where $g\left(\Sigma_{i=1}^n x_i; \lambda \right) = e^{-n\lambda} \lambda^{\Sigma_{i=1}^n x_i}$ and $h(x_1, ..., x_n) = \left(\Pi_{i=1}^n x_i! \right)^{-1}$ for all $x_1, ..., x_n \in \{0, 1, 2, ...\}$. So, the statistic $T = \Sigma_{i=1}^n X_i$ is sufficient for λ. Again, from Equation (6.2.8) one notes that $\mathbf{X}$ is sufficient too, but $\Sigma_{i=1}^n X_i$ is a significantly reduced summary. ▲

Example 6.2.3 Suppose $X_1, ..., X_n$ are iid $N(\mu, \sigma^2)$ with μ, σ^2 both unknown, $-\infty < \mu < \infty$, $0 < \sigma < \infty$. Denote $\boldsymbol{\theta} = (\mu, \sigma^2)$, $\mathcal{X} = \Re$, and $\Theta = \Re \times \Re^+$. Now,

$$L(\boldsymbol{\theta}) = \{\sigma\sqrt{2\pi}\}^{-n} exp\left\{-\tfrac{1}{2} \left(\Sigma_{i=1}^n x_i^2 - 2\mu\Sigma_{i=1}^n x_i + n\mu^2 \right)/\sigma^2 \right\} \qquad (6.2.9)$$

It matches with factorization (6.2.2) where

$$g(\Sigma_{i=1}^n x_i, \Sigma_{i=1}^n x_i^2; \boldsymbol{\theta}) = \sigma^{-n} exp\left\{-\tfrac{1}{2} \left(\Sigma_{i=1}^n x_i^2 - 2\mu\Sigma_{i=1}^n x_i + n\mu^2 \right)/\sigma^2 \right\}$$

$$\text{and } h(x_1, ..., x_n) = \{\sqrt{2\pi}\}^{-n}$$

for all $(x_1, ..., x_n) \in \Re^n$. So, $\mathbf{T} = \left(\Sigma_{i=1}^n X_i, \Sigma_{i=1}^n X_i^2 \right)$ is a *jointly* sufficient statistic for (μ, σ^2). ▲

> If $\mathbf{T}$ is a (jointly) sufficient statistic for $\boldsymbol{\theta}$, then any statistic $\mathbf{U}$ which is a *one-to-one* function of $\mathbf{T}$ is (jointly) sufficient for $\boldsymbol{\theta}$.

Example 6.2.4 (Example 6.2.3 Continued) We have $\overline{X} = n^{-1}\Sigma_{i=1}^n X_i$, $S^2 = (n-1)^{-1}\{\Sigma_{i=1}^n X_i^2 - n^{-1}(\Sigma_{i=1}^n X_i)^2\}$. It is clear that the transformation from $\mathbf{T} = \left(\Sigma_{i=1}^n X_i, \Sigma_{i=1}^n X_i^2 \right)$ to $\mathbf{U} = (\overline{X}, S^2)$ is *one-to-one*. So, we can claim that $(\overline{X}, S^2)$ is jointly sufficient for (μ, σ^2). ▲

> Let $\mathbf{T}$ be a sufficient statistic for $\boldsymbol{\theta}$. Consider a statistic $\mathbf{T}'$, a function of $\mathbf{T}$. Then, $\mathbf{T}'$ is not necessarily sufficient for $\boldsymbol{\theta}$.

An arbitrary function of a sufficient statistic $\mathbf{T}$ need not be sufficient for $\boldsymbol{\theta}$. Suppose that X is distributed as $N(\theta, 1)$ where $-\infty < \theta < \infty$ is an unknown parameter. Obviously, X is sufficient for θ. One may check that $T' = |X|$, a function of X, is *not* sufficient for θ.

> From joint sufficiency of a statistic $\mathbf{T} = (T_1, ..., T_p)$ for $\boldsymbol{\theta} = (\theta_1, ..., \theta_p)$, one should not claim that T_i is sufficient for θ_i, $i = 1, ..., p$. Note that $\mathbf{T}, \boldsymbol{\theta}$ may not even have the same dimension! See Example 6.2.5.

Example 6.2.5 (Example 6.2.4 Continued) We know that $(\overline{X}, S^2)$ is jointly sufficient for (μ, σ^2). So, $(S^2, \overline{X})$ is also jointly sufficient for (μ, σ^2). Should one claim that S^2 is sufficient for μ or $\overline{X}$ is sufficient for σ^2? Of course, not. ▲

Example 6.2.6 Suppose that $X_1, ..., X_n$ are iid Uniform$(0, \theta)$, and θ (> 0) is unknown. Here, $\mathcal{X} = (0, \theta)$ and $\Theta = \Re^+$. Now,

$$
\begin{aligned}
L(\theta) &= \Pi_{i=1}^n \{\theta^{-1} I(0 < x_i < \theta)\} \\
&= \theta^{-n} I(0 < x_{n:n} < \theta) I(0 < x_{n:1} < x_{n:n})
\end{aligned}
\tag{6.2.10}
$$

where $x_{n:1}, x_{n:n}$ are, respectively, the observed smallest and largest order statistics. The last step in Equation (6.2.10) matches with factorization (6.2.2) where $g(x_{n:n}; \theta) = \theta^{-n} I(0 < x_{n:n} < \theta)$ and $h(x_1, ..., x_n) = I(0 < x_{n:1} < x_{n:n})$ for all $x_1, ..., x_n \in (0, \theta), \theta > 0$. So, $X_{n:n}$ is sufficient for θ. ▲

It is *not* crucial that $X_1, ..., X_n$ be iid to apply Neyman factorization.

Example 6.2.7 Let X_1, X_2 be independent with

$$
f_1(x_1; \theta) = \theta e^{-\theta x_1}, f_2(x_2; \theta) = 2\theta e^{-2\theta x_2}
$$

as the respective pdfs where $\theta(> 0)$ is an unknown parameter and $0 < x_1, x_2 < \infty$. The likelihood function is the joint pdf:

$$
L(\theta) = f_1(x_1; \theta) f_2(x_2; \theta) = 2\theta^2 e^{-\theta(x_1 + 2x_2)}
\tag{6.2.11}
$$

for $0 < x_1, x_2 < \infty$. The step in Equation (6.2.11) matches with factorization (6.2.2) so that $T = X_1 + 2X_2$ is a sufficient statistic for θ. ▲

The next result shows a simple way to find sufficient statistics when a pmf or a pdf belongs to an *exponential family*. Recall Section 3.7. Its proof follows easily from Equation (6.2.2).

Theorem 6.2.2 (Sufficiency in an Exponential Family) *Let the random variables* $X_1, ..., X_n$ *be iid with a common pmf or pdf*

$$
f(x; \boldsymbol{\theta}) = a(\boldsymbol{\theta})g(x)exp\{\Sigma_{j=1}^k b_j(\boldsymbol{\theta})R_j(x)\}
$$

belonging to a k-parameter exponential family defined by Equation (3.7.1). *Denote the statistic* $T_j = \Sigma_{i=1}^n R_j(X_i), j = 1, ..., k$. *Then,* $\mathbf{T} = (T_1, ..., T_k)$ *is jointly sufficient for* $\boldsymbol{\theta}$.

Sufficient statistics derived earlier can also be found using Theorem 6.2.2. We leave these as exercises.

How to verify that a statistic is *not* sufficient for θ?
Discussions follow.

If T is *not* sufficient, then the conditional pmf or pdf of $X_1, ..., X_n$ given $T = t$ must involve θ, for some data $x_1, ..., x_n$ and t. We follow this route.

Example 6.2.8 (Example 6.2.1 Continued) Denote $U = X_1 X_2 + X_3$. The question is whether U is a sufficient statistic for p. Observe that

$$
\begin{aligned}
P(U = 0) \\
&= P\{X_1 X_2 = 0 \cap X_3 = 0\} \\
&= P\{[X_1 = 0 \cap X_2 = 0 \cap X_3 = 0] \\
&\quad \cup [X_1 = 0 \cap X_2 = 1 \cap X_3 = 0] \\
&\quad \cup [X_1 = 1 \cap X_2 = 0 \cap X_3 = 0]\} \\
&= (1-p)^3 + 2p(1-p)^2
\end{aligned}
\tag{6.2.12}
$$

which reduces to $(1-p)^2(1+p)$. Since $\{X_1 = 1 \cap X_2 = 0 \cap X_3 = 0\}$ is a subset of $\{U = 0\}$, we have

$$
\begin{aligned}
P(X_1 = 1 \cap X_2 = 0 \cap X_3 = 0 \mid U = 0) \\
&= P(X_1 = 1 \cap X_2 = 0 \cap X_3 = 0)/P(U = 0) \\
&= p(1-p)^2/\{(1-p)^2(1+p)\} = p/(1+p)
\end{aligned}
$$

which involves p. So, U is *not* sufficient for p. ▲

Example 6.2.9 Let X_1, X_2 be iid $N(\theta, 1)$ where θ is unknown, $-\infty < \theta < \infty$. Denote $T = X_1 + 2X_2$ and we verify that T is *not* sufficient for θ. The joint distribution of (X_1, T) is $N_2(\theta, 3\theta, 1, 5, 1/\sqrt{5})$, and so the conditional distribution of X_1 given $T = t$ is normal with mean $= \frac{1}{5}(t+2\theta)$ and variance $= \frac{4}{5}$ for $t \in \Re$. Since this conditional distribution involves θ, T is *not* sufficient for θ. ▲

Example 6.2.10 Suppose that X has its pdf $f(x) = \lambda e^{-\lambda x} I(x > 0)$ where $\lambda(> 0)$ is an unknown parameter. Instead of original data, suppose that we are told the observed value of the statistic, $T \equiv I(X > 2)$. Is T sufficient for λ? Express $P\{X > 3 \mid T = 1\}$ as

$$
P\{X > 3 \cap X > 2\}/P\{X > 2\} = P\{X > 3\}/P\{X > 2\} = e^{-\lambda}
$$

which involves λ. Hence, T is *not* sufficient for λ. ▲

6.3   Minimal Sufficiency

We noted earlier that $\mathbf{X}$ must always be sufficient for $\boldsymbol{\theta}$. But, we aim at reducing the data by means of summary statistics in lieu of considering $\mathbf{X}$. We found that Neyman factorization provided sufficient statistics which were substantially "reduced" compared with $\mathbf{X}$ in a number of examples. As a principle, one should use the "shortest sufficient" summary. Pertinent questions arise: How to define a "shortest sufficient" summary and how to get hold of such a summary?

Lehmann and Scheffé (1950) developed a mathematical formulation of *minimal sufficiency* and gave a technique to locate minimal sufficient statistics. Lehmann and Scheffé (1955, 1956) included important follow-ups.

Definition 6.3.1 *A statistic* **T** *is called minimal sufficient for unknown parameter* $\boldsymbol{\theta}$ *if and only if*

(*i*)   **T** *is sufficient for* $\boldsymbol{\theta}$, *and*

(*ii*)   **T** *is minimal or "shortest" in the sense that* **T** *is a function of any other sufficient statistic.*

Let us think about this concept for a moment. We want to summarize **X** by reducing it to some appropriate statistic such as $\overline{X}$, a median (M), or a histogram, and so on. Suppose that in a particular situation, a summary statistic $\mathbf{T} = (\overline{X}, M)$ is *minimal sufficient* for $\boldsymbol{\theta}$. Can we reduce this summary any further? Of course, we can. We may simply look at, for example, $T_1 = \overline{X}$ or $T_2 = M$ or $T_3 = \frac{1}{2}(\overline{X} + M)$. Can T_1 or T_2 or T_3 individually be sufficient for $\boldsymbol{\theta}$? The answer is no, none of these could be sufficient for $\boldsymbol{\theta}$. Because if, for example, T_1 was sufficient for $\boldsymbol{\theta}$, then $\mathbf{T} = (\overline{X}, M)$ would have to be a function of T_1. But, **T** cannot be a function of T_1 because we cannot uniquely specify **T** from the value of T_1 alone. A minimal sufficient summary **T** cannot be reduced any further to another sufficient summary statistic. In this sense, a minimal sufficient statistic **T** is the best sufficient statistic.

6.3.1   Lehmann-Scheffé Approach

The following theorem was proved in Lehmann and Scheffé (1950). This is an essential tool to locate minimal sufficient statistics. Its proof requires some understanding of a correspondence between a statistic and a *partition* it induces on a sample space.

Consider $\mathbf{X} \equiv (X_1, ..., X_n)$ with $\mathbf{x} = (x_1, ..., x_n) \in \mathcal{X}^n$. A statistic $\mathbf{T} \equiv \mathbf{T}(X_1, ..., X_n)$ is a mapping from $\mathcal{X}^n$ onto some space $\mathcal{T}$. For $\mathbf{t} \in \mathcal{T}$, let $\mathcal{X}_{\mathbf{t}} = \{\mathbf{x} : \mathbf{x} \in \mathcal{X}^n \text{ such that } \mathbf{T}(\mathbf{x}) = \mathbf{t}\}$. These are disjoint subsets of $\mathcal{X}^n$ and also $\mathcal{X}^n = \cup_{\mathbf{t} \in \mathcal{T}} \mathcal{X}_{\mathbf{t}}$. In other words, $\{\mathcal{X}_{\mathbf{t}} : \mathbf{t} \in \mathcal{T}\}$ forms a *partition* of the space $\mathcal{X}^n$ *induced* by the statistic **T**.

Theorem 6.3.1 (Minimal Sufficient Statistic) *Consider* $h(\mathbf{x}, \mathbf{y}; \boldsymbol{\theta}) = \Pi_{i=1}^n f(x_i; \boldsymbol{\theta})/\Pi_{i=1}^n f(y_i; \boldsymbol{\theta})$, *the ratio of the likelihood functions from Equation (6.2.1) at* **x** *and* **y**, $\mathbf{x}, \mathbf{y} \in \mathcal{X}^n$. *Suppose that there is a statistic* $\mathbf{T} \equiv \mathbf{T}(X_1, ..., X_n) = (T_1, ..., T_k)$ *such that the following holds:*

> *With arbitrary data points* $\mathbf{x} = (x_1, ..., x_n)$, $\mathbf{y} = (y_1, ..., y_n)$,
> *both from* $\mathcal{X}^n$, *the expression* $h(\mathbf{x}, \mathbf{y}; \boldsymbol{\theta})$ *does not involve* $\boldsymbol{\theta}$ (6.3.1)
> *if and only if* $T_i(\mathbf{x}) = T_i(\mathbf{y}), i = 1, ..., k.$

Then, **T** *is a minimal sufficient statistic for* $\boldsymbol{\theta}$.

Proof We first show that $\mathbf{T}$ is a sufficient statistic for $\boldsymbol{\theta}$ and then we verify that $\mathbf{T}$ is minimal. For simplicity, let us assume that $f(\mathbf{x}; \boldsymbol{\theta})$ is positive for all $\mathbf{x} \in \mathcal{X}^n$ and $\boldsymbol{\theta}$.

Sufficiency part: Start with $\{\mathcal{X}_t : \mathbf{t} \in \mathcal{T}\}$ which is a partition of $\mathcal{X}^n$ induced by $\mathbf{T}$. In $\mathcal{X}_t$, fix an element $\mathbf{x}_t$. If we look at an arbitrary element $\mathbf{x} \in \mathcal{X}^n$, then this element $\mathbf{x}$ belongs to $\mathcal{X}_t$ for some unique $\mathbf{t}$ so that both $\mathbf{x}$ and $\mathbf{x}_t$ belong to the same set $\mathcal{X}_t$. So, one has $\mathbf{T}(\mathbf{x}) = \mathbf{T}(\mathbf{x}_t)$. Thus, by invoking the "if part" of the statement in Equation (6.3.1), we can claim that $h(\mathbf{x}, \mathbf{x}_t; \boldsymbol{\theta})$ is free from $\boldsymbol{\theta}$. Denote $h(\mathbf{x}) \equiv h(\mathbf{x}, \mathbf{x}_t; \boldsymbol{\theta}), \mathbf{x} \in \mathcal{X}^n$. Hence, we write:

$$\Pi_{i=1}^n f(x_i; \boldsymbol{\theta}) = \Pi_{i=1}^n f(x_{ti}; \boldsymbol{\theta}) h(\mathbf{x}) = g(\mathbf{T}(\mathbf{x}); \boldsymbol{\theta}) h(\mathbf{x})$$

with $\mathbf{x}_t = (x_{t1}, ..., x_{tn})$. Using Neyman factorization, the statistic $\mathbf{T}$ is sufficient for $\boldsymbol{\theta}$. ◆

Minimal part: Suppose $\mathbf{U} = \mathbf{U}(\mathbf{X})$ is another sufficient statistic for $\boldsymbol{\theta}$. Then, by Neyman factorization, we write:

$$\Pi_{i=1}^n f(x_i; \boldsymbol{\theta}) = g_0(\mathbf{U}(\mathbf{x}); \boldsymbol{\theta}) h_0(\mathbf{x})$$

with some appropriate $g_0(.; \boldsymbol{\theta})$ and $h_0(.)$. Here, $h_0(.)$ does not involve $\boldsymbol{\theta}$. Now, for any two sample points $\mathbf{x} = (x_1, ..., x_n)$, $\mathbf{y} = (y_1, ..., y_n)$ from $\mathcal{X}^n$ such that $\mathbf{U}(\mathbf{x}) = \mathbf{U}(\mathbf{y})$, we obtain:

$$\begin{aligned} h(\mathbf{x}, \mathbf{y}; \boldsymbol{\theta}) \\ &= \Pi_{i=1}^n f(x_i; \boldsymbol{\theta}) / \Pi_{i=1}^n f(y_i; \boldsymbol{\theta}) \\ &= \{g_0(\mathbf{U}(\mathbf{x}); \boldsymbol{\theta}) h_0(\mathbf{x})\} / \{g_0(\mathbf{U}(\mathbf{y}); \boldsymbol{\theta}) h_0(\mathbf{y})\} \\ &= h_0(\mathbf{x}) / h_0(\mathbf{y}), \text{ since } g_0(\mathbf{U}(\mathbf{x}); \boldsymbol{\theta}) = g_0(\mathbf{U}(\mathbf{y}); \boldsymbol{\theta}) \end{aligned}$$

Thus, $h(\mathbf{x}, \mathbf{y}; \boldsymbol{\theta})$ is free from $\boldsymbol{\theta}$. Now, by invoking the "only if" part from Equation (6.3.1), we claim that $\mathbf{T}(\mathbf{x}) = \mathbf{T}(\mathbf{y})$. That is, $\mathbf{T}$ is a function of $\mathbf{U}$. Now, the proof is complete. ∎

Example 6.3.1 (Example 6.2.1 Continued) With $\mathbf{x} = (x_1, ..., x_n)$ and $\mathbf{y} = (y_1, ..., y_n)$, both data points from $\mathcal{X}$, we have:

$$\{\Pi_{i=1}^n f(x_i; \theta)\} / \{\Pi_{i=1}^n f(y_i; \theta)\} = \left(p(1-p)^{-1}\right)^{\{\Sigma_{i=1}^n x_i - \Sigma_{i=1}^n y_i\}} \quad (6.3.2)$$

From Equation (6.3.2), it is clear that $\left(p(1-p)^{-1}\right)^{\{\Sigma_{i=1}^n x_i - \Sigma_{i=1}^n y_i\}}$ would become free from p if and only if $\Sigma_{i=1}^n x_i - \Sigma_{i=1}^n y_i = 0$, that is, if and only if $\Sigma_{i=1}^n x_i = \Sigma_{i=1}^n y_i$. Hence, by the theorem of Lehmann-Scheffé, $T = \Sigma_{i=1}^n X_i$ is minimal sufficient for p. ▲

We have shown nonsufficiency of a statistic U in Example 6.2.8. One may arrive at the same conclusion by contrasting U with a minimal sufficient statistic. Look at Examples 6.3.2 and 6.3.5.

Example 6.3.2 (Example 6.3.1 Continued) With $n = 3$, $T = \Sigma_{i=1}^{3} X_i$ is minimal sufficient for p, and let $U = X_1 X_2 + X_3$. Assume that U is sufficient for p. Then, $\Sigma_{i=1}^{3} X_i$ must be a function of U, by the definition of minimal sufficiency. That is, knowing an observed value of U, we must be able to come up with a unique observed value of T. Now, the event $\{U = 0\}$ consists of a union of $\{X_1 = 0 \cap X_2 = 0 \cap X_3 = 0\}$, $\{X_1 = 0 \cap X_2 = 1 \cap X_3 = 0\}$, and $\{X_1 = 1 \cap X_2 = 0 \cap X_3 = 0\}$. If the event $\{U = 0\}$ is observed, we know then that either $T = 0$ or $T = 1$. But, we cannot be sure about a unique observed value of T. Thus, T cannot be a function of U and so U cannot be sufficient for p. ▲

Example 6.3.3 (Example 6.2.3 Continued) With $\mathbf{x} = (x_1, ..., x_n)$ and $\mathbf{y} = (y_1, ..., y_n)$, both data points from $\mathcal{X}$, we have:

$$\{\Pi_{i=1}^{n} f(x_i; \boldsymbol{\theta})\}/\{\Pi_{i=1}^{n} f(y_i; \boldsymbol{\theta})\}$$
$$= \; exp\left\{\mu \left(\Sigma_{i=1}^{n} x_i - \Sigma_{i=1}^{n} y_i\right)/\sigma^2\right\} exp\left\{-\tfrac{1}{2}\left(\Sigma_{i=1}^{n} x_i^2 - \Sigma_{i=1}^{n} y_i^2\right)/\sigma^2\right\}$$
(6.3.3)

From Equation (6.3.3), it is clear that the last expression would not involve $\boldsymbol{\theta} = (\mu, \sigma^2)$ if and only if $\Sigma_{i=1}^{n} x_i - \Sigma_{i=1}^{n} y_i = 0$ as well as $\Sigma_{i=1}^{n} x_i^2 - \Sigma_{i=1}^{n} y_i^2 = 0$, that is, if and only if $\Sigma_{i=1}^{n} x_i = \Sigma_{i=1}^{n} y_i$ and $\Sigma_{i=1}^{n} x_i^2 = \Sigma_{i=1}^{n} y_i^2$. Hence, by the theorem of Lehmann-Scheffé, $\mathbf{T} = (\Sigma_{i=1}^{n} X_i, \Sigma_{i=1}^{n} X_i^2)$ is minimal sufficient for (μ, σ^2). ▲

Example 6.3.4 (Example 6.2.6 Continued) With data $\mathbf{x} = (x_1, ..., x_n)$ and $\mathbf{y} = (y_1, ..., y_n)$, both from $\mathcal{X}$, we have:

$$\{\Pi_{i=1}^{n} f(x_i; \theta)\}/\{\Pi_{i=1}^{n} f(y_i; \theta)\} = I\left(0 < x_{n:n} < \theta\right)/I(0 < y_{n:n} < \theta)$$
(6.3.4)

Denote $a(\theta) = I\left(0 < x_{n:n} < \theta\right)/I(0 < y_{n:n} < \theta)$. Now, the question is this: does $a(\theta)$ become free from θ if and only if $x_{n:n} = y_{n:n}$? One may verify that the answer is "yes" and hence $X_{n:n}$, the largest order statistic, is minimal sufficient for θ. ▲

Theorem 6.3.2 *A statistic which is a one-to-one function of a minimal sufficient statistic is minimal sufficient.*

The following theorem provides a useful tool for finding minimal sufficient statistics within a rich class of statistical models, namely an *exponential family*. One may refer to Lehmann (1983, pp. 43-44) or Lehmann and Casella (1998) for details.

Theorem 6.3.3 (Minimal Sufficiency in an Exponential Family) *Let $X_1, ..., X_n$ be iid with a common pmf or pdf*

$$f(x; \boldsymbol{\theta}) = \; a(\boldsymbol{\theta})g(x)exp\{\Sigma_{j=1}^{k} b_j(\boldsymbol{\theta})R_j(x)\}$$
(6.3.5)

belonging to a k-parameter exponential family defined by Equation (3.7.1). Now, let us denote the statistic $T_j = \Sigma_{i=1}^{n} R_j(X_i), j = 1, ..., k$. Then, $\mathbf{T} = (T_1, ..., T_k)$ is (jointly) minimal sufficient for $\boldsymbol{\theta}$.

The following result provides the nature of the distribution of a minimal sufficient statistic when a common pmf or pdf comes from an exponential family. One may refer to Theorem 4.3 of Lehmann (1983) and Lemma 8 in Lehmann (1986). One may also review Barankin and Maitra (1963), Brown (1964), Hipp (1974), Barndorff-Nielsen (1978), and Lehmann and Casella (1998) to gain broader perspectives.

Theorem 6.3.4 (Distribution of a Minimal Sufficient Statistic in an Exponential Family) *Under the conditions of Theorem 6.3.3, the pmf or pdf of a minimal sufficient statistic* $(T_1, ..., T_k)$ *belongs to a k-parameter exponential family.*

6.4 Information

We remarked earlier that we would work with a sufficient or minimal sufficient statistic **T** because it would reduce the data *and* preserve *all* "information" about θ contained in the original data. But, how much *information* does one have in original data that one tries to preserve? A notion of information was introduced by F. Y. Edgeworth in a series of papers, published in *J. Roy. Statist. Soc.* during 1908-1909. See Lehmann (1986, p. 147) for details. Fisher truly articulated a systematic development of this concept in his 1922 paper. One is referred to Efron's (1998, p. 101) commentaries on information, or Fisher-information.

For simplicity, we include only a one-parameter situation. Suppose that X is an observable real valued random variable with its pmf or pdf $f(x; \theta)$ where the unknown parameter $\theta \in \Theta$, an open subinterval of $\Re$. The $\mathcal{X}$ space is *assumed* not to depend upon θ.

Assumptions: (i) The partial derivative $\frac{\partial}{\partial \theta} f(x; \theta)$ is finite for all $x \in \mathcal{X}$, $\theta \in \Theta$; (ii) we can interchange the derivative (with respect to θ) and the integral (with respect to x).

Definition 6.4.1 *The Fisher-information, or simply information about* θ, *contained in X is:*

$$\mathcal{I}_X(\theta) = E_\theta \left[\left\{ \frac{\partial}{\partial \theta} log f(X; \theta) \right\}^2 \right] \qquad (6.4.1)$$

Example 6.4.1 Suppose X is Poisson(λ), $\lambda > 0$. Now,

$$log f(x; \lambda) = -\lambda + x log(\lambda) - log(x!)$$

so that $\frac{\partial}{\partial \lambda} log f(x; \lambda) = -1 + x\lambda^{-1}$. Thus,

$$\mathcal{I}_X(\lambda) = E_\lambda \left[\left\{ \frac{\partial}{\partial \lambda} log f(X; \lambda) \right\}^2 \right] = E_\lambda \left[(X - \lambda)^2 / \lambda^2 \right] = \lambda^{-1} \qquad (6.4.2)$$

As we contemplate having larger values of λ, variability in X increases. Hence, it is natural that information about λ contained in X will go down as λ increases. ▲

Example 6.4.2 Suppose X is $N(\mu, \sigma^2)$ where $\mu \in (-\infty, \infty)$ is unknown, but $\sigma \in (0, \infty)$ is known. Now,

$$logf(x;\mu) = -\tfrac{1}{2}\left\{(x-\mu)^2/\sigma^2\right\} - log(\sigma\sqrt{2\pi})$$

so that $\frac{\partial}{\partial\mu}logf(x;\mu) = (x-\mu)/\sigma^2$. Thus,

$$\mathcal{I}_X(\mu) = E_\mu\left[\left\{\tfrac{\partial}{\partial\mu}logf(X;\mu)\right\}^2\right] = E_\mu\left[(X-\mu)^2/\sigma^4\right] = \sigma^{-2} \qquad (6.4.3)$$

Again, as we contemplate having larger values of σ, variability in X increases. So, it is natural that information about μ contained in X will go down as σ increases. ▲

The following result quantifies information about θ contained in a random sample $X_1, ..., X_n$. It is not hard to prove. Refer to Mukhopadhyay (2000, pp. 302-303).

Theorem 6.4.1 *Let $X_1, ..., X_n$ be iid with a common pmf or pdf $f(x; \theta)$. We denote $\mathcal{I}_{X_1}(\theta)$, the information contained in one observation, X_1. Then, the information $\mathcal{I}_X(\theta)$ contained in $\mathbf{X} = (X_1, ..., X_n)$ is:*

$$\mathcal{I}_\mathbf{X}(\theta) = n\mathcal{I}_{X_1}(\theta) \text{ for all } \theta \in \Theta \qquad (6.4.4)$$

Suppose that we evaluate $\mathcal{I}_\mathbf{X}(\theta)$ and $\mathcal{I}_T(\theta)$ where $T \equiv T(\mathbf{X})$ is a statistic. If we find that $\mathcal{I}_T(\theta) = \mathcal{I}_\mathbf{X}(\theta)$, can we claim that T is indeed sufficient for θ? The answer is "yes," we certainly can (Theorem 6.4.2). One may refer to Rao (1973, result [iii], p. 330). In a recent exchange of personal communications, C. R. Rao gave a simple way to look at this important result. Exercise 6.4.5 gives an outline of Rao's elegant proof.

Theorem 6.4.2 *Suppose $\mathbf{X}$ is the data and $T = T(\mathbf{X})$ is a statistic. Then, $\mathcal{I}_\mathbf{X}(\theta) \geq \mathcal{I}_T(\theta)$ for all $\theta \in \Theta$. The equality holds for all θ if and only if T is sufficient for θ.*

Example 6.4.3 (Example 6.4.1 Continued) One knows that $T = \Sigma_{i=1}^n X_i$ is distributed as Poisson$(n\lambda)$ and hence one may work with the pmf $g(t; \lambda)$ of T. Now, observe the following: $log\{g(t;\lambda)\} = -n\lambda + tlog(n\lambda) - log(t!)$ which implies that $\frac{\partial}{\partial\lambda}log\{g(t;\lambda)\} = -n + t\lambda^{-1}$. So, one can claim that $\mathcal{I}_T(\lambda) = E_\lambda\left[(T-n\lambda)^2/\lambda^2\right] = n\lambda^{-1}$.

Example 6.4.1 and Equation (6.4.4) imply $\mathcal{I}_\mathbf{X}(\lambda) = n\mathcal{I}_{X_1}(\lambda) = n\lambda^{-1} = \mathcal{I}_T(\lambda)$. That is, T is indeed sufficient for λ. ▲

Example 6.4.4 (Example 6.4.2 Continued) One knows that $T = \overline{X}$ is distributed as $N(\mu, n^{-1}\sigma^2)$. One can start with the pdf $g(t; \mu)$ of T and verify that $\mathcal{I}_T(\mu) = n\sigma^{-2}$ as follows: $log\{g(t;\mu)\} = -\tfrac{1}{2}\left\{n(t-\mu)^2/\sigma^2\right\} - log(\sigma\sqrt{2n^{-1}\pi})$ which implies that $\frac{\partial}{\partial\mu}log\{g(t;\mu)\} = n(t-\mu)/\sigma^2$. So, $\mathcal{I}_T(\mu) = E_\mu\left[n^2(T-\mu)^2/\sigma^4\right] = n\sigma^{-2}$.

Example 6.4.2 and Equation (6.4.4) imply $\mathcal{I}_{\mathbf{X}}(\mu) = n\mathcal{I}_{X_1}(\mu) = n\sigma^{-2} = \mathcal{I}_T(\mu)$ so that T is indeed sufficient for μ. ▲

Remark 6.4.1 Suppose that the pmf or pdf $f(x;\theta)$ is such that $\frac{\partial^2}{\partial\theta^2} f(x;\theta)$ is finite for all $x \in \mathcal{X}$ and $E_\theta \left[\frac{\partial^2}{\partial\theta^2} f(X;\theta) \right]$ is finite for all $\theta \in \Theta$. Then, *Fisher-information* can alternatively be found as follows:

$$\mathcal{I}_X(\theta) = -E_\theta \left[\frac{\partial^2}{\partial\theta^2} \log f(X;\theta) \right] \tag{6.4.5}$$

Its proof is left out as an exercise.

6.5   Ancillarity

A concept called *ancillarity* is apparently the furthest removed from sufficiency. A sufficient statistic $\mathbf{T}$ preserves *all* information about $\boldsymbol{\theta}$. In contrast, an ancillary statistic $\mathbf{T}$ provides *no* information about $\boldsymbol{\theta}$. This is not to imply that an ancillary statistic is necessarily bad or useless. The fixed sample size n seldom carries any information about $\boldsymbol{\theta}$, but n is a crucial input in statistical analysis! This concept evolved from Fisher (1925a) and later it blossomed into a vast area of *conditional inference*. In his 1956 book, Fisher emphasized many positive aspects of ancillarity in analyzing real data. For fuller discussions of *conditional inference* one may look at Basu (1964), Hinkley (1980), and Ghosh (1988). The interesting article of Reid (1995) provides an assessment of conditional inference procedures.

Consider real valued observations $X_1, ..., X_n$ from some population with a common pmf or pdf $f(x; \boldsymbol{\theta})$ with unknown parameter vector $\boldsymbol{\theta} \in \Theta \subseteq \Re^p$. Denote $\mathbf{X} \equiv (X_1, ..., X_n)$ and a vector valued statistic $\mathbf{T} \equiv \mathbf{T}(\mathbf{X})$.

Definition 6.5.1 *A statistic* $\mathbf{T}$ *is called ancillary for* $\boldsymbol{\theta}$ *or simply ancillary provided that its pmf or pdf* $g(\mathbf{t})$, $\mathbf{t} \in \mathcal{T}$, *does not involve* $\boldsymbol{\theta}$.

Example 6.5.1 Suppose $X_1, ..., X_n$ are iid $N(\theta, 1)$ where θ is an unknown parameter, $-\infty < \theta < \infty, n \geq 3$. A statistic $T_1 = X_1 - X_2$ is distributed as $N(0, 2)$ whatever be θ. Hence T_1 is ancillary for θ. The statistic $T_2 = X_1 + ... + X_{n-1} - (n-1)X_n$ is distributed as $N(0, n(n-1))$ whatever be θ. Hence, T_2 is also ancillary for θ. The sample variance S^2 is distributed as $(n-1)^{-1}\chi^2_{n-1}$ whatever be θ and hence S^2 is ancillary too for θ. Another statistic, $\mathbf{T} = (T_1, T_2)$, is distributed as $N_2(0, 0, 2, n(n-1), 0)$ whatever be θ, and hence $\mathbf{T}$ is ancillary too. ▲

Example 6.5.2 Suppose $X_1, ..., X_n$ are iid $N(\mu, \sigma^2)$, $\boldsymbol{\theta} = (\mu, \sigma^2)$, $-\infty < \mu < \infty, 0 < \sigma^2 < \infty, n \geq 2$. Both parameters μ, σ are unknown. Reconsider the statistics T_1 or T_2 from Example 6.5.1. T_1, T_2 are no longer ancillary statistics. But, consider another statistic $T_3 = (X_1 - X_2)/S$ where S^2 is the sample variance. Denoting $Y_i = (X_i - \mu)/\sigma$, observe that one can equivalently express T_3 as:

$$T_3 = \sqrt{(n-1)}(Y_1 - Y_2)/\sqrt{\Sigma_{i=1}^n (Y_i - \overline{Y})^2}$$

Since $Y_1, ..., Y_n$ are iid $N(0,1)$, one can show that the pdf of T_3 would not involve $\boldsymbol{\theta}$. So, T_3 is an ancillary *statistic*. Note that we do not need the explicit pdf of T_3 to conclude this. ▲

We remarked earlier that a statistic which is ancillary for $\boldsymbol{\theta}$ can play useful roles in the process of inference making. The following examples would clarify this point.

> It is possible to have two statistics T_1, T_2 such that (i) T_2 has some information about θ, but it is not sufficient for θ; (ii) T_1 is ancillary for θ, and yet (iii) (T_1, T_2) is jointly sufficient for θ.
> Look at Example 6.5.3.

$$(6.5.1)$$

Example 6.5.3 (Example 6.5.1 Continued) Fix $n = 2$. It is clear that $T_1 = X_1 - X_2$ is ancillary. Consider another statistic $T_2 = X_1$. Recall that $\mathcal{I}_{T_2}(\theta) = 1$ whereas $\mathcal{I}_{(X_1, X_2)}(\theta) = 2$ and so T_2 cannot be sufficient. But, if we are told the observed value of the statistic $\mathbf{T} = (T_1, T_2)$, we can reconstruct $\mathbf{X}$ *uniquely*. That is, $\mathbf{T}$ is *jointly sufficient*. ▲

> It is possible to have two statistics T_1, T_2 such that (i) T_1 has no information about θ, (ii) T_2 has no information about θ, and yet (iii) (T_1, T_2) is jointly minimal sufficient for θ.
> Look at Example 6.5.4.

$$(6.5.2)$$

Example 6.5.4 (Example Due to D. Basu) Suppose (X, Y) is distributed as $N_2(0, 0, 1, 1, \rho)$ with unknown parameter $\rho \in (-1, 1)$. Consider two statistics $T_1 = X, T_2 = Y$. Since T_1, T_2 individually have standard normal distributions, it follows that T_1 is ancillary for ρ and so is T_2. Note that $\mathbf{T} = (T_1, T_2)$, equivalent to (X, Y), is minimal sufficient for ρ. What is remarkable is that T_1 (or T_2) has no information, but the statistic (T_1, T_2) has *all* information. ▲

6.5.1 Location, Scale, and Location-Scale Families

We briefly introduce special families of distributions which are frequently encountered. We start with a pdf $g(x), x \in \mathcal{X} \subseteq \Re$ and construct the following families of distributions:

(i) Location: $\mathcal{F}_1 = \{f(x; \theta) = g(x - \theta) : \theta \in \Re, x \in \Re\}$;

(ii) Scale: $\mathcal{F}_2 = \{f(x; \delta) = \delta^{-1} g(x/\delta) : \delta \in \Re^+, x \in \Re\}$;

(iii) Location-Scale: $\mathcal{F}_3 = \{f(x; \theta, \delta) = \delta^{-1} g((x - \theta)/\delta) : \theta \in \Re,$

$$\delta \in \Re^+, x \in \Re\}.$$

(6.5.3)

The reader should check that the corresponding members $f(.)$ from families $\mathcal{F}_1, \mathcal{F}_2, \mathcal{F}_3$ are indeed pdfs themselves. In part (i) θ is called a *location parameter*, (ii) δ is called a *scale parameter*, and (iii) θ, δ are respectively called *location* and *scale parameters*.

For example, a $N(\mu, 1)$ distribution, with $\mu \in \Re$, forms a *location* family. The $N(0, \sigma^2)$ distribution, with $\sigma \in \Re^+$, forms a *scale* family. The $N(\mu, \sigma^2)$ distribution, with $\mu \in \Re, \sigma \in \Re^+$, forms a *location-scale* family.

In a location family, the role of a location parameter θ is felt in the "movement" of the pdf along the horizontal axis as different values of θ are contemplated. In a large factory, we may look at the monthly wage of each employee and postulate that the distribution of wages as $N(\mu, \sigma^2)$ where $\sigma = \$100$. After negotiation of a new contract, suppose that each employee receives a \$50 monthly raise. Then the distribution moves to the right with its new center of symmetry at $\mu + 50$. The intrinsic shape of the distribution does not change.

In a scale family, the role of a scale parameter δ is felt in "squeezing" or "expanding" the pdf along the horizontal axis as different values of δ are contemplated. The $N(0, 1)$ distribution has its center of symmetry at the point $x = 0$. The $N(0, \delta^2)$ distribution's center of symmetry stays put at the point $x = 0$, but depending on whether δ is larger or smaller than one, the shape of the density curve will become more flat or more peaked at the center, compared with the standard normal distribution. Suppose that we record heights (in inches) of individuals and we postulate the distribution of heights as $N(70, \sigma^2)$. If heights are measured in centimeters instead, then the distribution would *appear* more spread out around the new center. One needs to keep in mind that recording the heights in centimeters would amount to multiplying each original observation X measured in inches by 2.54.

In a location-scale family, one would notice movement of the distribution along the horizontal axis as well as the squeezing or expansion effect in its shape. We may be looking at data on weekly maximum temperature in a city, in Fahrenheit ($°F$) or Celsius ($°C$). If one postulates a normal distribution for temperatures, changing the unit of measurement from Fahrenheit to Celsius would amount to shifts in both origin and scale. One needs to recall the relationship $\frac{1}{5}C = \frac{1}{9}(F - 32)$ between Fahrenheit and Celsius.

What is the relevance of such special families of distributions in the context of ancillarity? Suppose that $X_1, ..., X_n$ are iid having a common pdf

$f(x)$, *indexed* by some appropriate parameter(s). Then, we may summarize the following:

> If a common pdf of $X_1, ..., X_n$ belongs to a *location* family $\mathcal{F}_1$ from Equation (6.5.3), the statistic $U = (X_1 - X_n, X_2 - X_n, ..., X_{n-1} - X_n)$ is ancillary.   (6.5.4)

> If a common pdf of $X_1, ..., X_n$ belongs to a *scale* family $\mathcal{F}_2$ from Equation (6.5.3), the statistic $V = \left(\dfrac{X_1}{X_n}, \dfrac{X_2}{X_n}, ..., \dfrac{X_{n-1}}{X_n} \right)$ is ancillary.   (6.5.5)

> If a common pdf of $X_1, ..., X_n$ belongs to a *location-scale* family $\mathcal{F}_3$ from Equation (6.5.3), the statistic $W = \left(\dfrac{X_1 - X_n}{S}, \dfrac{X_2 - X_n}{S}, ..., \dfrac{X_{n-1} - X_n}{S} \right)$ is ancillary where S^2 is the sample variance.   (6.5.6)

6.6   Completeness

Suppose that a real valued random variable X has its pmf or pdf $f(x; \theta)$ for $x \in \mathcal{X}$ and $\theta \in \Theta$. Let $T \equiv T(X)$ be a statistic with its pmf or pdf $g(t; \theta)$ for $t \in \mathcal{T}$ and $\theta \in \Theta$.

Definition 6.6.1 *The family $\{g(t; \theta): \theta \in \Theta\}$ is called the family of distributions induced by the statistic T.*

Definition 6.6.2 *The family $\{g(t; \theta): \theta \in \Theta\}$ is called complete if and only if the following condition holds: consider any real valued function $h(t)$ defined for $t \in \mathcal{T}$ with finite expectation, such that*

$$E_\theta [h(T)] = 0 \text{ for all } \theta \in \Theta \text{ implies } h(t) \equiv 0 \ \ w.p.1 \qquad (6.6.1)$$

Definition 6.6.3 *A statistic T is called complete if and only if $\{g(t; \theta): \theta \in \Theta\}$ is complete.*

In these definitions, observe that neither T nor θ has to be necessarily real valued. For vector valued statistic $\mathbf{T} = (T_1, T_2, ..., T_k)$ and $\boldsymbol{\theta}$, the requirement in Equation (6.6.1) would be obviously interpreted as follows:

$$\int \cdots \int_{\mathbf{t} \in \mathcal{T}} h(\mathbf{t}) g(\mathbf{t}; \boldsymbol{\theta}) \Pi_{i=1}^k dt_i = 0 \text{ for all } \boldsymbol{\theta} \in \Theta \text{ implies}$$
$$h(\mathbf{t}) \equiv 0 \ \ w.p.1 \qquad (6.6.2)$$

Here, $h(\mathbf{t})$ is a real valued function of $\mathbf{t} = (t_1, ..., t_k) \in \mathcal{T}$ and $g(\mathbf{t}; \boldsymbol{\theta})$ is a pmf or pdf of $\mathbf{T}$.

The concept was introduced by Lehmann and Scheffé (1950) and further explored measure-theoretically by Bahadur (1957).

Example 6.6.1 A *statistic* T is distributed as Bernoulli(p), $0 < p < 1$. Is T complete? The pmf induced by T is $g(t; p) = p^t(1 - p)^{1-t}$, $t = 0, 1$. Consider any real valued function $h(t)$ such that $E_p[h(T)] = 0$ for all $0 < p < 1$. Now, we write:

$$E_p[h(T)] = (1 - p)h(0) + ph(1) = p\{h(1) - h(0)\} + h(0) = 0,$$
$$\text{for all } p \in (0, 1) \tag{6.6.3}$$

The expression in Equation (6.6.3) is linear in p and so it *may* be zero for at most one value of p between zero and one. But, we demand that $p\{h(1) - h(0)\} + h(0)$ must be zero for infinitely many values of p in $(0, 1)$. Hence, this expression must be identically zero which means that the constant term as well as the coefficient of p must be individually both zero. That is, we must have $h(0) = 0$ and $h(1) - h(0) = 0$ so that $h(1) = 0$. In other words, we have $h(t) \equiv 0$ for $t = 0, 1$. Thus, T is complete. ▲

Example 6.6.2 Consider $g(t; \sigma) = \{\sigma\sqrt{2\pi}\}^{-1}\exp\left\{-\frac{1}{2}t^2/\sigma^2\right\}$, $-\infty < t < \infty$, $\sigma \in \Re^+$. Is the family $\{g(t; \sigma): \sigma > 0\}$ complete? The answer is "no," it is not complete. In order to prove this claim, consider $h(t) = t$ and then observe that $E_\sigma[h(T)] = E_\sigma[T] = 0$ for all $0 < \sigma < \infty$, but $h(t)$ is not identically zero. ▲

6.6.1 Complete Sufficient Statistics

The completeness of a statistic T is a mathematical property. From a statistical point of view, however, this concept can lead to important results when a *complete statistic* T is also *sufficient*.

Definition 6.6.4 A statistic $\mathbf{T}$ *is called complete sufficient for $\boldsymbol{\theta}$ if and only if* (i) $\mathbf{T}$ *is sufficient for $\boldsymbol{\theta}$ and* (ii) $\mathbf{T}$ *is complete.*

Example 6.6.3 Let $X_1, ..., X_n$ be iid Bernoulli(p), $0 < p < 1$. Then, $T = \Sigma_{i=1}^n X_i$ is sufficient for p. We verify that T is complete. The pmf induced by T is $g(t; p) = \binom{n}{t}p^t(1 - p)^{n-t}$, $t \in T = \{0, 1, ..., n\}$, $0 < p < 1$. Consider a real valued function $h(t)$ such that $E_p[h(T)] = 0$ for all $0 < p < 1$. Now, with $\gamma = p(1 - p)^{-1}$, we write:

$$E_p[h(T)] = \Sigma_{t=0}^n h(t)\binom{n}{t}p^t(1 - p)^{n-t} = (1 - p)^n\Sigma_{t=0}^n \binom{n}{t}h(t)\gamma^t \tag{6.6.4}$$

Observe that $E_p[h(T)]$ has been expressed as a polynomial of the nth degree in $\gamma \in (0, \infty)$. An nth-degree polynomial in γ may be equal to zero for at most n values of $\gamma \in (0, \infty)$. If we assume:

$$E_p[h(T)] = 0 \text{ for all } p \in (0, 1)$$

then $\Sigma_{t=0}^n \binom{n}{t}h(t)\gamma^t \equiv 0$ for all $\gamma \in (0, \infty)$, that is, $\binom{n}{t}h(t) \equiv 0$ for all $t = 0, 1, ..., n$. Hence, $h(t) \equiv 0$ for all $t = 0, 1, ..., n$, proving completeness of the sufficient statistic T. ▲

Example 6.6.4 Let $X_1, ..., X_n$ be iid Poisson$(\lambda), 0 < \lambda < \infty$. Then, $T = \Sigma_{i=1}^n X_i$ is sufficient for λ. Is T complete? The pmf induced by T is $g(t; \lambda) = e^{-n\lambda}(n\lambda)^t/t!, t \in \mathcal{T} = \{0, 1, 2, ...\}, 0 < \lambda < \infty$. Consider a real valued function $h(t)$ such that $E_\lambda[h(T)] = 0$ for all $0 < \lambda < \infty$. Now, with $k(t) = h(t)n^t/t!$, we write:

$$E_\lambda[h(T)] = \Sigma_{t=0}^\infty h(t)e^{-n\lambda}(n\lambda)^t/t! = e^{-n\lambda}\Sigma_{t=0}^\infty k(t)\lambda^t \qquad (6.6.5)$$

Note that $E_\lambda[h(T)]$ is expressed as a power series in λ. Such power series forms a vector space with $\mathcal{G} = \{1, \lambda, \lambda^2, \lambda^3, ..., \lambda^t, ...\}$ as its minimal generator. That is, the vectors in $\mathcal{G}$ are *linearly independent* so that if we assume

$$E_\lambda[h(T)] = 0 \text{ for all } \lambda \in (0, \infty)$$

then $\Sigma_{t=0}^\infty k(t)\lambda^t \equiv 0$, that is, $k(t) \equiv 0$ for all $t = 0, 1, 2, ...$. Hence, $h(t) \equiv 0$ for all $t = 0, 1, 2, ...$, proving completeness of T. ▲

> A sufficient or minimal sufficient statistic **T** for an unknown parameter $\boldsymbol{\theta}$ may not necessarily be complete.
> Look at Example 6.6.5.

Example 6.6.5 Suppose $X_1, ..., X_n$ are iid Normal(θ, θ^2) with an unknown parameter $\theta(>0)$. Then, $\mathbf{T} = (\overline{X}, S^2)$ is a minimal sufficient statistic for θ, and note that $E_\theta[n(n+1)^{-1}\overline{X}^2] = \theta^2 = E_\theta(S^2)$ for all $\theta > 0$. That is, $E_\theta[n(n+1)^{-1}\overline{X}^2 - S^2] = 0$ for all $\theta > 0$. Let $h(\mathbf{T}) = n(n+1)^{-1}\overline{X}^2 - S^2$ so that $E_\theta[h(\mathbf{T})] \equiv 0$ for all $\theta > 0$, but $h(\mathbf{t}) = n(n+1)^{-1}\overline{x}^2 - s^2$ is not identically zero. Thus, **T** cannot be complete. ▲

Theorem 6.6.1 *Suppose that a statistic* $\mathbf{T} = (T_1, ..., T_k)$ *is complete. Let* $\mathbf{U} = (U_1, ..., U_k)$ *be another statistic with* $\mathbf{U} = g(\mathbf{T})$ *where* $g : \mathcal{T} \to \mathcal{U}$ *is one-to-one. Then,* **U** *is complete.*

> Are complete sufficient statistics also minimal sufficient? The answer is "yes." From Example 6.6.5, it should be clear, however, that the converse is not necessarily true. Refer to Lehmann and Scheffé (1950) and Bahadur (1957).   (6.6.6)

Now, we state a remarkably general result in the case of an exponential family of distributions. One may refer to Lehmann (1986, pp. 142-143) for its proof.

Theorem 6.6.2 (Completeness of Minimal Sufficient Statistics in an Exponential Family) *Let* $X_1, ..., X_n$ *be iid with a common pmf or pdf*

$$f(x; \boldsymbol{\theta}) = q(\boldsymbol{\theta})p(x)exp\{\Sigma_{j=1}^k \theta_j R_j(x)\} \qquad (6.6.7)$$

belonging to a k-parameter exponential family defined by Equation (3.7.1). Denote the statistic $T_j = \Sigma_{i=1}^n R_j(X_i), j = 1, ..., k$. *Then, the (jointly) minimal sufficient statistic* $\mathbf{T} = (T_1, ..., T_k)$ *for* $\boldsymbol{\theta}$ *is complete.*

Example 6.6.6 Suppose $X_1, ..., X_n$ are iid $N(\mu, \sigma^2)$ with $(\mu, \sigma^2) \in \Theta = \Re \times \Re^+$ where μ, σ^2 are both unknown. The common pdf belongs to a two-parameter exponential family. So, the minimal sufficient statistic is $\mathbf{T} = (T_1(\mathbf{X}), T_2(\mathbf{X}))$ with $T_1(\mathbf{X}) = \Sigma_{i=1}^n X_i$ (or $\overline{X}$) and $T_2(\mathbf{X}) = \Sigma_{i=1}^n X_i^2$ (or S^2). In view of Theorem 6.6.2, $(\overline{X}, S^2)$ is complete. ▲

Theorem 6.6.2 covers a lot of ground by helping to prove the completeness property of sufficient statistics. But it fails to reach out to *nonexponential* families. The case in point will become clear from Example 6.6.7.

Example 6.6.7 Suppose $X_1, ..., X_n$ are iid Uniform$(0, \theta)$, $\theta(> 0)$ being an unknown parameter. Then, $T(\mathbf{X}) \equiv X_{n:n}$ is a minimal sufficient statistic for θ and its pdf is $g(t; \theta) = nt^{n-1}\theta^{-n} I(0 < t < \theta)$. This does not belong to exponential family (6.6.7) with $k = 1$. But, one may show directly that T is complete. Let $h(t)$, $t > 0$ be an arbitrary real valued function such that $E_\theta[h(T)] = 0$ for all $\theta > 0$ and write:

$$0 \equiv \tfrac{d}{d\theta} E_\theta[h(T)] = \tfrac{d}{d\theta} \int_0^\theta h(t) nt^{n-1}\theta^{-n} dt = n\theta^{-1} h(\theta)$$

which proves that $h(\theta) \equiv 0$ for all $\theta > 0$. Hence, T is complete. ▲

6.6.2   Basu's Theorem

Suppose that $\mathbf{X} = (X_1, ..., X_n)$ has a joint pmf or pdf indexed with some unknown parameter $\boldsymbol{\theta}$. It is not essential to assume that $X_1, ..., X_n$ are iid. Let $\mathbf{U} \equiv \mathbf{U}(\mathbf{X})$ and $\mathbf{W} \equiv \mathbf{W}(\mathbf{X})$ be two *statistics*. In general, showing that two statistics $\mathbf{U}, \mathbf{W}$ are independent is a fairly tedious process. The following result, known as *Basu's Theorem*, provides an elegant tool (Basu, 1955a) to prove *independence* of two appropriate statistics painlessly. One may refer to Basu's (1958) sequel for more insight.

Theorem 6.6.3 (Basu's Theorem) *Suppose that* $\mathbf{U} \equiv \mathbf{U}(\mathbf{X})$ *is a complete sufficient statistic for* $\boldsymbol{\theta}$ *and* $\mathbf{W} \equiv \mathbf{W}(\mathbf{X})$ *is an ancillary statistic. Then,* $\mathbf{U}$ *and* $\mathbf{W}$ *are independently distributed.*

Proof For simplicity, we supply a proof in a discrete case only. Denote the domain spaces for $\mathbf{U}$ and $\mathbf{W}$ by $\mathcal{U}$ and $\mathcal{W}$, respectively. To prove that $\mathbf{U}$ and $\mathbf{W}$ are independently distributed, we need to show:

$$P_{\boldsymbol{\theta}}\{\mathbf{W} = \mathbf{w} \mid \mathbf{U} = \mathbf{u}\} = P_{\boldsymbol{\theta}}(\mathbf{W} = \mathbf{w}) \text{ for all } \mathbf{w} \in \mathcal{W},$$
$$\mathbf{u} \in \mathcal{U}, \text{ and } \boldsymbol{\theta} \in \Theta \tag{6.6.8}$$

For $\mathbf{w} \in \mathcal{W}$, we denote $P_{\boldsymbol{\theta}}(\mathbf{W} = \mathbf{w}) = h(\mathbf{w})$. Obviously, $h(\mathbf{w})$ is free from $\boldsymbol{\theta}$ since $\mathbf{W}$ is ancillary. Also, observe that $P_{\boldsymbol{\theta}}\{\mathbf{W} = \mathbf{w} \mid \mathbf{U} = \mathbf{u}\}$ must be free from $\boldsymbol{\theta}$ since $\mathbf{U}$ is a sufficient statistic for $\boldsymbol{\theta}$. So, write $g(\mathbf{u}) = P_{\boldsymbol{\theta}}\{\mathbf{W} =$

$\mathbf{w} \mid \mathbf{U} = \mathbf{u}$. Now, $E_{\boldsymbol{\theta}}[g(\mathbf{U})] = P_{\boldsymbol{\theta}}(\mathbf{W} = \mathbf{w})$ which was denoted earlier by $h(\mathbf{w})$. So, we claim that $g(\mathbf{U}) - h(\mathbf{W})$ is a genuine *statistic*.

Now, note that $E_{\boldsymbol{\theta}}[g(\mathbf{U}) - h(\mathbf{w})] \equiv 0$ for all $\boldsymbol{\theta}$ and the fact that $\mathbf{U}$ is complete. Thus, by Definition 6.6.4, we must have $g(\mathbf{u}) - h(\mathbf{w}) \equiv 0$ w.p.1, that is, $g(\mathbf{u}) \equiv h(\mathbf{w})$ for all $\mathbf{w} \in \mathcal{W}, \mathbf{u} \in \mathcal{U}$. We have shown the validity of Equation (6.6.8). ∎

Example 6.6.8 Suppose that $X_1, ..., X_n$ are iid $N(\mu, \sigma^2)$ with $n \geq 2$, $(\mu, \sigma^2) \in \Re \times \Re^+$ where μ is unknown, but σ^2 is known. Let $U \equiv \overline{X}$ which is a complete sufficient statistic for μ. Observe that $W \equiv S^2$ is an ancillary statistic for μ. By Basu's Theorem, the two *statistics* $\overline{X}$ and S^2 are independently distributed. ▲

Example 6.6.9 (Example 6.6.8 Continued) Denote $V = X_{n:n} - X_{n:1}$, the sample range. Then, $\overline{X}$ and S/V are independently distributed. Also, $\overline{X}$ and $(X_{n:n} - \overline{X})$ are independent. In the same spirit, $\overline{X}$ and $(X_{n:1} - \overline{X})^2$ are independent. Is $\overline{X}$ independent of $log(\mid X_{n:n} - \overline{X} \mid /S)$? The ancillarity of the relevant statistics may be verified by appealing to a location family of distributions. We leave out some details as exercises. ▲

Example 6.6.10 Suppose that $X_1, ..., X_n$ are iid $N(\mu, \sigma^2)$ with $n \geq 2$, $(\mu, \sigma^2) \in \Re \times \Re^+$ where σ^2 is unknown, but μ is known. Now, $U^2 \equiv \Sigma_{i=1}^{n}(X_i - \mu)^2$ is a complete sufficient statistic for σ^2. Observe that $W = (X_1 - \overline{X})/U$ is ancillary for σ^2. Immediately we can claim that $\tanh(\mid W \mid)$ and $\sin^{-1}\left(\Sigma_{i=1}^{n}(X_i - \mu)^2\right)$ are independently distributed. Also, $\Sigma_{i=1}^{n}(X_i - \mu)^2$ and $(X_{n:1} - \overline{X})/S$ are independent where S^2 is the customary sample variance. The ancillarity of the relevant statistics may be verified by appealing to a scale family of distributions. We leave out some details as exercises. ▲

6.7 Exercises and Complements

6.2.1 Let $X_1, ..., X_n$ be iid Geometric(p) with a common pmf $f(x; p) = p(1 - p)^x$, $x = 0, 1, 2, ...$ where $0 < p < 1$ is an unknown parameter. Show that $\Sigma_{i=1}^{n} X_i$ is sufficient for p.

6.2.2 Let $X_1, ..., X_m$ be iid Bernoulli(p), $Y_1, ..., Y_n$ be iid Bernoulli(q), and X, Y be independent, where $0 < p < 1$ is an unknown parameter, $q = 1 - p$. Show that $\Sigma_{i=1}^{m} X_i - \Sigma_{j=1}^{n} Y_j$ is sufficient for p. {*Hint*: Can one justify looking at data $(X_1, ..., X_m, 1 - Y_1, ..., 1 - Y_n)$?}

6.2.3 Suppose that X is $N(\theta, 1)$ where $-\infty < \theta < \infty$ is an unknown parameter. Show that $\mid X \mid$ cannot be sufficient for θ.

6.2.4 Let $X_1, ..., X_m$ be iid Poisson(λ), $Y_1, ..., Y_n$ be iid Poisson(2λ), and that X, Y be independent, where $0 < \lambda < \infty$ is an unknown parameter. Show that $\Sigma_{i=1}^{m} X_i + \Sigma_{j=1}^{n} Y_j$ is sufficient for λ.

6.2.5 Let $X_1, ..., X_4$ be iid Bernoulli(p) where $0 < p < 1$ is an unknown parameter. Denote $U = X_1(X_3 + X_4) + X_2$. Show that U is not a sufficient statistic for p.

6.2.6 Suppose that $X_1, ..., X_n$ are iid Beta(α, β) involving parameters $\alpha > 0$ and $\beta > 0$. Show that

(i) $\Pi_{i=1}^n X_i$ is sufficient for α if β is known;

(ii) $\Pi_{i=1}^n (1 - X_i)$ is sufficient for β if α is known;

(iii) $(\Pi_{i=1}^n X_i, \Pi_{i=1}^n (1 - X_i))$ is jointly sufficient for (α, β) if both parameters are unknown.

6.2.7 Suppose $X_1, ..., X_n$ are iid with a common Rayleigh pdf:

$$f(x; \theta) = 2\theta^{-1} x \, exp(-x^2/\theta) I(x > 0)$$

where $\theta(> 0)$ is an unknown parameter. Show that $\Sigma_{i=1}^n X_i^2$ is a sufficient statistic for θ.

6.3.1 Suppose $X_1, ..., X_n$ are iid with a common pdf:

$$\sigma^{-1} exp\{-(x - \mu)/\sigma\} I(x > \mu) \text{ where } -\infty < \mu < \infty, 0 < \sigma < \infty$$

Show that

(i) $X_{n:1}$ is minimal sufficient for μ if σ is known;

(ii) $n^{-1}\Sigma_{i=1}^n (X_i - \mu)$ is minimal sufficient for σ if μ is known;

(iii) $(X_{n:1}, \Sigma_{i=1}^n (X_i - X_{n:1}))$ is minimal sufficient for (μ, σ) if both parameters are unknown.

6.3.2 Show that the pmf or pdf corresponding to distributions such as Binomial(n, p), Poisson(λ), Gamma(α, β), $N(\mu, \sigma^2)$, or Beta(α, β) belong to the exponential family defined in Equation (6.3.5) when

(i) $0 < p < 1$ is unknown;

(ii) $\lambda \in \Re^+$ is unknown;

(iii) $\mu \in \Re$ is unknown but $\sigma \in \Re^+$ is known;

(iv) $\sigma \in \Re^+$ is unknown but $\mu \in \Re$ is known;

(v) $\mu \in \Re$ and $\sigma \in \Re^+$ are both unknown;

(vi) $\alpha \in \Re^+$ is known but $\beta \in \Re^+$ is unknown;

(vii) $\alpha \in \Re^+, \beta \in \Re^+$ are both unknown.

In each case, obtain the minimal sufficient statistic(s) for the associated unknown parameter(s).

6.3.3 (Exercise 6.2.1 Continued) Show that this pmf belongs to an exponential family defined in Equation (6.3.5). Hence, show that $\Sigma_{i=1}^n X_i$ is minimal sufficient for p.

6.3.4 Suppose that $X_1, ..., X_n$ are iid with a Uniform($\theta - \frac{1}{2}, \theta + \frac{1}{2}$) distribution where $\theta(> 0)$ is an unknown parameter. Show that $(X_{n:1}, X_{n:n})$ is a jointly minimal sufficient statistic for θ.

6.3.5 Suppose $X_1, ..., X_n$ are iid $N(\theta, \theta^2)$ where $0 < \theta < \infty$ is an unknown parameter. Does the common pdf belong to exponential family (6.3.5)? Derive a minimal sufficient statistic for θ.

6.3.6 Suppose that $X_1, ..., X_n$ are iid $N(\theta, \theta)$ where $0 < \theta < \infty$ is an unknown parameter. Does the common pdf belong to exponential family (6.3.5)? Derive a minimal sufficient statistic for θ.

6.3.7 Let $X_1, ..., X_n$ be iid Uniform$(-\theta, \theta)$ where $0 < \theta < \infty$ is an unknown parameter. Derive a minimal sufficient statistic for θ.

6.3.8 Let $X_1, ..., X_m$ be iid $N(\mu_1, \sigma^2)$, $Y_1, ..., Y_n$ be iid $N(\mu_2, \sigma^2)$, and X, Y be independent where $-\infty < \mu_1, \mu_2 < \infty$, $0 < \sigma^2 < \infty$ are unknown parameters. Derive minimal sufficient statistics for (μ_1, μ_2, σ^2).

6.4.1 Let $X_1, ..., X_n$ be iid Bernoulli(p) where $0 < p < 1$ is an unknown parameter. Evaluate $\mathcal{I}_{\mathbf{X}}(p), \mathcal{I}_{\overline{X}}(p)$ and compare $\mathcal{I}_{\mathbf{X}}(p)$ with $\mathcal{I}_{\overline{X}}(p)$. Can Theorem 6.4.2 be used to claim that $\overline{X}$ is sufficient for p?

6.4.2 In a $N(\mu, \sigma^2)$ distribution with $-\infty < \mu < \infty$, $0 < \sigma^2 < \infty$, let only μ be known. Evaluate $\mathcal{I}_{U^2}(\sigma^2), \mathcal{I}_{S^2}(\sigma^2)$ and then show that $\mathcal{I}_{U^2}(\sigma^2) > \mathcal{I}_{S^2}(\sigma^2)$ where

$$U^2 = n^{-1}\Sigma_{i=1}^n (X_i - \mu)^2, S^2 = (n-1)^{-1}\Sigma_{i=1}^n (X_i - \overline{X})^2, n \geq 2.$$

6.4.3 (Exercise 6.2.7 Continued) Denote a statistic $T = \Sigma_{i=1}^n X_i^2$. Evaluate $\mathcal{I}_{\mathbf{X}}(\theta)$ and $\mathcal{I}_T(\theta)$. Are they same? If so, what conclusion can one draw from this?

6.4.4 Suppose that $X_1, ..., X_n$ are iid with a Weibull distribution, that is, the common pdf is:

$$f(x; \alpha) = \alpha^{-1}\beta x^{\beta-1}\exp(-x^\beta/\alpha)I(x > 0)$$

where $\alpha(> 0)$ is an unknown parameter, but $\beta(> 0)$ is assumed known. Denote $T = \Sigma_{i=1}^n X_i^\beta$. Evaluate $\mathcal{I}_{\mathbf{X}}(\alpha), \mathcal{I}_T(\alpha)$, and compare them. What conclusion can one draw from this comparison?

6.4.5 Prove Theorem 6.4.2. {*Hint*: This interesting idea was included in a personal communication from C. R. Rao to N. Mukhopadhyay. Note that the likelihood function $f(\mathbf{x}; \theta)$ can be written as $g(t; \theta)h(\mathbf{x} \mid T = t; \theta)$ where $g(t; \theta)$ is a pdf or pmf of T and $h(\mathbf{x} \mid T = t; \theta)$ is a conditional pdf or pmf of $\mathbf{X}$ given that $T = t$. From this equality, first derive the identity: $\mathcal{I}_{\mathbf{X}}(\theta) = \mathcal{I}_T(\theta) + \mathcal{I}_{\mathbf{X}|T}(\theta)$ for all $\theta \in \Theta$. This implies that $\mathcal{I}_{\mathbf{X}}(\theta) \geq \mathcal{I}_T(\theta)$ for all $\theta \in \Theta$. One will have $\mathcal{I}_{\mathbf{X}}(\theta) = \mathcal{I}_T(\theta)$ if and only if $\mathcal{I}_{\mathbf{X}|T}(\theta) = 0$, that is, $h(\mathbf{x} \mid T = t; \theta)$ must be free from θ. It then follows from the definition of sufficiency that T is sufficient for θ.}

6.5.1 Let X_1, X_2, X_3 be iid $N(\theta, 1)$ where θ is an unknown parameter, $-\infty < \theta < \infty$. Denote $T_1 = X_1 - X_2, T_2 = X_1 + X_2 - 2X_3$, and $\mathbf{T} = (T_1, T_2)$. Is $\mathbf{T}$ ancillary for θ? Are T_1, T_2 independent?

6.5.2 Let $X_1, ..., X_n$ be iid $N(\mu, \sigma^2), \infty < \mu < \infty, 0 < \sigma^2 < \infty, n \geq 4$. Both μ and σ^2 are unknown. Let S^2 be the sample variance and denote:

$$T_1 = (X_1 - X_3)/S, T_2 = (X_1 + X_3 - 2X_2)/S$$
$$T_3 = (X_1 - X_3 + 2X_2 - 2X_4)/S$$

and $\mathbf{T} = (T_1, T_2, T_3)$. Show that $\mathbf{T}$ is ancillary for $\boldsymbol{\theta} = (\mu, \sigma^2)$.

6.5.3 (Exercise 6.3.4 Continued) Show that $X_{n:n} - X_{n:1}$ is ancillary for θ. Is $(X_{n:n} - X_{n:1})/(\overline{X} - X_{n:1})$ ancillary?

6.5.4 (Exercise 6.3.7 Continued) Show that $X_{n:n}/X_{n:1}$ is ancillary for θ. Is $X_{n:n}^2/(X_{n:n} - X_{n:1})^2$ ancillary?

6.5.5 (**Curved Exponential Family**) Suppose that (X, Y) has a *curved exponential family* of distribution with its joint pdf:

$$f(x, y; \theta) = \begin{cases} exp\{-\theta x - \theta^{-1}y\} & \text{if } 0 < x, y < \infty \\ 0 & \text{elsewhere} \end{cases}$$

where $\theta(> 0)$ is an unknown parameter. This distribution was discussed by Fisher (1934, 1956) in the context of his famous "Nile" example. Denote $U = XY, V = X/Y$. Show that: U is ancillary for θ, (U, V) is minimal sufficient for θ, whereas V is not sufficient for θ.

6.6.1 Let X_1, X_2 be iid Poisson(λ) where $\lambda(> 0)$ is unknown. Is the family of distributions induced by the statistic $\mathbf{T} = (X_1, X_2)$ complete?

6.6.2 (Exercise 6.2.1 Continued) Is $\Sigma_{i=1}^n X_i$ complete sufficient for p?

6.6.3 Let $X_1, ..., X_n$ be iid $N(2\theta, 5\theta^2)$ where $0 < \theta < \infty$ is an unknown parameter. Is the minimal sufficient statistic complete?

6.6.4 Let $X_1, ..., X_n$ be iid $N(\theta, \theta)$ where $0 < \theta < \infty$ is an unknown parameter. Is the minimal sufficient statistic complete?

6.6.5 (Exercise 6.3.7 Continued) Is the minimal sufficient statistic T complete?

6.6.6 (Exercise 6.3.8 Continued) Is the minimal sufficient statistic $\mathbf{T}$ complete?

6.6.7 Suppose that $X_1, ..., X_m$ are iid Gamma(α, β), $Y_1, ..., Y_n$ are iid Gamma($\alpha, k\beta$), and X, Y are independent, with $0 < \alpha, \beta < \infty$ where β is the *only* unknown parameter. Assume that $k (> 0)$ is known. Is minimal sufficient statistic T complete?

6.6.8 Let $X_1, ..., X_n$ be iid having a Beta distribution with parameters $\alpha = \beta = \theta (> 0)$ unknown. Is the minimal sufficient statistic T complete?

6.6.9 Suppose that $X_1, ..., X_n$ are iid having a Uniform($-\theta, \theta$) distribution with $\theta(> 0)$ unknown. Denote:

$$T = \max_{1 \leq i \leq n} |X_i|, U_1 = |X_{n:1}|/X_{n:n}, \text{ and } U_2 = (X_1 - X_2)^2/\{X_{n:1}X_{n:n}\}$$

Is T distributed independently of $\mathbf{U} \equiv (U_1, U_2)$?

6.6.10 (Exercise 6.3.8 Continued) Check whether the following two-dimensional statistics

$$\mathbf{U} = \left(\{\overline{X} - \overline{Y}\}^3, \{\overline{X} - \overline{Y}\}^2/T \right), \text{ and } \mathbf{V} = (V_1, \ V_2)$$

are distributed independently where

$$T = \Sigma_{i=1}^m (X_i - \overline{X})^2 + \Sigma_{i=1}^n (Y_i - \overline{Y})^2, V_1 = \{\overline{X} - Y_{n:n} - X_1 + Y_2\}^2/T$$
$$\text{and } V_2 = \{\overline{X} - \overline{Y} - X_2 + Y_{n:1}\}^3 / |X_{m:m} - X_{m:1}|^3$$

6.6.11 Suppose that $X_1, ..., X_n$ are iid having a common pdf:

$$\sigma^{-1} exp\{-(x - \theta)/\sigma\} I(x > \theta) \text{ with } \theta, \sigma \text{ unknown}$$

and $-\infty < \theta < \infty, 0 < \sigma < \infty$. Now, argue that

$$X_{n:1} \text{ and } \Sigma_{i=1}^n (X_i - X_{n:1})$$

are distributed independently.

6.6.12 Suppose that $X_1, ..., X_4$ are iid $N(\mu, \sigma^2)$ with $-\infty < \mu < \infty$ and $0 < \sigma < \infty$ both unknown. Check whether

$$\frac{X_1 - X_4}{|\overline{X} - X_3|} \text{ and } \frac{|\overline{X}|}{S}$$

are distributed independently.

7

Point Estimation

7.1 Introduction

We begin with *independent and identically distributed* (iid) observations $X_1, ..., X_n$ having a common *probability mass function* (pmf) or *probability density function* (pdf) $f(x; \boldsymbol{\theta})$, $x \in \mathcal{X}$. It is not essential for X to be real valued or iid. The sample size n is assumed known. We suppose that $\boldsymbol{\theta}$ is fixed but unknown and $\boldsymbol{\theta} \in \Theta \subseteq \Re^k$. This chapter develops *point estimation* techniques only. Section 7.2 introduces the *method of maximum likelihood* that was pioneered by Fisher (1922, 1925a, 1934). Since one may encounter many estimators for $\boldsymbol{\theta}$, some criteria to compare their performances are addressed in Section 7.3. We show ways to find the *best estimator* among all unbiased estimators. Sections 7.4 and 7.5 give a number of fundamental tools, for example, the *Rao-Blackwell Theorem*, the *Cramér-Rao Inequality*, and the *Lehmann-Scheffé Theorems*. Section 7.6 discusses a large-sample criterion called *consistency* due to Fisher (1922).

7.2 Maximum Likelihood Estimator

Definition 7.2.1 *A point estimator of an unknown parameter $\boldsymbol{\theta}$ is a function* $\mathbf{T} \equiv \mathbf{T}(X_1, ..., X_n)$ *involving only the observations* $X_1, ..., X_n$. *Once* $\mathbf{X} = \mathbf{x}$ *is observed, the value* $t \equiv \mathbf{T}(\mathbf{x})$ *is called an estimate of $\boldsymbol{\theta}$.*

An arbitrary estimator T of θ, for example, can be practically any function of $X_1, ..., X_n$. We may think of $X_1, X_1^2 - \overline{X}, X_{n:n} - S, \overline{X}, S^2$ and so on as competing estimators.

R. A. Fisher was keenly aware of Karl Pearson's way of finding estimators by the *method of moments*. Fisher (1912) was critical of Pearson's approach of curve fitting and wrote that "The method of moments ... though its arbitrary nature is apparent" and went on to formulate the *method of maximum likelihood* in the same paper. Fisher's preliminary ideas took concrete shapes in a path-breaking article appearing in 1922 and was followed by more elaborate discussions laid out in Fisher (1925a,b, 1934).

Consider $X_1, ..., X_n$ which are iid with a common pmf or pdf $f(x; \boldsymbol{\theta})$ where $x \in \mathcal{X} \subseteq \Re$ and $\boldsymbol{\theta} = (\theta_1, ..., \theta_k) \in \Theta \subseteq \Re^k$. Here, $\theta_1, ..., \theta_k$ are unknown parameters. Recall a *likelihood function* from Equation (6.2.1). Having data $\mathbf{X} = \mathbf{x}$, we write:

$$L(\boldsymbol{\theta}) = \Pi_{i=1}^n f(x_i; \boldsymbol{\theta}) \qquad (7.2.1)$$

> Throughout this chapter and the ones that follow, we often pay
> attention to a likelihood function when it is positive.

Definition 7.2.2 *A maximum likelihood estimate of θ is a value $\widehat{\theta} \equiv \widehat{\theta}(\mathbf{x})$ for which $L(\widehat{\theta}) = \max_{\theta \in \Theta} L(\theta)$. The maximum likelihood estimator (MLE) of θ is $\widehat{\theta}(\mathbf{X})$. From the context, it should be clear whether the notation $\widehat{\theta}$ is used as an estimator or an estimate.*

Intuitively, an MLE is interpreted as that value of θ which maximizes the "chance" of observing a particular dataset $\mathbf{x}$. Note that there is no dictum regarding a specific mathematical tool to be used for locating where $L(\theta)$ attains its maximum. If $L(\theta)$ is twice differentiable, then one finds $\widehat{\theta}$ by simply using calculus. In regular cases, we equivalently maximize $log L(\theta)$ by solving simultaneously:

$$\frac{\partial}{\partial \theta} log L(\theta) = 0$$

which is called the *likelihood equation.* One may, however, need special numerical optimization techniques in some problems that are mathematically intractable.

In the sequel, we temporarily write c to denote a *generic* constant not involving θ.

Example 7.2.1 Let $X_1, ..., X_n$ be iid $N(\mu, \sigma^2)$ where μ is unknown but σ^2 is known, $-\infty < \mu < \infty$, $0 < \sigma < \infty$, and $\mathcal{X} = \Theta = \Re$. Then,

$$L(\mu) = \{\sigma\sqrt{2\pi}\}^{-n} exp\{-\tfrac{1}{2\sigma^2}\Sigma_{i=1}^n (x_i - \mu)^2\}$$

which is to be maximized with respect to μ. It is *equivalent* to maximizing $log L(\mu)$. Now,

$$log L(\mu) = c - (2\sigma^2)^{-1}\Sigma_{i=1}^n (x_i - \mu)^2$$

and

$$\frac{d}{d\mu} log L(\mu) = \Sigma_{i=1}^n (x_i - \mu)/\sigma^2$$

Next, we equate $\frac{d}{d\mu} log L(\mu)$ to zero and solve for μ. This implies $\mu = \overline{x}$, and so $\widehat{\mu} = \overline{X}$. At this point, one would also check whether $\mu = \overline{x}$ maximizes $log L(\mu)$. Observe that

$$\frac{d^2}{d\mu^2} log L(\mu) \mid_{\mu=\overline{x}} = -n/\sigma^2$$

which is negative so that $L(\mu)$ is maximized at $\mu = \overline{x}$. Thus, the MLE for μ is $\overline{X}$, the sample mean. ▲

Example 7.2.2 Suppose $X_1, ..., X_n$ are iid $N(\mu, \sigma^2)$ where μ and σ^2 are both unknown, $\theta = (\mu, \sigma^2)$, $-\infty < \mu < \infty$, $0 < \sigma < \infty$, $n \geq 2$, $\mathcal{X} = \Re$, and $\Theta = \Re \times \Re^+$. We wish to find the MLE for θ. Now,

$$L(\mu, \sigma^2) = \{2\pi\}^{-n/2}\{\sigma^2\}^{-n/2} exp\{-\tfrac{1}{2\sigma^2}\Sigma_{i=1}^n (x_i - \mu)^2\}$$

which is to be maximized with respect to both μ, σ^2. It is *equivalent* to maximizing $\log L(\mu, \sigma^2)$. One has

$$\log L(\mu, \sigma^2) = c - \tfrac{n}{2}\log(\sigma^2) - \tfrac{1}{2\sigma^2}\Sigma_{i=1}^{n}(x_i - \mu)^2$$

which leads to

$$\frac{\partial}{\partial \mu}\log L(\mu, \sigma^2) = \tfrac{1}{\sigma^2}\Sigma_{i=1}^{n}(x_i - \mu), \text{ and}$$

$$\frac{\partial}{\partial \sigma^2}\log L(\mu, \sigma^2) = -\tfrac{n}{2\sigma^2} + \tfrac{1}{2\sigma^4}\Sigma_{i=1}^{n}(x_i - \mu)^2$$

We equate both partial derivatives to zero and solve the resulting equations simultaneously for μ and σ^2. But, $\frac{\partial}{\partial \mu}\log L(\mu, \sigma^2) = 0$ and $\frac{\partial}{\partial \sigma^2}\log L(\mu, \sigma^2) = 0$ imply $\mu = \overline{x}$ as well as $-\tfrac{n}{2}\sigma^{-2} + \tfrac{1}{2}\sigma^{-4}\Sigma_{i=1}^{n}(x_i - \overline{x})^2 = 0$, thereby leading to $\sigma^2 = n^{-1}\Sigma_{i=1}^{n}(x_i - \overline{x})^2 = u$, say. At this point, one would also check whether $L(\mu, \sigma^2)$ is maximized at $(\mu, \sigma^2) = (\overline{x}, u)$. Obtain the matrix H of second-order partial derivatives of $\log L(\mu, \sigma^2)$ and show that H evaluated at $(\mu, \sigma^2) = (\overline{x}, u)$ is *negative definite* (n.d.). We have:

$$H = \begin{pmatrix} \frac{\partial^2}{\partial \mu^2}\log L(\mu, \sigma^2) & \frac{\partial^2}{\partial \sigma^2 \partial \mu}\log L(\mu, \sigma^2) \\[2mm] \frac{\partial^2}{\partial \mu \partial \sigma^2}\log L(\mu, \sigma^2) & \frac{\partial^2}{\partial (\sigma^2)^2}\log L(\mu, \sigma^2) \end{pmatrix}$$

$$= \begin{pmatrix} -n\sigma^{-2} & -\Sigma_{i=1}^{n}(x_i - \mu)\sigma^{-4} \\[2mm] -\Sigma_{i=1}^{n}(x_i - \mu)\sigma^{-4} & \tfrac{n}{2}\sigma^{-4} - \Sigma_{i=1}^{n}(x_i - \mu)^2\sigma^{-6} \end{pmatrix}$$

which evaluated at $(\mu, \sigma^2) = (\overline{x}, u)$ reduces to

$$G = \begin{pmatrix} -nu^{-1} & 0 \\ 0 & -\tfrac{1}{2}nu^{-2} \end{pmatrix}$$

The matrix G would be n.d. if and only if its odd-order principal minors are negative and all even-order principal minors are positive. Refer to Equation (4.7.5). The first diagonal of G is $-nu^{-1}$, which is negative, and $det(G) = \tfrac{1}{2}n^2u^{-3}$, which is positive. So, G is a n.d. matrix. Thus, $L(\mu, \sigma^2)$ is maximized at $(\mu, \sigma^2) = (\overline{x}, u)$. Hence, the MLE of μ, σ^2 are respectively $\widehat{\mu} = \overline{X}$ and $\widehat{\sigma}^2 = n^{-1}\Sigma_{i=1}^{n}(X_i - \overline{X})^2$. ▲

Next, we give some examples to highlight the point that $L(\theta)$ may *not* be a differentiable function of θ. In such situations, the method of finding a MLE proceeds on a case-by-case basis.

Example 7.2.3 Suppose that a single observation X is distributed as Bernoulli(p) with $0 \leq p \leq 1$ unknown. Write

$$L(p) = \begin{cases} 1 - p & \text{if } x = 0 \\ p & \text{if } x = 1 \end{cases}$$

Whether we observe $x = 0$ or 1, the resulting likelihood function $L(p)$ is not differentiable at the end points. By drawing a picture of $L(p)$ one can verify that (i) when $x = 0$, $L(p)$ is maximized if p is the smallest, that is, if $p = 0$ or (ii) when $x = 1$, $L(p)$ is maximized if p is the largest, that is, if $p = 1$. Hence, the MLE of p is $\hat{p} = X$ when $\Theta = [0, 1]$.

But, if the parameter space happens to be $\Theta = [\frac{1}{3}, \frac{2}{3}]$ instead, what will be the MLE of p? Again, $L(p)$ is maximized at the end points where $L(p)$ is not differentiable. By examining a picture (see Figure 7.2.1) of $L(p)$, it will become clear (i) when $x = 0$, $L(p)$ is maximized if p is the smallest, that is, if $p = \frac{1}{3}$ or (ii) when $x = 1$, $L(p)$ is maximized if p is the largest, that is, if $p = \frac{2}{3}$. Hence, the MLE of p is $\hat{p} = \frac{1}{3}(X + 1)$. ▲

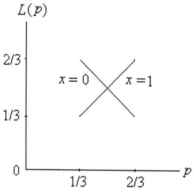

Figure 7.2.1. $L(p)$ when $\Theta = [\frac{1}{3}, \frac{2}{3}]$, $x = 0$ or $x = 1$.

Example 7.2.4 Suppose that $X_1, ..., X_n$ that are iid Uniform$(0, \theta)$ where $0 < \theta < \infty$ is an unknown parameter, $\mathcal{X} = (0, \theta)$, and $\Theta = \mathfrak{R}^+$. Write

$$L(\theta) = \theta^{-n} I(0 < x_{n:n} < \theta) I(0 < x_{n:1} < x_{n:n})$$

which is maximized at an *end point*. See Figure 7.2.2. From this picture of $L(\theta)$, it should be apparent that $L(\theta)$ is maximized when $\theta = x_{n:n}$. So, the MLE of θ is $\hat{\theta} = X_{n:n}$. ▲

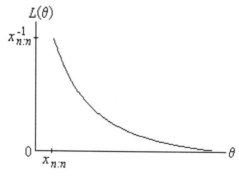

Figure 7.2.2. The likelihood function $L(\theta)$ in Example 7.2.4.

Remark 7.2.1 In Examples 7.2.1 through 7.2.4, an MLE of $\boldsymbol{\theta}$ turned out to be a (minimal) sufficient statistic. This may not be entirely surprising. Neyman factorization rewrites $L(\boldsymbol{\theta})$ as $g(\mathbf{T};\boldsymbol{\theta})h(\mathbf{x})$ where $\mathbf{T}$ is a (minimal) sufficient statistic for $\boldsymbol{\theta}$. Any maximizing technique will reduce to a problem of maximizing $g(\mathbf{T};\boldsymbol{\theta})$ with respect to $\boldsymbol{\theta}$! Thus, an MLE of $\boldsymbol{\theta}$ will necessarily be a function of the (minimal) sufficient statistic $\mathbf{T}$.

An MLE is quite versatile. It has a remarkable feature which is known as its *invariance* property. This is useful to obtain the MLE of parametric functions of $\boldsymbol{\theta}$ if one already has the MLE $\widehat{\boldsymbol{\theta}}$. This interesting result was proved by Zehna (1966).

Theorem 7.2.1 (Invariance of MLE) *Suppose that an MLE $\widehat{\boldsymbol{\theta}}$ of $\boldsymbol{\theta}$ ($\in \Theta \subseteq \Re^k$) exists. Let $g(.)$ be a function, not necessarily one-to-one, from Θ to a subset of $\Re^m$. Then, an MLE of a parametric function $g(\boldsymbol{\theta})$ is $g(\widehat{\boldsymbol{\theta}})$.*

Example 7.2.5 (Example 7.2.2 Continued) The MLEs of μ, σ^2 were respectively $\widehat{\mu} = \overline{X}$ and $\widehat{\sigma}^2 = U = n^{-1}\Sigma_{i=1}^n(X_i - \overline{X})^2$. In view of the invariance property, the MLE of (i) σ will be $\sqrt{U}$, (ii) $\mu+\sigma$ will be $\overline{X}+\sqrt{U}$, and (iii) μ^2/σ^2 will be $\overline{X}^2/U$. ▲

Example 7.2.6 (Example 7.2.4 Continued) An MLE of θ was $X_{n:n}$. In view of the invariance property, the MLE of (i) θ^2 will be $X_{n:n}^2$, (ii) $1/\theta$ will be $1/X_{n:n}$, (iii) $(\theta - 1)\sqrt{\theta + 1}$ will be $(X_{n:n} - 1)\sqrt{X_{n:n} + 1}$, and (iv) $\sin^{-1}(\theta)$ will be $\sin^{-1}(X_{n:n})$. ▲

7.3 Criteria to Compare Estimators

After observing $\mathbf{X} \equiv (X_1, ..., X_n)$, we may have competing estimators for a real valued parametric function of interest, $\tau(\boldsymbol{\theta})$. How should we compare performances of rival estimators and decide which one is the "best"? Some basic ideas are introduced in this section.

7.3.1 Unbiasedness, Variance, and Mean Squared Error

Definition 7.3.1 *A real valued statistic $T \equiv T(X_1, ..., X_n)$ is called an unbiased estimator of $\tau(\boldsymbol{\theta})$ if and only if $E_{\boldsymbol{\theta}}(T) = \tau(\boldsymbol{\theta})$ for all $\boldsymbol{\theta} \in \Theta$. A statistic T is called a biased estimator of $\tau(\boldsymbol{\theta})$ if and only if it is not unbiased for $\tau(\boldsymbol{\theta})$.*

Definition 7.3.2 *The bias of a real valued estimator T of $\tau(\boldsymbol{\theta})$ is given by*

$$B_{\boldsymbol{\theta}}(T) \equiv E_{\boldsymbol{\theta}}(T) - \tau(\boldsymbol{\theta}), \ \boldsymbol{\theta} \in \Theta$$

Gauss (1821) originally introduced the concept of an unbiased estimator in the context of least squares. Intuitively speaking, an unbiased estimator of $\tau(\boldsymbol{\theta})$ hits its target on an average and hence unbiasedness is an attractive property. A class of all unbiased estimators can be fairly rich and we may restrict ourselves to consider only unbiased estimators.

In order to clarify ideas, consider $N(\mu, \sigma^2)$ population where $\mu \in \Re$ is unknown but $\sigma \in \Re^+$ is assumed known with $\mathcal{X} = \Re$ and $\Theta = \Re$. The problem is one of estimating μ. Data consist of iid observations $X_1, ..., X_4$. Define rival estimators of μ as follows:

$$T_1 = X_1 + X_4 \qquad T_2 = \tfrac{1}{2}(X_1 + X_3) \qquad T_3 = \overline{X}$$
$$T_4 = \tfrac{1}{3}(X_1 + X_3) \quad T_5 = X_1 + T_2 - X_4 \quad T_6 = \tfrac{1}{10}\Sigma_{i=1}^4 iX_i \qquad (7.3.1)$$

Surely, one may form other estimators for μ. Observe that $E_\mu(T_1) = 2\mu$, $E_\mu(T_2) = \mu$, $E_\mu(T_3) = \mu$, $E_\mu(T_4) = \tfrac{2}{3}\mu$, $E_\mu(T_5) = \mu$, and $E_\mu(T_6) = \mu$. Thus, T_1 and T_4 are both biased estimators of μ, but T_2, T_3, T_5, and T_6 are unbiased estimators of μ. If we wish to estimate μ unbiasedly, then among T_1 through T_6, we should include only T_2, T_3, T_5, and T_6 for further evaluations.

Definition 7.3.3 *Let* $T \equiv T(X_1, ..., X_n)$ *be an estimator of* $\tau(\boldsymbol{\theta})$. *The mean squared error (MSE) of* T, *often denoted by* MSE_T, *is given by* $E_{\boldsymbol{\theta}}[(T - \tau(\boldsymbol{\theta}))^2]$. *If* T *is an unbiased estimator of* $\tau(\boldsymbol{\theta})$, *then* MSE_T *is the variance of* T, *denoted by* $V_{\boldsymbol{\theta}}(T)$.

We obtain

$$V_\mu(T_2) = \tfrac{1}{4}V_\mu(X_1 + X_3) = \tfrac{1}{4}[V_\mu(X_1) + V_\mu(X_3)] = \tfrac{1}{2}\sigma^2$$
$$V_\mu(T_3) = V_\mu(\overline{X}) = \tfrac{1}{4}\sigma^2$$
$$V_\mu(T_5) = V_\mu(\tfrac{3}{2}X_1) + V_\mu(\tfrac{1}{2}X_3) + V_\mu(X_4) = [\tfrac{9}{4} + \tfrac{1}{4} + 1]\sigma^2 = \tfrac{7}{2}\sigma^2$$
$$V_\mu(T_6) = \tfrac{1}{100}\Sigma_{i=1}^4 V_\mu(iX_i) = \tfrac{1}{100}\sigma^2\Sigma_{i=1}^4 i^2 = \tfrac{3}{10}\sigma^2$$
$$(7.3.2)$$

so that these are the associated MSEs. How would one find the MSE of T_1 or T_4? The following result will help. Its proof is left as an exercise!

Theorem 7.3.1 MSE_T, *the MSE associated with* T, *is:*

$$E_{\boldsymbol{\theta}}[(T - \tau(\boldsymbol{\theta}))^2] = V_{\boldsymbol{\theta}}(T) + [E_{\boldsymbol{\theta}}(T) - \tau(\boldsymbol{\theta})]^2$$

That is, MSE_T is the variance of T plus the square of the bias of T.

Now, $MSE_{T_1} = 2\sigma^2 + (2\mu - \mu)^2 = \mu^2 + 2\sigma^2$ and $MSE_{T_4} = (\mu^2 + 2\sigma^2)/9$. The evaluation of MSE_{T_4} is left as an exercise.

> Sometimes it is possible to have T_1, T_2 which are respectively biased and unbiased estimators of $\tau(\boldsymbol{\theta})$, but $MSE_{T_1} < MSE_{T_2}$. Consider Example 7.3.1.

Example 7.3.1 Suppose $X_1, ..., X_n$ are iid $N(\mu, \sigma^2)$ where μ, σ^2 are unknown, $\boldsymbol{\theta} = (\mu, \sigma^2)$, $-\infty < \mu < \infty$, $0 < \sigma < \infty, n \geq 2, \mathcal{X} = \Re$, and

$\Theta = \Re \times \Re^+$. Our goal is estimation of $\tau(\boldsymbol{\theta}) = \sigma^2$, the population variance. Consider the sample variance, $S^2 = (n-1)^{-1}\Sigma_{i=1}^n(X_i - \overline{X})^2$, which is unbiased for σ^2. Now, $(n-1)S^2/\sigma^2$ is distributed as χ_{n-1}^2 and $V_{\boldsymbol{\theta}}(S^2) = 2\sigma^4(n-1)^{-1}$. Consider another estimator,

$$T = (n+1)^{-1}\Sigma_{i=1}^n(X_i - \overline{X})^2$$

which is rewritten as $(n-1)(n+1)^{-1}S^2$. Thus,

$$E_{\boldsymbol{\theta}}(T) = (n-1)(n+1)^{-1}\sigma^2 \neq \sigma^2$$

so that T is a *biased estimator* of σ^2. Next,

$$V_{\boldsymbol{\theta}}(T) = (n+1)^{-2}\sigma^4 V_{\boldsymbol{\theta}}[(n-1)S^2/\sigma^2] = 2(n-1)(n+1)^{-2}\sigma^4$$

and we express MSE_T as:

$$2(n-1)(n+1)^{-2}\sigma^4 + \sigma^4\left[(n-1)(n+1)^{-1} - 1\right]^2 = 2(n+1)^{-1}\sigma^4$$

This is smaller than $V_{\boldsymbol{\theta}}(S^2)$. That is, S^2 is unbiased for σ^2, T is biased for σ^2, but MSE_T is smaller than MSE_{S^2} for all $\boldsymbol{\theta}$. ▲

7.3.2   Best Unbiased Estimator

There are situations where a parametric function of interest has no unbiased estimator. In such situations, we do *not* seek out the best unbiased estimator. Consider Example 7.3.2.

Example 7.3.2 Let $X_1, ..., X_n$ be iid Bernoulli(p), with $p \in (0,1)$ unknown, and $\theta = p$. $T = \Sigma_{i=1}^n X_i$ is sufficient for p with its distribution Binomial(n, p), $0 < p < 1$. In this case, there is *no* unbiased estimator of $\tau(\theta) = 1/p$. To prove this result, *assume* that $h(T)$ is an unbiased estimator of $\tau(\theta)$ and write, for all $0 < p < 1$:

$$p^{-1} \equiv E_p[h(T)] = \Sigma_{t=0}^n h(t)\binom{n}{t}p^t(1-p)^{n-t}$$

which is rewritten as

$$\Sigma_{t=0}^n h(t)\binom{n}{t}p^{t+1}(1-p)^{n-t} - 1 = 0 \tag{7.3.3}$$

But, the *left-hand side* (lhs) in Equation (7.3.3) is a polynomial in p of degree $n+1$, and it must be zero for all $p \in (0,1)$. So, the lhs in Equation (7.3.3) must be identically zero, that is, we must have $-1 \equiv 0$. Contradiction! Hence, there is no unbiased estimator of $1/p$. ▲

How should one compare performances of two unbiased estimators of $\tau(\boldsymbol{\theta})$? If T_1, T_2 are unbiased estimators of $\tau(\boldsymbol{\theta})$, then T_1 is *preferable* to (or *better* than) T_2 if $V_{\boldsymbol{\theta}}(T_1) \leq V_{\boldsymbol{\theta}}(T_2)$ for all $\boldsymbol{\theta} \in \Theta$ but $V_{\boldsymbol{\theta}}(T_1) < V_{\boldsymbol{\theta}}(T_2)$ for some $\boldsymbol{\theta} \in \Theta$.

Definition 7.3.4 *Assume that there is at least one unbiased estimator of* $\tau(\boldsymbol{\theta})$. *Consider* $\mathcal{C}$, *a class of all unbiased estimators of* $\tau(\boldsymbol{\theta})$. *An estimator* $T \in \mathcal{C}$ *is called the best unbiased estimator or a uniformly minimum variance unbiased estimator (UMVUE) of* $\tau(\boldsymbol{\theta})$ *if and only if for all estimators* $T^* \in \mathcal{C}$, *we have*

$$V_{\boldsymbol{\theta}}(T) \leq V_{\boldsymbol{\theta}}(T^*) \text{ for all } \boldsymbol{\theta} \in \Theta \qquad (7.3.4)$$

7.4   Improved Unbiased Estimators via Sufficiency

It appears that one of the earliest examples of UMVUE was found by Aitken and Silverstone (1942). The UMVUEs as suitable and unique functions of sufficient statistics were investigated by Halmos (1946), Kolmogorov (1950a), and more generally by Rao (1947).

7.4.1   Rao-Blackwell Theorem

A method for improving upon an initial unbiased estimator of $\tau(\boldsymbol{\theta})$ is called *Rao-Blackwellization*. C. R. Rao and D. Blackwell independently discovered this path-breaking idea in their fundamental papers that appeared in 1945 and 1947, respectively.

Theorem 7.4.1 (Rao-Blackwell Theorem) *Suppose that T is an unbiased estimator of a real valued parametric function* $\tau(\boldsymbol{\theta})$, $\boldsymbol{\theta} \in \Theta \subset \Re^k$. **U** *is a (jointly) sufficient statistic for* $\boldsymbol{\theta}$. *Define* $g(\mathbf{u}) = E_{\boldsymbol{\theta}}[T \mid \mathbf{U} = \mathbf{u}]$, *for* $\mathbf{u}$ *belonging to* $\mathcal{U}$, *the domain space of* **U**. *Then, one has:*

(i) W *is an unbiased estimator of* $\tau(\boldsymbol{\theta})$ *with* $W \equiv g(\mathbf{U})$;

(ii) $V_{\boldsymbol{\theta}}[W] \leq V_{\boldsymbol{\theta}}[T]$ *for all* $\boldsymbol{\theta} \in \Theta$, *with equality holding if and only if T is the same as W with probability (w.p.) 1.*

Proof (i) Since **U** is sufficient for $\boldsymbol{\theta}$, the conditional distribution of T given $\mathbf{U} = \mathbf{u}$ cannot involve $\boldsymbol{\theta}$. This is true for all $\mathbf{u} \in \mathcal{U}$. It follows from the definition of sufficiency. Hence, $W = g(\mathbf{U})$ is indeed a real valued statistic. Using Theorem 3.2.1, part (i), we have for all $\boldsymbol{\theta} \in \Theta$,

$$\tau(\boldsymbol{\theta}) = E_{\boldsymbol{\theta}}[T] = E_{\mathbf{U}}[E\{T \mid \mathbf{U}\}] = E_{\boldsymbol{\theta}}[g(\mathbf{U})] = E_{\boldsymbol{\theta}}[W] \qquad (7.4.1)$$

which shows part (i). ◆

(ii) We express $V_{\theta}[T]$ as follows for all $\theta \in \Theta$:

$$E_{\theta}[\{T - \tau(\theta)\}^2]$$
$$= E_{\theta}[\{(T - W) + (W - \tau(\theta))\}^2]$$
$$= E_{\theta}[\{W - \tau(\theta)\}^2] + E_{\theta}[\{T - W\}^2] + 2E_{\theta}[(T - W)(W - \tau(\theta))]$$
$$= V_{\theta}[W] + E_{\theta}[\{T - W\}^2] \tag{7.4.2}$$

since

$$E_{\theta}[(T - W)(W - \tau(\theta))] = E_{\mathbf{U}}[\{g(\mathbf{U}) - \tau(\theta)\}E\{T - g(\mathbf{U}) \mid \mathbf{U}\}]$$
$$= E_{\mathbf{U}}[\{g(\mathbf{U}) - \tau(\theta)\}\{g(\mathbf{U}) - g(\mathbf{U})\}] = 0$$

From Equation (7.4.2), the first conclusion in part (ii) is obvious since $\{T - W\}^2$ is nonnegative w.p.1 so that $E_{\theta}[\{T - W\}^2] \geq 0$ for all $\theta \in \Theta$. For the second conclusion in part (ii), notice again from Equation (7.4.2) that $V_{\theta}[W] = V_{\theta}[T]$ for all $\theta \in \Theta$ *if and only if* $E_{\theta}[\{T - W\}^2] = 0$ for all $\theta \in \Theta$, that is, *if and only if* $T = W$ w.p.1. Now, the proof is complete. ∎

Example 7.4.1 Suppose $X_1, ..., X_n$ are iid Bernoulli(p) where $0 < p < 1$ is unknown. We wish to estimate $\tau(p) = p$ unbiasedly. Consider initially $T \equiv X_1$, an unbiased estimator of p. The values of T are 0 or 1. Consider $U \equiv \Sigma_{i=1}^{n}X_i$ which is sufficient for p. The domain space for U is $\mathcal{U} = \{0, 1, 2, ..., n\}$. We write for $u \in \mathcal{U}$:

$$E_p[T \mid U = u]$$
$$= [1 \times P_p\{T = 1 \mid U = u\}] + [0 \times P_p\{T = 0 \mid U = u\}]$$
$$= P_p\{X_1 = 1 \mid U = u\} \tag{7.4.3}$$
$$= P_p\{X_1 = 1 \cap U = u\}/P_p\{U = u\}$$
$$= P_p\{X_1 = 1 \cap \Sigma_{i=2}^{n}X_i = u - 1\}/P_p\{\Sigma_{i=1}^{n}X_i = u\}$$

Observe that $\Sigma_{i=1}^{n}X_i$ is Binomial(n, p), $\Sigma_{i=2}^{n}X_i$ is Binomial$(n - 1, p)$, and X_1 and $\Sigma_{i=2}^{n}X_i$ are independently distributed. We rewrite Equation (7.4.3) as:

$$E_p[X_1 \mid U = u]$$
$$= P_p\{X_1 = 1\}P_p\{\Sigma_{i=2}^{n}X_i = u - 1\}/P_p\{\Sigma_{i=1}^{n}X_i = u\}$$
$$= p\binom{n-1}{u-1}p^{u-1}(1 - p)^{n-u}/\{\binom{n}{u}p^u(1 - p)^{n-u}\}$$
$$= \binom{n-1}{u-1}/\binom{n}{u} = u/n = \overline{x}$$

That is, the Rao-Blackwellized version of an initial unbiased estimator X_1 is $\overline{X}$, the sample mean, even though T was indeed a naive and practically useless estimator of p. When $n = 1$, the sufficient statistic is X_1 so that if one starts with $T \equiv X_1$, then a final estimator obtained through Rao-Blackwellization will remain X_1. ▲

Example 7.4.2 (Example 7.4.1 Continued) We wish to estimate $\tau(p) = p(1 - p)$ unbiasedly. Consider initially $T \equiv X_1(1 - X_2)$, an unbiased estimator of $\tau(p)$. Consider $U = \Sigma_{i=1}^n X_i$, which is sufficient for p. Denote $h_p(u) = P_p\{\Sigma_{i=1}^n X_i = u\}$ and write for $u \in \mathcal{U}$:

$$E_p[X_1(1 - X_2) \mid U = u] = P_p\{X_1(1 - X_2) = 1 \mid U = u\}$$

since $X_1(1-X_2)$ takes the value 0 or 1 only. Thus, we can express $E_p[X_1(1 - X_2) \mid U = u]$ as:

$$P_p\{X_1 = 1 \cap X_2 = 0 \cap U = u\}/P_p\{U = u\}$$
$$= P_p\{X_1 = 1 \cap X_2 = 0 \cap \Sigma_{i=3}^n X_i = u - 1\}/h_p(u) \quad (7.4.4)$$

Observe that $\Sigma_{i=1}^n X_i$ is Binomial(n, p), $\Sigma_{i=3}^n X_i$ is Binomial$(n - 2, p)$, and that $X_1, X_2, \Sigma_{i=3}^n X_i$ are independent. Rewrite Equation (7.4.4):

$$E_p[X_1(1 - X_2) \mid U = u]$$
$$= P_p\{X_1 = 1\}P_p\{X_2 = 0\}P_p\{\Sigma_{i=3}^n X_i = u - 1\}/h_p(u)$$
$$= p(1 - p)\binom{n-2}{u-1}p^{u-1}(1 - p)^{n-u-1}/\{\binom{n}{u}p^u(1 - p)^{n-u}\}$$
$$= \binom{n-2}{u-1}/\binom{n}{u} = u(n - u)/\{n(n - 1)\}$$

which is $n(n - 1)^{-1}\overline{x}(1 - \overline{x})$. That is, the Rao-Blackwellized version of the initial unbiased estimator $X_1(1 - X_2)$ turns out to be

$$n(n - 1)^{-1}\overline{X}(1 - \overline{X})$$

For the Bernoulli random samples, since each observation is either 0 or 1, observe that the sample variance can be rewritten as:

$$S^2 = (n - 1)^{-1}\{\Sigma_{i=1}^n X_i^2 - n\overline{X}^2\} = (n - 1)^{-1}\{\Sigma_{i=1}^n X_i - n\overline{X}^2\}$$
$$= (n - 1)^{-1}\{n\overline{X} - n\overline{X}^2\} = n(n - 1)^{-1}\overline{X}(1 - \overline{X})$$

which coincides with the Rao-Blackwellized version! ▲

Example 7.4.3 Let $X_1, ..., X_n$ be iid $N(\mu, \sigma^2)$ where μ is unknown but σ^2 is known with $-\infty < \mu < \infty$, $0 < \sigma^2 < \infty$, and $\mathcal{X} = \Re$. We wish to estimate $\tau(\mu) = \mu$ unbiasedly. Consider initially $T \equiv X_1$, an unbiased estimator of μ. Consider $U \equiv n^{-1}\Sigma_{i=1}^n X_i$, sufficient for μ. The domain space for U is $\mathcal{U} = \Re$. For $u \in \mathcal{U}$, conditionally given $U = u$, the distribution of T is $N(u, \sigma^2(1 - n^{-1}))$. Now, $E_\mu[T \mid U = u] = u = \overline{x}$ so that the Rao-Blackwellized version of the initial unbiased estimator T is $\overline{X}$. ▲

Example 7.4.4 (Example 7.4.3 Continued) We wish to estimate $\tau(\mu) = P_\mu\{X_1 > a\}$ unbiasedly where a is some fixed and *known* real number. Consider initially $T \equiv I(X_1 > a)$, an unbiased estimator of $\tau(\mu)$. Let

$U = n^{-1}\Sigma_{i=1}^{n}X_i$, a sufficient statistic for μ. Conditionally, given $U = u$, X_1 is $N\left(u, \sigma^2(1 - n^{-1})\right)$. We have:

$$E_{\mu}[T \mid U = u] = P_{\mu}\{X_1 > a \mid U = u\}$$
$$= 1 - \Phi\left([a - u]/\{\sigma^2[1 - n^{-1}]\}^{1/2}\right)$$

So, the Rao-Blackwellized version of the initial unbiased estimator T is $W \equiv 1 - \Phi\left([a - \overline{X}]/\{\sigma^2[1 - n^{-1}]\}^{1/2}\right)$. ▲

Example 7.4.5 (Example 7.4.4 Continued) We wish to estimate $\tau(\mu) = \mu^2$ unbiasedly. Consider initially $T \equiv X_1^2 - \sigma^2$, an unbiased estimator of $\tau(\mu)$. Let $U = n^{-1}\Sigma_{i=1}^{n}X_i$, a sufficient statistic μ. Conditionally, given $U = u$, X_1 is $N\left(u, \sigma^2(1 - n^{-1})\right)$. We have:

$$E_{\mu}\{X_1^2 - \sigma^2 \mid U = u\} = \sigma^2(1 - n^{-1}) + u^2 - \sigma^2 = u^2 - n^{-1}\sigma^2$$

So, the Rao-Blackwellized version of the initial unbiased estimator T is $W \equiv \overline{X}^2 - n^{-1}\sigma^2$. ▲

Remark 7.4.1 In Example 7.4.5, $W = \overline{X}^2 - n^{-1}\sigma^2$ was the final unbiased estimator of μ^2. One may feel uneasy to use this estimator since $P_{\mu}\{W < 0\}$ is positive whatever be n, μ and σ. The parametric function μ^2 is nonnegative, but W can be negative with positive probability! At times, unbiasedness criterion may create an awkward situation like this.

7.5 Uniformly Minimum Variance Unbiased Estimator

This section gives methodologies for finding a *uniformly minimum variance unbiased estimator* (UMVUE) of a parametric function. The first approach relies upon the celebrated *Cramér-Rao Inequality* (Theorem 7.5.1). C. R. Rao and H. Cramér independently discovered, under mild regularity conditions, a lower bound for the variance of an unbiased estimator $\tau(\theta)$ in their classic papers, Rao (1945) and Cramér (1946b). Then, we introduce another fundamental result (Theorem 7.5.2) due to Lehmann and Scheffé (1950) which jumped out of the Rao-Blackwell Theorem. The Lehmann-Scheffé (1950, 1955, 1956) papers are invaluable in this regard.

7.5.1 Cramér-Rao Inequality

Lehmann (1983) referred to this as "information inequality," a name which was suggested by Savage (1954). Lehmann (1983, p. 145, Section 9) wrote, "The first version of the information inequality appears to have been given by Fréchet (1943). Early extensions and rediscoveries are due to Darmois

(1945), Rao (1945), and Cramér (1946b)." We will continue to refer to this inequality by its commonly used name, the *Cramér-Rao Inequality*, for ease of cross-referencing and searching other sources.

The variance bound, called the *Cramér-Rao lower bound* (CRLB), for unbiased estimators of $\tau(\theta)$ is appreciated where one can (i) derive an explicit expression of CRLB, and (ii) easily locate an unbiased estimator of $\tau(\theta)$ whose variance coincides with CRLB. In these situations, one has then found the UMVUE for $\tau(\theta)$!

Consider iid real valued observations $X_1, ..., X_n$ with a common pmf or pdf $f(x; \theta)$ where the unknown parameter $\theta \in \Theta \subseteq \Re$ and $x \in \mathcal{X} \subseteq \Re$. Denote $\mathbf{X} \equiv (X_1, ..., X_n)$. We *pretend* working with a pdf and hence expectations would be written as appropriate (multiple) integrals. In a discrete case, one would simply replace the integrals by summations.

Standing Assumptions: The support $\mathcal{X}$ does not involve θ and the first partial derivative of $f(x; \theta)$ with respect to θ and integrals with respect to $\mathbf{x} = (x_1, ..., x_n)$ are interchangeable.

Theorem 7.5.1 (Cramér-Rao Inequality) *Suppose that $T = T(\mathbf{X})$ is an unbiased estimator of a real valued parametric function $\tau(\theta)$. Assume that $\frac{d}{d\theta}\tau(\theta)$, denoted by $\tau'(\theta)$, is finite for all $\theta \in \Theta$. Then, for all $\theta \in \Theta$, under the standing assumptions, we have:*

$$V_\theta(T) \geq \frac{\{\tau'(\theta)\}^2}{nE_\theta\left[\left\{\frac{\partial}{\partial\theta}[\log f(X_1; \theta)]\right\}^2\right]} \tag{7.5.1}$$

The expression on the right-hand side (rhs) of this inequality is the Cramér-Rao lower bound.

Proof Without any loss of generality, assume that $0 < V_\theta(T) < \infty$. We have

$$\tau(\theta) = \int_\mathcal{X} \cdots \int_\mathcal{X} T(x_1, ..., x_n)\Pi_{i=1}^n f(x_i; \theta)\Pi_{i=1}^n dx_i \tag{7.5.2}$$

which implies that $\tau'(\theta)$ is:

$$\begin{aligned}&\frac{d}{d\theta}\left[\int_\mathcal{X} \cdots \int_\mathcal{X} T(x_1, ..., x_n)\Pi_{i=1}^n f(x_i; \theta)\Pi_{i=1}^n dx_i\right]\\&= \int_\mathcal{X} \cdots \int_\mathcal{X} T(x_1, ..., x_n)\left[\frac{\partial}{\partial\theta}\Pi_{i=1}^n f(x_i; \theta)\right]\Pi_{i=1}^n dx_i\end{aligned} \tag{7.5.3}$$

Observe that

$$\Pi_{i=1}^n f(x_i; \theta) = exp\{\Sigma_{i=1}^n \log f(x_i; \theta)\} \text{ for } x \in \mathcal{X} \tag{7.5.4}$$

so that, we get:

$$\begin{aligned}&\frac{\partial}{\partial\theta}\left[\Pi_{i=1}^n f(x_i; \theta)\right]\\&= exp\{\Sigma_{i=1}^n \log f(x_i; \theta)\}\frac{\partial}{\partial\theta}\{\Sigma_{i=1}^n \log f(x_i; \theta)\}\\&= \{\Sigma_{i=1}^n \frac{\partial}{\partial\theta}[\log f(x_i; \theta)]\}\Pi_{i=1}^n f(x_i; \theta)\end{aligned} \tag{7.5.5}$$

Denote $Y = \Sigma_{i=1}^{n} \frac{\partial}{\partial \theta}[log f(X_i; \theta)]$ and combine Equation (7.5.3) with Equation (7.5.5) to rewrite

$$\tau'(\theta)$$
$$= \int_{\mathcal{X}} \cdots \int_{\mathcal{X}} T(x_1, ..., x_n)\{\Sigma_{i=1}^{n} \frac{\partial}{\partial \theta}[log f(x_i; \theta)]\}\Pi_{i=1}^{n} f(x_i; \theta)\Pi_{i=1}^{n} dx_i$$
$$= E_\theta\{TY\}$$

(7.5.6)

One obviously has $\int_{\mathcal{X}} f(x; \theta)dx = 1$ so that

$$0 \equiv \frac{d}{d\theta} \int_{\mathcal{X}} f(x; \theta)dx = \int_{\mathcal{X}} \left[\frac{\partial}{\partial \theta} f(x; \theta)\right] dx = \int_{\mathcal{X}} \left[\frac{\partial}{\partial \theta} log f(x; \theta)\right] f(x; \theta)dx$$

(7.5 .7)

Hence, we write

$$E_\theta[Y] = E_\theta \left\{\Sigma_{i=1}^{n} \frac{\partial}{\partial \theta}[log f(X_i; \theta)]\right\} = \Sigma_{i=1}^{n} E_\theta \left\{\frac{\partial}{\partial \theta}[log f(X_i; \theta)]\right\} = 0$$

for all $\theta \in \Theta$, since $X_1, ..., X_n$ have identical distributions. Thus, Equation (7.5.6) gives:

$$\tau'(\theta) = E_\theta\{TY\} = Cov_\theta(T, Y)$$

which is rewritten as

$$\{\tau'(\theta)\}^2 = Cov_\theta^2(T, Y) \leq V_\theta(T)V_\theta(Y)$$

(7.5.8)

by virtue of the Cauchy-Schwartz Inequality from Equation (3.8.4) or Covariance Inequality from Equation (3.8.5). Recall that Y is a sum of n iid random variables and thus, in view of Equation (7.5.7), we obtain:

$$V_\theta(Y) = nV_\theta \left\{\frac{\partial}{\partial \theta}[log f(X_1; \theta)]\right\} = nE_\theta \left[\left\{\frac{\partial}{\partial \theta}[log f(X_1; \theta)]\right\}^2\right]$$

(7.5.9)

Now, the inequality from Equation (7.5.1) follows by combining Equations (7.5.8) and (7.5.9). ∎

Remark 7.5.1 One can see that CRLB will be attained by the variance of an unbiased estimator T of $\tau(\theta)$ for all $\theta \in \Theta$ *if and only if* the equality in Equation (7.5.8) holds, that is, *if and only if* T and Y are linearly related w.p.1. Hence, CRLB will be attained by the variance of T *if and only if*

$$T - a(\theta) = b(\theta)Y \text{ w.p.1, for all } \theta \in \Theta$$

(7.5.10)

with some fixed real valued functions $a(.)$ and $b(.)$.

Remark 7.5.2 Combining CRLB and $\mathcal{I}_{X_1}(\theta)$ from Equation (6.4.1), we can immediately restate the Cramér-Rao Inequality as follows:

$$V_\theta(T) \geq \frac{\{\tau'(\theta)\}^2}{n\mathcal{I}_{X_1}(\theta)}$$

(7.5.11)

We will interchangeably use the Cramér-Rao Inequality given by Equations (7.5.1) or (7.5.11).

Example 7.5.1 Suppose $X_1, ..., X_n$ are iid Poisson(λ) where $\lambda(> 0)$ is an unknown parameter. Let us consider $\tau(\lambda) = \lambda$ so that $\tau'(\lambda) = 1$. Now, $\overline{X}$ is an unbiased estimator of λ and $V_\lambda(\overline{X}) = n^{-1}\lambda$. Can we claim that $\overline{X}$ is the UMVUE of λ? In Equation (6.4.2) we found $\mathcal{I}_{X_1}(\lambda) = \lambda^{-1}$ and from Equation (7.5.11) we see that CRLB $= 1/(n\lambda^{-1}) = \lambda/n$. This coincides with $V_\lambda(\overline{X})$ for all $\lambda > 0$. That is, $\overline{X}$ must be the UMVUE of λ. ▲

Example 7.5.2 (Example 7.4.3 Continued) We wish to estimate $\tau(\mu) = \mu$ unbiasedly. Consider $\overline{X}$, an unbiased estimator of μ. Is $\overline{X}$ the UMVUE of μ? In Equation (6.4.3) we found $\mathcal{I}_{X_1}(\mu) = \sigma^{-2}$ and from Equation (7.5.11) we see that CRLB $= 1/(n\,\sigma^{-2}) = \sigma^2/n$. This coincides with $V_\mu(\overline{X})$ for all $\mu \in \Re$. Then, $\overline{X}$ must be the UMVUE of μ. ▲

In these examples, we thought of "natural" unbiased estimators of $\tau(\theta)$ and these estimators' variances coincided with CRLB. So, we could claim that the estimators we started with were in fact the UMVUEs of $\tau(\theta)$. The reader may also note that these UMVUEs agreed with the Rao-Blackwellized versions of naive initial unbiased estimators!

Example 7.5.3 Suppose $X_1, ..., X_n$ are iid Poisson(λ) where $0 < \lambda < \infty$ is unknown with $n \geq 2$. We wish to estimate $\tau(\lambda) = e^{-\lambda}$ unbiasedly. One may use an initial unbiased estimator, $T \equiv I(X_1 = 0)$, and find that its Rao-Blackwellized version is:

$$W \equiv (1 - n^{-1})^{\Sigma_{i=1}^n X_i}$$

Does $V_\lambda(W)$ attain the CRLB? Recall that $\Sigma_{i=1}^n X_i$ is Poisson($n\lambda$) and its *moment generating function* (mgf) is

$$E_\lambda[e^{s\Sigma_{i=1}^n X_i}] = exp\{n\lambda(e^s - 1)\} \qquad (7.5.12)$$

Use Equation (7.5.12) with $s = 2log(1 - n^{-1})$ to claim:

$$E_\lambda[W^2] = exp\{n\lambda[(\tfrac{n-1}{n})^2 - 1]\} = exp\{-2\lambda + n^{-1}\lambda\}$$

Hence, we have:

$$V_\lambda[W] = exp\{-2\lambda + n^{-1}\lambda\} - exp(-2\lambda) = e^{-2\lambda}(e^{\lambda/n} - 1)$$

But, CRLB $= \{\tau'(\lambda)\}^2/(n\lambda^{-1}) = n^{-1}\lambda e^{-2\lambda}$. Since $e^x > 1 + x$ for $x > 0$, we claim:

$$V_\lambda[W] = e^{-2\lambda}(e^{\lambda/n} - 1) > e^{-2\lambda}(1 + \tfrac{\lambda}{n} - 1) = e^{-2\lambda}\tfrac{\lambda}{n} = \text{CRLB}$$

so that $V_\lambda[W]$ does *not* attain CRLB, for any λ. At this point, it is not clear whether W is UMVUE of $\tau(\lambda)$. ▲

Example 7.5.4 (Example 7.4.5 Continued) We wish to estimate $\tau(\mu) = \mu^2$ unbiasedly. In Example 7.4.5, we found the Rao-Blackwellized unbiased estimator $W \equiv \overline{X}^2 - n^{-1}\sigma^2$ of $\tau(\mu)$. One may check that

$$V_\mu[W] = 2n^{-2}\sigma^4 + 4n^{-1}\mu^2\sigma^2$$

But, CRLB $= 4n^{-1}\mu^2\sigma^2$ and it is clear that $V_\mu[W] >$ CRLB, for all μ. That is, the CRLB *is not* attained. At this point, it is not clear whether W is UMVUE of $\tau(\mu)$. ▲

7.5.2   Lehmann-Scheffé Theorems

In situations encountered in Examples 7.5.3 and 7.5.4, neither the Rao-Blackwell Theorem nor the Cramér-Rao Inequality helps in deciding whether W is the UMVUE of $\tau(\theta)$. An alternative approach is needed.

Theorem 7.5.2 (Lehmann-Scheffé Theorem-I) *Suppose that T is an unbiased estimator of a real valued parametric function $\tau(\boldsymbol{\theta})$, $\boldsymbol{\theta} \in \Theta \subseteq \Re^k$. Let $\mathbf{U}$ be a complete (jointly) sufficient statistic for $\boldsymbol{\theta}$. Define $g(\mathbf{u}) \equiv E_{\boldsymbol{\theta}}[T \mid \mathbf{U} = \mathbf{u}]$ for $\mathbf{u} \in \mathcal{U}$. Then, the statistic $W \equiv g(\mathbf{U})$ is a unique (w.p.1) UMVUE of $\tau(\boldsymbol{\theta})$.*

Proof   The difference between the Rao-Blackwell Theorem and this theorem is that now $\mathbf{U}$ is also assumed complete! The Rao-Blackwell Theorem assures us that in order to search for the best unbiased estimator of $\tau(\boldsymbol{\theta})$, we need to focus on unbiased estimators which are functions of $\mathbf{U}$ alone. We already know that (i) W is a function of $\mathbf{U}$, and (ii) W is an unbiased estimator of $\tau(\boldsymbol{\theta})$. Suppose that there is another unbiased estimator W^* of $\tau(\boldsymbol{\theta})$ where W^* is also a function of $\mathbf{U}$. Define $h(\mathbf{U}) = W - W^*$ and then we have:

$$E_{\boldsymbol{\theta}}[h(\mathbf{U})] = E_{\boldsymbol{\theta}}[W - W^*] = \tau(\boldsymbol{\theta}) - \tau(\boldsymbol{\theta}) \equiv 0 \text{ for all } \theta \in \Theta \qquad (7.5.13)$$

Now, use Definition 6.6.1 of completeness of a statistic. Since $\mathbf{U}$ is a complete statistic, it follows from Equation (7.5.13) that $h(\mathbf{U}) \equiv 0$ w.p.1. So, $W = W^*$ w.p.1. ∎

In our quest for finding the UMVUE of $\tau(\boldsymbol{\theta})$, we may not always go through conditioning with respect to a complete sufficient statistic $\mathbf{U}$. In many problems, the following alternate result may be directly applicable. Its proof can be easily constructed.

Theorem 7.5.3 (Lehmann-Scheffé Theorem-II) *Suppose that $\mathbf{U}$ is a complete sufficient statistic for $\boldsymbol{\theta} \in \Theta \subseteq \Re^k$. Also, suppose that a statistic $W \equiv g(\mathbf{U})$ is an unbiased estimator of a real valued parametric function $\tau(\boldsymbol{\theta})$. Then, W is a unique (w.p.1) UMVUE of $\tau(\boldsymbol{\theta})$.*

> In Examples 7.5.5 and 7.5.6, neither the Rao-Blackwell Theorem nor the Cramér-Rao Inequality helped in identifying the UMVUE. But, the Lehmann-Scheffé approach is right on the money.

Example 7.5.5 (Example 7.5.3 Continued) We wished to estimate a parametric function, $\tau(\lambda) = e^{-\lambda}$, unbiasedly and found:

$$W \equiv (1 - n^{-1})^{\Sigma_{i=1}^n X_i}$$

But, W depended on a complete sufficient statistic, $U \equiv \Sigma_{i=1}^{n} X_i$, only. So, in view of the Lehmann-Scheffé Theorems, W is a unique (w.p.1) UMVUE for the parametric function $\tau(\lambda)$. ▲

Example 7.5.6 (Example 7.5.4 Continued) We estimated $\tau(\mu) = \mu^2$ unbiasedly and found $W \equiv \overline{X}^2 - n^{-1}\sigma^2$. But, W depended on a complete sufficient statistic, $U \equiv \Sigma_{i=1}^{n} X_i$, only. So, in view of the Lehmann-Scheffé Theorems, W is a unique (w.p.1) UMVUE for $\tau(\mu)$. ▲

> Next we give examples where the Cramér-Rao Inequality is not applicable, but we can conclude the UMVUE property of a natural unbiased estimator via the Lehmann-Scheffé Theorems.

Example 7.5.7 Suppose $X_1, ..., X_n$ are iid Uniform$(0, \theta)$ where $\theta (> 0)$ is an unknown parameter. Here, the domain space $\mathcal{X}$ depends on θ itself so that an approach via the Cramér-Rao Inequality is not feasible. Now, $U \equiv X_{n:n}$ is complete sufficient for θ. We wish to estimate $\tau(\theta) = \theta$ unbiasedly. The pdf of U is:

$$nu^{n-1}\theta^{-n}I(0 < u < \theta)$$

So, $E_\theta[X_{n:n}] = n(n+1)^{-1}\theta$. Hence, $E_\theta[(n+1)n^{-1}X_{n:n}] = \theta$ so that $W \equiv (n+1)n^{-1}X_{n:n}$ is an unbiased estimator of θ. Thus, by the Lehmann-Scheffé Theorems, $(n+1)n^{-1}X_{n:n}$ is the UMVUE for θ. ▲

Example 7.5.8 Suppose that $X_1, ..., X_n$ are iid $N(\mu, \sigma^2)$ where μ, σ^2 are both unknown, $-\infty < \mu < \infty$, $0 < \sigma^2 < \infty$, and $\mathcal{X} = \Re$. Let us write $\boldsymbol{\theta} = (\mu, \sigma^2) \in \Theta = \Re \times \Re^+$ and we wish to estimate $\tau(\boldsymbol{\theta}) = \mu$ unbiasedly. $\mathbf{U} \equiv (\overline{X}, S^2)$ is a complete sufficient statistic for $\boldsymbol{\theta}$. Obviously, $\overline{X}$ is an unbiased estimator of $\tau(\boldsymbol{\theta})$ and $\overline{X}$ is a function of $\mathbf{U}$ only. So, by the Lehmann-Scheffé Theorems, $\overline{X}$ is the UMVUE for μ. Similarly, S^2 is the UMVUE for σ^2. ▲

Example 7.5.9 (Example 7.5.8 Continued) We wish to estimate $\tau(\boldsymbol{\theta}) = \mu + \sigma$ unbiasedly. $\mathbf{U} = (\overline{X}, S^2)$ is a complete sufficient statistic for $\boldsymbol{\theta}$. Note that $E_{\boldsymbol{\theta}}[S] = a_n\sigma$ where

$$a_n = \sqrt{2}\Gamma(\frac{n}{2})\{\Gamma(\frac{n-1}{2})\sqrt{n-1}\}^{-1}$$

Thus, $W = \overline{X} + a_n^{-1}S$ is an unbiased estimator of $\tau(\boldsymbol{\theta})$ and W depends only on $\mathbf{U}$. So, by the Lehmann-Scheffé Theorems,

$$\overline{X} + a_n^{-1}S$$

is the UMVUE for $\mu + \sigma$. ▲

7.6   Consistent Estimator

Consistency is a large sample property of an estimator. Fisher introduced a notion of the *consistency* property in his 1922 paper.

Definition 7.6.1 *Suppose that* $\{T_n \equiv T_n(X_1, ..., X_n); n \geq 1\}$ *is a sequence of estimators for an unknown real valued parametric function* $\tau(\boldsymbol{\theta})$, $\boldsymbol{\theta} \in \Theta \subseteq \Re^k$. *Then,* T_n *is called consistent for* $\tau(\boldsymbol{\theta})$ *if and only if* $T_n \xrightarrow{P} \tau(\boldsymbol{\theta})$ *as* $n \to \infty$. *Also,* T_n *is called inconsistent if it is not consistent.*

> One may verify that estimators found in this chapter are indeed all consistent for associated parameters of interest.

It may be noted that Fisher never suggested that one should pick an estimator because of its consistency property alone. Common sense dictates that the consistency property has to take a back seat when considered in conjunction with the sufficiency property. It so happens that in many problems, the usual estimators such as the MLE or UMVUE are frequently consistent for parameter(s) of interest.

We may point out that in many cases, $\widehat{\theta}_{\text{MLE}}$ is consistent for θ. In Chapter 12, under mild regularity conditions, we will quote a result showing that an MLE $\widehat{\theta}$ of θ is indeed consistent. But, in general, $\widehat{\theta}_{\text{MLE}}$ may not be consistent for θ. Examples of inconsistent MLEs were constructed by Neyman and Scott (1948), Basu (1955b), and Bahadur (1958).

7.7   Exercises and Complements

7.2.1 Let $X_1, ..., X_n$ be iid Geometric(p) with a common pmf $f(x; p) = p(1-p)^x$, $x = 0, 1, 2, ...$, and $0 < p < 1$ is an unknown parameter. Find the MLE of p. Is the MLE sufficient for p?

7.2.2 Suppose $X_1, ..., X_n$ are iid Bernoulli(p) random variables, and $0 \leq p \leq 1$ is an unknown parameter.
- (i)   Derive the MLE for p^2;
- (ii)   Derive the MLE for p/q where $q = 1 - p$;
- (iii)   Derive the MLE for $p(1-p)$.

7.2.3 Suppose that $X_1, ..., X_n$ are iid $N(\mu, \sigma^2)$ where μ is known but σ^2 is unknown, $-\infty < \mu < \infty$, $0 < \sigma < \infty$, $n \geq 2$.
- (i)   Show that the MLE of σ^2 is $n^{-1}\sum_{i=1}^{n}(X_i - \mu)^2$;
- (ii)   Is the MLE in part (i) sufficient for σ^2?
- (iii)   Derive the MLE for $1/\sigma$;
- (iv)   Derive the MLE for $(\sigma + \sigma^{-1})^{1/2}$.

7.2.4 Suppose that $X_1, ..., X_n$ are iid with a common Weibull pdf:

$$f(x; \alpha) = \alpha^{-1} \beta x^{\beta-1} exp(-x^\beta/\alpha) I(x > 0)$$

where $\alpha(> 0)$ is the unknown parameter, but $\beta(> 0)$ is assumed known. Find the MLE for α.

7.2.5 Suppose that $X_1, ..., X_n$ are iid with a common pdf:

$$\sigma^{-1}exp\{-(x-\mu)/\sigma\}I(x > \mu)$$

where μ and σ are both unknown, $-\infty < \mu < \infty, 0 < \sigma < \infty, n \geq 2$. Show that the MLE for μ and σ are respectively $X_{n:1}$ and $n^{-1}\Sigma_{i=1}^{n}(X_i - X_{n:1})$. Derive the MLEs for $\mu/\sigma, \mu/\sigma^2$ and $\mu + \sigma$.

7.2.6 Let $X_1, ..., X_n$ be iid Uniform$(-\theta, \theta)$ where $0 < \theta < \infty$ is an unknown parameter. Derive the MLEs for θ, θ^2 and θ^{-2}.

7.2.7 Suppose $X_1, ..., X_m$ are iid $N(\mu_1, \sigma^2)$, $Y_1, ..., Y_n$ are iid $N(\mu_2, \sigma^2)$, and X, Y are independent where $-\infty < \mu_1, \mu_2 < \infty$, $0 < \sigma < \infty$ are unknown parameters. Derive the MLE for (μ_1, μ_2, σ^2). Also, derive the MLE for $(\mu_1 - \mu_2)/\sigma$.

7.2.8 Consider the following *linear regression* problem: let $Y_1, ..., Y_n$ be independent random variables where Y_i is distributed as $N(\beta_0 + \beta_1 x_i, \sigma^2)$ with unknown parameters β_0, β_1. Here, $\sigma(> 0)$ is assumed known and $x_1, ..., x_n$ are fixed real numbers with $\boldsymbol{\theta} = (\beta_0, \beta_1) \in \Re^2, n \geq 2$. Denote $a = \Sigma_{i=1}^{n}(x_i - \overline{x})^2$ with $\overline{x} = n^{-1}\Sigma_{i=1}^{n}x_i$ and assume that $a > 0$. Suppose that the MLEs for β_0, β_1 are respectively denoted by $\widehat{\beta}_0, \widehat{\beta}_1$.

(i)   Write down the likelihood function of $Y_1, ..., Y_n$;

(ii)   Show that $\widehat{\beta}_{1MLE} \equiv a^{-1}\Sigma_{i=1}^{n}(x_i - \overline{x})Y_i$ and $\widehat{\beta}_{1MLE}$ is normally distributed with mean β_1 and variance σ^2/a;

(iii)   Show that $\widehat{\beta}_{0MLE} \equiv \overline{Y} - \widehat{\beta}_{1MLE}\overline{x}$ and $\widehat{\beta}_{0MLE}$ is normally

distributed with mean β_0 and variance $\sigma^2 \left\{ \dfrac{1}{n} + \dfrac{\overline{x}^2}{a} \right\}$.

7.3.1 Suppose $X_1, ..., X_4$ are iid $N(0, \sigma^2)$ where $0 < \sigma < \infty$ is an unknown parameter. Consider the following estimators:

$$T_1 = X_1^2 - X_2 + X_4, T_2 = \tfrac{1}{3}(X_1^2 + X_2^2 + X_4^2), T_3 = \tfrac{1}{4}\Sigma_{i=1}^{4}X_i^2$$

$$\text{and } T_4 = \tfrac{1}{3}\Sigma_{i=1}^{4}(X_i - \overline{X})^2, T_5 = \tfrac{1}{2}|X_1 - X_2|$$

(i)   Is T_i unbiased for σ^2, $i = 1, ..., 4$?

(ii)   Among estimators T_1, T_2, T_3, and T_4 for σ^2, which one has the smallest MSE?

(iii)   Is T_5 unbiased for σ? If not, find a suitable multiple of T_5 which is unbiased for σ. Evaluate the MSE of T_5.

7.3.2 (Exercise 7.2.5 Continued) Denote $U = \Sigma_{i=1}^{n}(X_i - X_{n:1})$ and let $V = cU$ be an estimator of σ where $c(> 0)$ is a constant.

(i)   Find the MSE of V. Then, minimize this MSE with respect to c. Call this latter estimator W which has the smallest MSE among the estimators of σ which are multiples of U;

(ii)   How do estimators W and $(n-1)^{-1}\Sigma_{i=1}^{n}(X_i - X_{n:1})$ compare relative to their respective bias and the MSE?

7.3.3 (Example 7.3.2 Continued) Show that there is no unbiased estimator for the parametric function (i) $\tau(p) = p^{-1}(1-p)^{-1}$, (ii) $\tau(p) = p^{-1}(1-p)^{-2}$.

7.4.1 Let $X_1, ..., X_n$ be iid Bernoulli(p) where $0 < p < 1$ is an unknown parameter with $n \geq 2$. Consider $\tau(p) = p^2$ and start with the unbiased estimator $T = X_1 X_2$. Derive the Rao-Blackwellized version of T.

7.4.2 (Example 7.4.2 Continued) Consider $\tau(p) = p^2(1-p)$ and $n \geq 3$. Start with the unbiased estimator $T = X_1 X_2 (1 - X_3)$ and derive its Rao-Blackwellized version.

7.4.3 Let $X_1, ..., X_n$ be iid Poisson(λ) where $0 < \lambda < \infty$ is unknown, and $n \geq 2$. We wish to estimate $\tau(\lambda) = \lambda$ and consider $T = \frac{1}{2}(X_1 + X_2)$ as an initial unbiased estimator of λ. Derive its Rao-Blackwellized version.

7.4.4 (Exercise 7.4.3 Continued) We wish to estimate $\tau(\lambda) = e^{-\lambda}$ unbiasedly. Consider $T = I(X_1 = 0)$ as an initial unbiased estimator of $\tau(\lambda)$. Show that $(1 - n^{-1})^{\sum_{i=1}^n X_i}$ is its Rao-Blackwellized version.

7.4.5 (Exercise 7.4.4 Continued) We wish to estimate $\lambda e^{-\lambda}$ unbiasedly. Consider $T = I(X_1 = 1)$ as an initial unbiased estimator of $\tau(\lambda)$. Derive its Rao-Blackwellized version.

7.4.6 (Example 7.4.4 Continued) Let $\tau(\mu) = P_\mu\{|X_1| \leq a\}$ where a is some known positive real number. Find an initial unbiased estimator T for $\tau(\mu)$. Next, derive its Rao-Blackwellized version.

7.4.7 (Example 7.4.5 Continued) Let $\tau(\mu) = \mu^3$. Find an initial unbiased estimator T for $\tau(\mu)$. Next, derive its Rao-Blackwellized version. {*Hint:* Start with the fact that $E_\mu[(X_1 - \mu)^3] = 0$ and hence show that $T = X_1^3 - 3\sigma^2 X_1$ is an unbiased estimator for $\tau(\mu)$. Next, work with the third moment of the conditional distribution of X_1 given that $\overline{X} = \overline{x}$.}

7.5.1 Suppose $X_1, ..., X_n$ are iid with a common Rayleigh pdf:

$$f(x; \theta) = 2\theta^{-1} x \, exp(-x^2/\theta) I(x > 0)$$

with $\theta(> 0)$ unknown. Find the UMVUE for (i) θ, (ii) θ^2, and (iii) θ^{-1}.

7.5.2 (Exercise 7.5.1 Continued) Find the CRLB for the variance of unbiased estimators of (i) θ, (ii) θ^2, and (iii) θ^{-1}. Is CRLB attained by the variance of the respective UMVUE obtained in Exercise 7.5.1?

7.5.3 Suppose $X_1, ..., X_n$ are iid $N(\mu, \sigma^2)$ where μ, σ are both unknown, $\mu \in \Re, \sigma \in \Re^+, n \geq 2$. Let $\boldsymbol{\theta} = (\mu, \sigma)$ and $\tau(\boldsymbol{\theta}) = \mu\sigma^k$ where k is a known and fixed real number. Derive the UMVUE for $\tau(\boldsymbol{\theta})$. Pay particular attention to any minimum sample size requirement! {*Hint:* Use independence between $\overline{X}$ and S to derive the expectation of $\overline{X}S^k$ where S^2 is the sample variance. Then, make some adjustments.}

7.5.4 Let $X_1, ..., X_n$ be iid Uniform$(-\theta, \theta)$ where θ is an unknown parameter, and $\theta \in \Re^+$. Let $\tau(\theta) = \theta^k$ where k is a known fixed positive real number. Derive the UMVUE for $\tau(\theta)$. {*Hint: $U \equiv \max_{1 \leq i \leq n} |X_i|$ is complete sufficient for θ. Use the pdf of U/θ in order to derive the expectation of U^k. Then, make some adjustments.*}

7.5.5 Let $X_1, ..., X_n$ be iid Bernoulli(p) where $0 < p < 1$ is an unknown parameter. Consider a parametric function $\tau(p) = p + (1-p)e^2$.

(*i*)   Find a suitable unbiased estimator T for $\tau(p)$;

(*ii*)   Since the complete sufficient statistic is $U = \sum_{i=1}^{n} X_i$, use Lehmann-Scheffé Theorems and evaluate the conditional expectation, $E_p[T \mid U = u]$;

(*iii*)   Hence, derive the UMVUE for $\tau(p)$.

7.5.6 (Exercise 7.5.5 Continued) Consider $\tau(p) = (p + qe^3)^2$ where $q = 1 - p$.

(*i*)   Find a suitable unbiased estimator T for $\tau(p)$;

(*ii*)   Since the complete sufficient statistic is $U = \sum_{i=1}^{n} X_i$, use Lehmann-Scheffé Theorems and evaluate the conditional expectation, $E_p[T \mid U = u]$;

(*iii*)   Hence, derive the UMVUE for $\tau(p)$.

7.5.7 (Example 7.5.3 Continued) Let $\tau(\mu) = e^{2\mu}$. Derive the UMVUE for $\tau(\mu)$.

7.5.8 Let $X_1, ..., X_n$ be iid Bernoulli(p) with $0 < p < 1$ and $n \geq 4$. Find the UMVUE for $\tau(p) = p(1-p)^2$. {*Hint: Start with the unbiased estimator $T = X_1(1 - X_2)(1 - X_3)$ for $\tau(p)$. Then Rao-Blackwellize T given the complete and sufficient statistic $\sum_{i=1}^{n} X_i$.*}

7.5.9 Let $X_1, ..., X_n$ be iid Bernoulli(p) with $0 < p < 1$ and $n \geq 3$. Find the UMVUE for $\tau(p) = p(1 - 2p)$. {*Hint: Note that $\tau(p) = p - 2p^2$. The sample mean $\overline{X}$ estimates p unbiasedly. Next, in order to estimate p^2, one may proceed as in the Exercise 7.4.1. Otherwise, one may start with the unbiased estimator $X_2 X_3$ for p^2. Consider the complete and sufficient statistic $\sum_{i=1}^{n} X_i$. Then Rao-Blackwellize $X_2 X_3$ given $\sum_{i=1}^{n} X_i$. Next, one would combine the two parts and use Lehmann-Scheffé Theorems.*}

7.5.10 Let $X_1, ..., X_n$ be iid Uniform($0, \theta$) with $\theta > 0$. Find the UMVUE for $\tau(\theta) = (\theta^{1/4} + 2\theta^{-1/4})^2$. {*Hint: Verify that all one needs are the UMVUEs for both $\theta^{1/2}$ and $\theta^{-1/2}$. Start with the complete and sufficient statistic $T = X_{n:n}$ and find functions of T which are unbiased estimators for $\theta^{1/2}$ and $\theta^{-1/2}$, respectively. Next, one would combine these parts and use Lehmann-Scheffé Theorems.*}

7.5.11 Let $X_1, ..., X_n$ be iid Uniform($-\theta, \theta$) where θ is an unknown parameter, and $\theta \in \Re^+$. Find the UMVUE for $\tau(\theta) = (\theta^{1/2} - 2\theta^{-1/2})^2$. {*Hint: Verify that all one needs are the UMVUEs for both θ and θ^{-1}. Start with the complete and sufficient statistic $T = X_{n:n}$ and find functions of T which are unbiased estimators for θ and θ^{-1}, respectively. Next, one would combine these parts and use Lehmann-Scheffé Theorems.*}

7.6.1 Suppose $X_1, ..., X_n$ are iid with a common Rayleigh pdf:

$$f(x; \theta) = 2\theta^{-1} x \, exp(-x^2/\theta) I(x > 0)$$

with $\theta(> 0)$ unknown. Show that the MLEs and UMVUEs for θ, θ^2 and θ^{-1} are all consistent.

7.6.2 Suppose $X_1, ..., X_n$ are iid with a common Weibull pdf:

$$f(x; \alpha) = \alpha^{-1} \beta x^{\beta-1} exp(-x^\beta/\alpha) I(x > 0)$$

with $\alpha(> 0)$ unknown, but $\beta(> 0)$ is assumed known. Show that the MLEs and UMVUEs for α, α^2 and α^{-1} are all consistent.

7.6.3 Suppose $X_1, ..., X_n$ are iid with a common pdf:

$$f(x; \theta) = \begin{cases} (\theta + 1)x^\theta & \text{if } 0 < x < 1 \\ 0 & \text{elsewhere} \end{cases}$$

with $\theta(> 0)$ unknown. Show that the MLE for θ is consistent.

7.6.4 Suppose $X_1, ..., X_n$ are iid whose common pdf is:

$$f(x) = \begin{cases} \{\sigma\sqrt{2\pi}\}^{-1} x^{-1} exp\{-\frac{1}{2\sigma^2}(log(x) - \mu)^2\} & \text{if } x > 0 \\ 0 & \text{elsewhere} \end{cases}$$

with μ, σ both unknown, $-\infty < \mu < \infty$, $0 < \sigma < \infty$, and $\boldsymbol{\theta} = (\mu, \sigma)$.

(i)   Evaluate the expression for $E_\theta[X_1^k]$, denoted by the parametric function $\tau(\boldsymbol{\theta})$, for any fixed $k(> 0)$;

(ii)   Derive the MLE, denoted by T_n, for $\tau(\boldsymbol{\theta})$;

(iii) Show that T_n is consistent for $\tau(\boldsymbol{\theta})$.

7.6.5 (Exercise 7.5.8 Continued) Derive the MLE U_n for $\tau(p) = p(1-p)^2$. Show that U_n is consistent for $\tau(p)$.

7.6.6 (Exercise 7.5.9 Continued) Derive the MLE U_n for $\tau(p) = p(1-2p)$. Show that U_n is consistent for $\tau(p)$.

7.6.7 (Exercise 7.5.10 Continued) Derive the MLE U_n for $\tau(\theta) = (\theta^{1/4} + 2\theta^{-1/4})^2$. Show that U_n is consistent for $\tau(\theta)$.

7.6.8 (Exercise 7.5.11 Continued) Derive the MLE U_n for $\tau(\theta) = (\theta^{1/2} - 2\theta^{-1/2})^2$. Show that U_n is consistent for $\tau(\theta)$.

7.6.9 Suppose $X_1, ..., X_n$ are iid $N(\mu, \sigma^2)$ where $-\infty < \mu < \infty$ and $0 < \sigma < \infty$ are unknown parameters. Denote

$$T_n = \frac{2X_1 + 4X_2 + 6X_3 + ... + 2nX_n}{n(n+1)}, n \geq 1$$

and address the following problems.

(i)   Evaluate the expressions for $E_{\mu,\sigma^2}[T_n]$ and $V_{\mu,\sigma^2}[T_n]$;

(ii)   Show that T_n is consistent for μ;

(iii)   Is $\max\{0, T_n\}$ consistent for μ?

7.6.10 Consider the setup from Exercise 7.6.9, but without the assumption of normality. That is, suppose $X_1, ..., X_n$ are iid having some common pdf $f(x)$ with mean μ and variance σ^2 where $-\infty < \mu < \infty$ and $0 < \sigma < \infty$ are unknown parameters. Denote

$$T_n = \frac{2X_1 + 4X_2 + 6X_3 + ... + 2nX_n}{n(n+1)}, n \geq 1$$

and address the following problems.

(i)   Evaluate the expressions for $E_f[T_n]$ and $V_f[T_n]$;

(ii)   Show that T_n is consistent for μ;

(iii)   Is $\max\{0, T_n\}$ consistent for μ?

8

Tests of Hypotheses

8.1 Introduction

Suppose that a population *probability mass function* (pmf) or *probability density function* (pdf) is $f(x; \theta)$ with $x \in \mathcal{X} \subseteq \Re$ where θ is an unknown parameter, $\theta \in \Theta \subseteq \Re$.

Definition 8.1.1 *A hypothesis is a statement under consideration about an unknown parameter θ.*

Jerzy Neyman and Egon S. Pearson discovered a fundamental approach to formulate a test for statistical hypotheses. The Neyman-Pearson collaboration (1928a,b) first emerged with formulations and constructions of tests through comparisons of likelihood functions which blossomed into other landmark papers (Neyman and Pearson, 1933a,b).

Suppose that one has to choose between two hypotheses:

$$H_0 : \theta \in \Theta_0 \text{ vs. } H_1 : \theta \in \Theta_1$$

where $\Theta_0 \subset \Theta, \Theta_1 \subset \Theta$, and $\Theta_0 \cap \Theta_1 = \varnothing$, an empty set. Based on the evidence collected via random samples $X_1, ..., X_n$, a statistical problem is to select one of the hypotheses which seems more reasonable. The basic question is this: given the sample evidence, if one must decide in favor of H_0 or H_1, which hypothesis is it going to be?

We refer to H_0 and H_1 respectively as the *null* and *alternative* hypotheses. A hypothesis is called *simple* if it pinpoints a specific pmf or pdf. That is, $H_0 : \theta = \theta_0$ and $H_1 : \theta = \theta_1$, with $\theta_0 \neq \theta_1$ known, would be both *simple*. A hypothesis which is not simple is called *composite*, for example, $H_0 : \theta > \theta_0, H_0 : \theta < \theta_0, H_1 : \theta \geq \theta_1, H_1 : \theta \leq \theta_1$ are *composite* hypotheses. A hypothesis, $H_1 : \theta \neq \theta_0$, is called a *two-sided* hypothesis. In Section 8.2, we formulate two types of errors in decision making. Section 8.3 develops a *most powerful* (MP) test for choosing between simple null vs. simple alternative hypotheses. In Section 8.4, the idea of a *uniformly most powerful* (UMP) test is pursued when H_0 is simple but H_1 is one-sided. Section 8.5 gives examples of two-sided tests and briefly touches upon *unbiased* and *uniformly most powerful unbiased* (UMPU) tests.

8.2 Error Probabilities and Power Function

We gather observations $X_1, ..., X_n$ and learn about unknown θ. One would favor H_0 (or H_1) if an estimator $\hat{\theta} \equiv \hat{\theta}(X_1, ..., X_n)$ seems more likely under

H_0 (or H_1). But, one may make a mistake by favoring a wrong hypothesis simply because one's decision is based on the evidence from a random sample.

The entries shown in the Table 8.2.1 may help in one's understanding of the kinds of errors that may be committed by choosing one hypothesis over the other:

Table 8.2.1. Type I and Type II Errors

Test Result or Decision	Nature's Choice	
	H_0 True	H_1 True
Accept H_0	No Error	Type II Error
Accept H_1	Type I Error	No Error

The entries reveal that there can be one of two possible kinds of errors. We may reject H_0 while H_0 is true or accept H_0 while H_1 is true.

What do we mean by a *test* of H_0 vs. H_1? Having observed the data $\mathbf{X} = (X_1, ..., X_n)$, a test will guide us *unambiguously* to either reject H_0 or accept H_0. This is accomplished by *partitioning* a sample space into two parts, $\mathcal{R}$ and $\mathcal{R}^c$, corresponding to respective final decisions "reject H_0" and "accept H_0."

> We reject H_0 whenever $\mathbf{x} \in \mathcal{R}$. The subset $\mathcal{R}$ is called a *rejection region* or a *critical region*.

Whenever H_0, H_1 are simple hypotheses, we respectively write α, β for the Type I and II error probabilities:

$$
\begin{aligned}
\alpha = P\{\text{Type I error}\} \quad &= P\{\text{Rejecting } H_0 \text{ when } H_0 \text{ is true}\} \\
&= P\{\mathbf{X} \in \mathcal{R} \text{ when } H_0 \text{ is true}\} \\
\beta = P\{\text{Type II error}\} \quad &= P\{\text{Accepting } H_0 \text{ when } H_1 \text{ is true}\} \\
&= P\{\mathbf{X} \in \mathcal{R}^c \text{ when } H_1 \text{ is true}\}
\end{aligned}
\tag{8.2.1}
$$

Example 8.2.1 Consider a population distribution $N(\theta, 1)$ where $\theta \in \Re$ is unknown. We postulate $H_0 : \theta = 5.5$ and $H_1 : \theta = 8$. Suppose that we have observed the data, $\mathbf{X} = (X_1, ..., X_9)$. Denote $\overline{X} = \frac{1}{9}\Sigma_{i=1}^{9}X_i$. Some illustrations of "tests" are given below:

$$
\begin{aligned}
&\text{Test \#1:} \quad \text{Reject } H_0 \text{ if and only if } X_1 > 7 \\
&\text{Test \#2:} \quad \text{Reject } H_0 \text{ if and only if } \tfrac{1}{2}(X_1 + X_2) > 7 \\
&\text{Test \#3:} \quad \text{Reject } H_0 \text{ if and only if } \overline{X} > 6 \\
&\text{Test \#4:} \quad \text{Reject } H_0 \text{ if and only if } \overline{X} > 7.5
\end{aligned}
\tag{8.2.2}
$$

We may summarize these tests in another way:

Test #1: $\mathcal{R}_1 = \{\mathbf{x} = (x_1, ..., x_9) : x_1 > 7\}$

Test #2: $\mathcal{R}_2 = \{\mathbf{x} = (x_1, ..., x_9) : \frac{1}{2}(x_1 + x_2) > 7\}$ (8.2.3)

Test #3: $\mathcal{R}_3 = \{\mathbf{x} = (x_1, ..., x_9) : \overline{x} > 6\}$

Test #4: $\mathcal{R}_4 = \{\mathbf{x} = (x_1, ..., x_9) : \overline{x} > 7.5\}$

Here, $\mathcal{R}_i$ is that part of the sample space $\Re^9$ where H_0 is rejected by means of the Test #$i, i = 1, ..., 4$.

Denote Z for a standard normal variable. For Test #1, we have:

$$\alpha = P\{X_1 > 7 \text{ when } \theta = 5.5\} = P\{Z > 1.5\} = 0.06681$$
$$\text{and } \beta = P\{X_1 \leq 7 \text{ when } \theta = 8\} = P\{Z \leq -1\} = 0.15866$$

Next, α, β associated with these tests are summarized below in Table 8.2.2.

Table 8.2.2. Values of α and β for Tests #1 through #4
from Equation (8.2.2)

Test #1 $\mathcal{R}_1$	Test #2 $\mathcal{R}_2$	Test #3 $\mathcal{R}_3$	Test #4 $\mathcal{R}_4$
$\alpha = 0.06681$ $\beta = 0.15866$	$\alpha = 0.01696$ $\beta = 0.07865$	$\alpha = 0.06681$ $\beta = 0.00000$	$\alpha = 0.00000$ $\beta = 0.06681$

We immediately conclude a few things: between Tests #1 and #2, the Test #2 appears better because both its error probabilities are smaller. Comparing Tests #1 and #3, we may say that Test #3 performs much better. Comparing Tests #1 through #3, we feel that Test #1 should not be in the running, and no clear-cut choice emerges between Tests #2 and #3. One of these has a smaller α but a larger β. If we must pick between Tests #2 and #3, we have to consider the consequences of committing either error in practice. It is clear that an experimenter may not be able to accomplish this just by looking at α or β. Tests #3 and #4 point out a slightly different story. By downsizing the rejection region $\mathcal{R}$ for Test #4 in comparison with that of Test #3, we are able to make α for Test #4 practically zero, but this happens at the expense of a sharp rise in β. ▲

> All tests may not be comparable among themselves such as Tests #2 and #3. By suitably adjusting a rejection region $\mathcal{R}$, we may make α (or β) as small as we please, but then β (or α) will be on the rise as the sample size n is held fixed.

Definition 8.2.1 *The power or power function of a test is*

$$Q(\theta) \equiv P_\theta(\mathcal{R}) \text{ for all } \theta \in \Theta \qquad (8.2.4)$$

In a simple null vs. simple alternative testing situation, one obviously has $Q(\theta_0) = \alpha$ and $Q(\theta_1) = 1 - \beta$.

Example 8.2.2 (Example 8.2.1 Continued) Verify that for Test #4, the power function is $Q(\theta) = 1 - \Phi(22.5 - 3\theta)$, $\theta \in \Re$. ▲

A test may also be identified by what is called a *critical function* or a *test function*.

Definition 8.2.2 *A function* $\psi(.) : \mathcal{X}^n \to [0, 1]$ *is called a critical function or a test function where* $\psi(\mathbf{x})$ *denotes the probability with which* H_0 *is rejected when the data* $\mathbf{X} = \mathbf{x}$ *have been observed.*

We may rewrite the power function from Equation (8.2.4) as follows:

$$Q(\theta) \equiv E_\theta\{\psi(\mathbf{X})\} \text{ for } \theta \in \Theta \qquad (8.2.5)$$

8.2.1   Best Test

How should one define a "best" test for H_0 vs. H_1? We consider testing $H_0 : \theta \in \Theta_0$ vs. $H_1 : \theta \in \Theta_1$ where $\Theta_0 \subset \Theta, \Theta_1 \subset \Theta$, and $\Theta_0 \cap \Theta_1 = \varnothing$.

Definition 8.2.3 *Fix a number* $\alpha \in (0, 1)$. *A test for* $H_0 : \theta \in \Theta_0$ *vs.* $H_1 : \theta \in \Theta_1$ *with its power function* $Q(\theta)$ *is called size (or level)* α *according as* $\max_{\theta \in \Theta_0} Q(\theta) = (or \leq) \alpha$.

In defining a size or level α test, $\max_{\theta \in \Theta_0} Q(\theta)$ is interpreted as the worst possible Type I error probability. Then, we may restrict attention to a class of *test functions* such that $\max_{\theta \in \Theta_0} E_\theta\{\psi(\mathbf{X})\} = \alpha$ or $\max_{\theta \in \Theta_0} E_\theta\{\psi(\mathbf{X})\} \leq \alpha$. Obviously, a size α test is also a level α test.

The idea is to compare all tests for H_0 vs. H_1, each having a basic common property to begin with. In defining a "best" test, we compare only *level* α tests.

Definition 8.2.4 *Consider* C, *the collection of all level* α *tests for* $H_0 : \theta \in \Theta_0$ *vs.* $H_1 : \theta \in \Theta_1$. *A test from* C *with its power function* $Q(\theta)$ *is called the best or uniformly most powerful (UMP) level* α *if and only if* $Q(\theta) \geq Q^*(\theta)$ *for all* $\theta \in \Theta_1$ *where* $Q^*(\theta)$ *is a power function of any other test from* C. *If* Θ_1 *is a singleton set, then the best test is called the most powerful (MP) level* α *test.*

Example 8.2.3 Consider the pdf $N(\theta, 1)$ with $\theta \in \Re$ unknown. We test $H_0 : \theta = 5.5$ vs. $H_1 : \theta = 8$. A random sample $\mathbf{X} = (X_1, ..., X_9)$ is collected. Now, let us consider the following tests:

   Test #1: Reject H_0 if and only if $X_1 > 7.1449$

   Test #2: Reject H_0 if and only if $\frac{1}{4}(X_1 + ... + X_4) > 6.32245$

   Test #3: Reject H_0 if and only if $\overline{X} > 6.0483$

Let us write $Q_i(\theta)$ for the power function of Test #$i, i = 1, 2, 3$. Using MAPLE, we noted that $Q_1(5.5) = Q_2(5.5) = Q_3(5.5) = 0.049995$. That is, these are level α tests where $\alpha = 0.049995$. In Figure 8.2.1, we have plotted three power functions. It is clear from this plot that among the three tests under investigation, Test #3 has the maximum power at each $\theta > 5.5$.

We may add that the respective power at $\theta = 8$ is: $Q_1(8) = 0.80375, Q_2(8) = 0.9996$, and $Q_3(8) = 1.0$. ▲

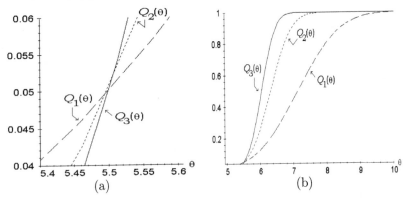

Figure 8.2.1. Power functions for Tests #1, #2, and #3 from Example 8.2.3.

8.3   Simple Null vs. Simple Alternative

We introduce the celebrated Neyman-Pearson Lemma which was originally formulated and proved in Neyman and Pearson (1933a). This is an indispensable tool for finding a MP level α test for a simple null hypothesis against a simple alternative hypothesis.

8.3.1   Neyman-Pearson Lemma

Suppose that $X_1, ..., X_n$ are *independent and identically distributed* (iid) real valued random variables with a common distribution $f(x; \theta), \theta \in \Theta \subseteq \Re$. Denote the likelihood function, $L(\mathbf{x}; \theta) \equiv \Pi_{i=1}^{n} f(x_i; \theta)$ under θ. We wish to test:

$$H_0 : \theta = \theta_0 \text{ vs. } H_1 : \theta = \theta_1 \qquad (8.3.1)$$

where $\theta_0 \neq \theta_1, \theta_0, \theta_1 \in \Theta$. Under $H_i : \theta = \theta_i$, the datum $\mathbf{x}$ has a completely specified likelihood function $L(\mathbf{x}; \theta_i), i = 0, 1$.

   Theorem 8.3.1 (Neyman-Pearson Lemma) *Consider a test of H_0 vs. H_1 with the rejection ($\mathcal{R}$) and acceptance ($\mathcal{R}^c$) regions for the null hypothesis defined as follows:*

$$\begin{aligned} \mathbf{x} \in \mathcal{R} & \quad \textit{if } L(\mathbf{x}; \theta_1) > kL(\mathbf{x}; \theta_0) \\ \mathbf{x} \in \mathcal{R}^c & \quad \textit{if } L(\mathbf{x}; \theta_1) < kL(\mathbf{x}; \theta_0) \end{aligned}$$

or equivalently, having a test function

$$\psi(\mathbf{x}) = \begin{cases} 1 & \textit{if } L(\mathbf{x}; \theta_1) > kL(\mathbf{x}; \theta_0) \\ 0 & \textit{if } L(\mathbf{x}; \theta_1) < kL(\mathbf{x}; \theta_0) \end{cases} \qquad (8.3.2)$$

The constant $k(\geq 0)$ is so determined that

$$E_{\theta_0}\{\psi(\mathbf{X})\} = \alpha \qquad (8.3.3)$$

Any test satisfying Equations (8.3.2) and (8.3.3) is a MP level α test.

Proof Let $X_1, ..., X_n$ be continuous random variables. A discrete case may be disposed of by replacing integrals with sums. Note that a test function $\psi(\mathbf{x})$ that satisfies Equation (8.3.3) has size α and hence it has level α too!

Let $\psi^*(\mathbf{x})$ be another level α test function. Let $Q(\theta), Q^*(\theta)$ be the power functions associated with ψ, ψ^*, respectively. Now, we verify:

$$\{\psi(\mathbf{x}) - \psi^*(\mathbf{x})\}\{L(\mathbf{x}; \theta_1) - kL(\mathbf{x}; \theta_0)\} \geq 0 \text{ for all } \mathbf{x} \in \mathcal{X}^n \qquad (8.3.4)$$

First, let $\mathbf{x} \in \mathcal{X}^n$ but $\psi(\mathbf{x}) = 1$ so that $L(\mathbf{x}; \theta_1) - kL(\mathbf{x}; \theta_0) > 0$, by definition of ψ in Equation (8.3.2). For such $\mathbf{x}$, one obviously has $\psi(\mathbf{x}) - \psi^*(\mathbf{x}) \geq 0$ since $\psi^*(\mathbf{x}) \in (0, 1)$ so that Equation (8.3.4) holds.

Next, let $\mathbf{x} \in \mathcal{X}^n$ but $\psi(\mathbf{x}) = 0$ so that $L(\mathbf{x}; \theta_1) - kL(\mathbf{x}; \theta_0) < 0$, by definition of ψ in Equation (8.3.2). For such $\mathbf{x}$, one obviously has $\psi(\mathbf{x}) - \psi^*(\mathbf{x}) \leq 0$ since $\psi^*(\mathbf{x}) \in (0, 1)$ so that Equation (8.3.4) holds. Now, if $\mathbf{x} \in \mathcal{X}^n$ is such that $0 < \psi(\mathbf{x}) < 1$, then from Equation (8.3.2) we must have $L(\mathbf{x}; \theta_1) - kL(\mathbf{x}; \theta_0) = 0$, and so Equation (8.3.4) is validated. That is, Equation (8.3.4) holds for all $\mathbf{x} \in \mathcal{X}^n$. Hence,

$$\begin{aligned}
0 \quad &\leq \int \cdots \int_{\mathcal{X}^n} \{\psi(\mathbf{x}) - \psi^*(\mathbf{x})\}\{L(\mathbf{x}; \theta_1) - kL(\mathbf{x}; \theta_0)\}\Pi_{i=1}^n dx_i \\
&= \int \cdots \int_{\mathcal{X}^n} \psi(\mathbf{x})\{L(\mathbf{x}; \theta_1) - kL(\mathbf{x}; \theta_0)\}\Pi_{i=1}^n dx_i \\
&\quad - \int \cdots \int_{\mathcal{X}^n} \psi^*(\mathbf{x})\{L(\mathbf{x}; \theta_1) - kL(\mathbf{x}; \theta_0)\}\Pi_{i=1}^n dx_i \\
&= \{E_{\theta_1}[\psi(\mathbf{X})] - kE_{\theta_0}[\psi(\mathbf{X})]\} - \{E_{\theta_1}[\psi^*(\mathbf{X})] - kE_{\theta_0}[\psi^*(\mathbf{X})]\} \\
&= \{Q(\theta_1) - Q^*(\theta_1)\} - k\{Q(\theta_0) - Q^*(\theta_0)\}.
\end{aligned}$$

$$(8.3.5)$$

Now, $Q(\theta_0)$ is the Type I error probability associated with ψ and thus $Q(\theta_0) = \alpha$ from Equation (8.3.3). $Q^*(\theta_0)$ is a similar entity associated with ψ^* which has level α, that is $Q^*(\theta_0) \leq \alpha$. Thus, $Q(\theta_0) - Q^*(\theta_0) \geq 0$ and we may rewrite Equation (8.3.5) as:

$$Q(\theta_1) - Q^*(\theta_1) \geq k\{Q(\theta_0) - Q^*(\theta_0)\} \geq 0$$

So, $Q(\theta_1) \geq Q^*(\theta_1)$. Hence, the test function ψ is at least as powerful as ψ^*. The proof is complete. ∎

Remark 8.3.1 Observe that the Neyman-Pearson Lemma rejects H_0 in favor of accepting H_1 provided that the ratio of the likelihoods under H_1 and H_0 is sufficiently large ($> k$) for some suitable $k(\geq 0)$.

> *Convention:* The ratio $\frac{c}{0}$ is interpreted as infinity if $c > 0$
> and one if $c = 0$.

Remark 8.3.2 In the Neyman-Pearson Lemma, note that nothing has been specified about $\mathbf{x}$, which may satisfy the equation $L(\mathbf{x}; \theta_1) = kL(\mathbf{x}; \theta_0)$. Now, if $X_1, ..., X_n$ are continuous variables, then a set of such points would have a zero probability. For a discrete data $\mathbf{x}$, however, one would *randomize* if $L(\mathbf{x}; \theta_1) = kL(\mathbf{x}; \theta_0)$ holds. Consider Example 8.3.5.

> A final MP or UMP level α test is given in the simplest *implementable* form. That is, having fixed α, the cut-off point of such a test defining regions $\mathcal{R}, \mathcal{R}^c$ must be explicitly found analytically or from a standard statistical table.

Remark 8.3.3 For all practical purposes, a MP level α test given by the Neyman-Pearson Lemma is unique. That is, if one finds another MP level α test with a test function ψ^*, then ψ^* and ψ will coincide on the sets $\{\mathbf{x} \in \mathcal{X}^n : L(\mathbf{x}; \theta_1) > kL(\mathbf{x}; \theta_0)\}$ and $\{\mathbf{x} \in \mathcal{X}^n : L(\mathbf{x}; \theta_1) < kL(\mathbf{x}; \theta_0)\}$.

> *Convention:* k is a *generic* and *nonstochastic* constant. k may not remain the same from one step to another.

Example 8.3.1 Suppose $X_1, ..., X_n$ are iid $N(\mu, \sigma^2)$ with unknown $\mu \in \Re$, but $\sigma^2 \in \Re^+$ is known. With preassigned $\alpha \in (0, 1)$, we wish to find a MP level α test for $H_0 : \mu = \mu_0$ vs. $H_1 : \mu = \mu_1(> \mu_0)$ with known real numbers μ_0, μ_1. Since H_0, H_1 are simple hypotheses, the Neyman-Pearson Lemma applies. The likelihood function is:

$$L(\mathbf{x}; \mu) = \{\sigma^2 2\pi\}^{-n/2} exp\{-(2\sigma^2)^{-1}\Sigma_{i=1}^n (x_i - \mu)^2\}, \ \mu \in \Re$$

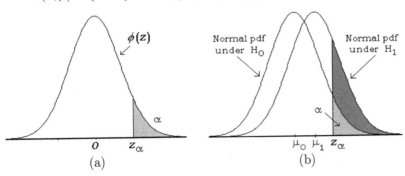

Figure 8.3.1. (a) Standard normal pdf: upper $100\alpha\%$ point; (b) probability on the right of z_α is larger under H_1 (darker plus lighter shaded areas) than under H_0 (lighter shaded area).

A MP test has the form:

$$\text{Reject } H_0 \text{ if and only if } L(\mathbf{x}; \mu_1)/L(\mathbf{x}; \mu_0) > k$$

That is, we will reject H_0 if and only if

$$exp\{\sigma^{-2}(\mu_1 - \mu_0)\Sigma_{i=1}^n x_i\} \text{ is "large" } (> k) \qquad (8.3.6)$$

Since $\mu_1 > \mu_0$, the Equation (8.3.6) may be equivalently expressed as:

Reject H_0 if and only if $\Sigma_{i=1}^n X_i$, or equivalently

the sample mean $\overline{X}$, is "large" $(> k)$ (8.3.7)

Since $E_\mu[X] = \mu$, it does make sense to reject H_0 when $\overline{X}$ is "large" $(> k)$ because the alternative hypothesis postulates a value μ_1 which is larger than μ_0. But, the MP test given in Equation (8.3.7) is *not* in an implementable form. Since the test must have size α, we rewrite Equation (8.3.7) as:

Reject H_0 if and only if $\sqrt{n}(\overline{X} - \mu_0)/\sigma > z_\alpha$ (8.3.8)

where z_α is the upper $100\alpha\%$ point of the standard normal distribution. See Figure 8.3.1. Under H_0, observe that $\sqrt{n}(\overline{X} - \mu_0)/\sigma$ is a statistic, referred to as a *test statistic*, which is distributed as a standard normal random variable.

Here, the critical region $\mathcal{R} \equiv \{\mathbf{x} = (x_1, ..., x_n) \in \Re^n : \sqrt{n}(\overline{x} - \mu_0)/\sigma > z_\alpha\}$. Now, we have:

Type I error probability

$= P_{\mu_0}\{\sqrt{n}(\overline{X} - \mu_0)/\sigma > z_\alpha\} = \alpha$, by the choice of z_α

So, we have a MP level α test. ▲

Example 8.3.2 (Example 8.3.1 Continued) With preassigned $\alpha \in (0, 1)$, we wish to find a MP level α test for $H_0 : \mu = \mu_0$ vs. $H_1 : \mu = \mu_1(< \mu_0)$ with known real numbers μ_0, μ_1. See Figure 8.3.2 for the position of $-z_\alpha$.

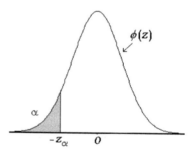

Figure 8.3.2. Standard normal pdf: lower $100\alpha\%$ point.

One may check that a MP level α test will have its critical region $\mathcal{R} = \{\mathbf{x} = (x_1, ..., x_n) \in \Re^n : \sqrt{n}(\overline{x} - \mu_0)/\sigma < -z_\alpha\}$. ▲

Example 8.3.3 Suppose $X_1, ..., X_n$ are iid Uniform$(0, \theta)$ with $\theta(> 0)$ unknown. With preassigned $\alpha \in (0, 1)$, we wish to obtain a MP level α test for $H_0 : \theta = \theta_0$ vs. $H_1 : \theta = \theta_1(> \theta_0)$ with known positive numbers θ_0, θ_1. Since H_0, H_1 are simple hypotheses, the Neyman-Pearson Lemma applies. The likelihood function is:

$$L(\mathbf{x}; \theta) = \theta^{-n} I(0 < x_{n:n} < \theta) I(0 < x_{n:1} < x_{n:n}) , \ \theta \in \Re^+$$

A MP test has the form:

Reject H_0 if and only if $L(\mathbf{x}; \theta_1)/L(\mathbf{x}; \theta_0) > k$

That is, one will reject H_0 if and only if

$$X_{n:n} \text{ is "large" } (> k) \tag{8.3.9}$$

But, a MP test given by Equation (8.3.9) is *not* in an implementable form. Under H_0, the pdf of $T = X_{n:n}$ is $nt^{n-1}/\theta_0^n I(0 < t < \theta_0)$. Hence, we determine k as follows:

Type I error probability

$$= P_{\theta_0}\{X_{n:n} > k\} = \int_k^{\theta_0} nt^{n-1}\theta_0^{-n}dt = (\theta_0^n - k^n)/\theta_0^n = \alpha$$

so that $k = \theta_0(1 - \alpha)^{1/n}$. A MP level α test will have its critical region $\mathcal{R} \equiv \{\mathbf{x} = (x_1, ..., x_n) \in \Re^{+n} : x_{n:n} > \theta_0(1 - \alpha)^{1/n}\}$. ▲

> In the Neyman-Pearson Lemma, we assumed that θ was real valued. This assumption was not crucial. The unknown parameter could be vector valued. It was crucial that the likelihood function involved no unknown component of $\boldsymbol{\theta}$ under either hypothesis $H_i, i = 0, 1$ if $\boldsymbol{\theta}$ were vector valued. Consider Example 8.3.4.

Example 8.3.4 Suppose $X_1, ..., X_n$ are iid with a common pdf:

$$b^{-\delta}[\Gamma(\delta)]^{-1}x^{\delta-1}exp(-x/b)$$

involving two unknown parameters $(\delta, b) \in \Re^+ \times \Re^+, x \in \Re^+$. With pre-assigned $\alpha \in (0,1)$, we wish to obtain a MP level α test for $H_0 : (b = b_0, \delta = \delta^*)$ vs. $H_1 : (b = b_1, \delta = \delta^*)$ where $b_1 > b_0$ are positive numbers and δ^* is also a positive number. Since H_0 and H_1 are simple hypotheses, the Neyman-Pearson Lemma applies. The likelihood function is:

$$L(\mathbf{x}; \delta, b) = b^{-n\delta}[\Gamma(\delta)]^{-n}exp\{-b^{-1}\Sigma_{i=1}^n x_i\}\{\Pi_{i=1}^n x_i\}^{\delta-1},$$
$$\text{for } (\delta, b) \in \Re^+ \times \Re^+$$

A MP test has the form:

Reject H_0 if and only if $L(\mathbf{x}; \delta^*, b_1)/L(\mathbf{x}; \delta^*, b_0) > k$

That is, one will reject H_0 if and only if

$$exp\{(b_0^{-1} - b_1^{-1})\Sigma_{i=1}^n x_i\} \text{ is "large" } (> k) \tag{8.3.10}$$

Observe that, under H_0, the statistic $\Sigma_{i=1}^n X_i$ has a Gamma$(n\delta^*, b_0)$ distribution which is completely specified for fixed values of n, δ^*, b_0. Let us equivalently express the test as:

Reject H_0 if and only if $\Sigma_{i=1}^n X_i > g_{n,\delta^*,b_0,\alpha}$ \tag{8.3.11} where $g_{n,\delta^*,b_0,\alpha}$ is the upper $100\alpha\%$ point of Gamma$(n\delta^*, b_0)$ distribution. In Table 8.3.1, $g_{n,\delta^*,b_0,\alpha}$ values are given for $\alpha = 0.01, 0.05, b_0 = 1, n = 2, 5, 6$ and $\delta^* = 2, 3$. ▲

Table 8.3.1. Selected Values of $g_{n,\delta^*,b_0,\alpha}$ with $b_0 = 1$

	$\alpha = 0.05$			$\alpha = 0.01$		
	$n = 2$	$n = 5$	$n = 6$	$n = 2$	$n = 5$	$n = 6$
$\delta^* = 2$	7.7537	15.7050	18.2080	10.0450	18.7830	21.4900
$\delta^* = 3$	10.5130	21.8860	25.4990	13.1080	25.4460	21.3100

> In a discrete case, one applies the Neyman-Pearson Lemma, but employs *randomization*. Consider Example 8.3.5.

Example 8.3.5 Let $X_1, ..., X_n$ be iid Bernoulli(p) where $p \in (0,1)$ is unknown. With preassigned $\alpha \in (0,1)$, we wish to derive a MP level α test for $H_0 : p = p_0$ vs. $H_1 : p = p_1(> p_0)$ with two numbers p_0, p_1 in $(0,1)$. Since H_0 and H_1 are simple hypotheses, the Neyman-Pearson Lemma applies. Writing $T = \Sigma_{i=1}^n X_i$, $t = \Sigma_{i=1}^n x_i$, the likelihood function is:

$$L(\mathbf{x}; p) = p^t (1-p)^{n-t}, \ p \in (0,1)$$

A MP test has the form:

Reject H_0 if and only if $L(\mathbf{x}; p_1)/L(\mathbf{x}; p_0) > k$

That is, we will reject H_0 if

$$\{[p_1(1-p_0)]/[p_0(1-p_1)]\}^t \{(1-p_1)/(1-p_0)\}^n \text{ is "large" } (> k) \quad (8.3.12)$$

Since $p_1 > p_0$, we have $[p_1(1-p_0)]/[p_0(1-p_1)] > 1$. So, "large" values of the *left-hand side* (lhs) in Equation (8.3.12) will correspond to "large" values of $\Sigma_{i=1}^n X_i$. So, a MP test may be expressed as:

$$\psi(\mathbf{X}) = \begin{cases} 1 & \text{if} \quad \Sigma_{i=1}^n X_i > k \\ \gamma & \text{if} \quad \Sigma_{i=1}^n X_i = k \\ 0 & \text{if} \quad \Sigma_{i=1}^n X_i < k \end{cases} \quad (8.3.13)$$

where a positive integer k and $\gamma \in (0,1)$ are to be chosen so that this test has size α. Note that $\Sigma_{i=1}^n X_i$ has Binomial(n, p_0) distribution under H_0. Now, one determines the *smallest integer* k such that $P_{p_0}\{\Sigma_{i=1}^n X_i > k\} < \alpha$ and lets

$$\gamma = [\alpha - P_{p_0}\{\Sigma_{i=1}^n X_i > k\}]/P_{p_0}\{\Sigma_{i=1}^n X_i = k\} \quad (8.3.14)$$

where one has

$$P_{p_0}\{\Sigma_{i=1}^n X_i = k\} = \binom{n}{k} p_0^k (1-p_0)^{n-k},$$
$$P_{p_0}\{\Sigma_{i=1}^n X_i > k\} = \Sigma_{u=k+1}^n \binom{n}{u} p_0^u (1-p_0)^{n-u} \quad (8.3.15)$$

Now, with k and γ defined by Equation (8.3.14), one may check that the associated Type I error probability is:

$$\gamma P_{p_0}\{\Sigma_{i=1}^n X_i = k\} + P_{p_0}\{\Sigma_{i=1}^n X_i > k\} = \alpha$$

So, we have a MP level α test. If $\Sigma_{i=1}^n X_i = k$, then one rejects H_0 with probability γ. For example, if $\gamma = 0.135$, then consider three-digit random numbers $000, 001, ..., 134, 135, ..., 999$ and look at a random number table to draw one three-digit number. If we come up with one of the numbers $000, 001, ...$ or 134, then and only then H_0 will be rejected. This is what is called *randomization*. Table 8.3.2 provides k and γ for some specific choices of $n, p_0,$ and α.

Table 8.3.2. Values of k and γ in the Bernoulli Case

$n = 10$	$\alpha = 0.10$		$n = 10$	$\alpha = 0.05$	
p_0	k	γ	p_0	k	γ
0.2	4	0.763	0.2	4	0.195
0.4	6	0.406	0.6	8	0.030
$n = 20$	$\alpha = 0.05$		$n = 25$	$\alpha = 0.10$	
p_0	k	γ	p_0	k	γ
0.3	9	0.031	0.5	16	0.757
0.5	14	0.792	0.6	18	0.330

A reader should verify some of these entries. ▲

> A MP level α test always depends on (jointly) sufficient statistics.

In each example, the reader has noticed that a MP level α test always depended only on (jointly) sufficient statistics. This is no coincidence. Let $\mathbf{T} \equiv \mathbf{T}(X_1, ..., X_n)$ be a (jointly) sufficient statistic for $\boldsymbol{\theta}$. By the Neyman factorization (Theorem 6.2.1), one can split a likelihood function as follows:

$$L(\mathbf{x}; \boldsymbol{\theta}) = g(\mathbf{T}(\mathbf{x}); \boldsymbol{\theta})h(\mathbf{x}) \text{ for all } \mathbf{x} \in \mathcal{X} \qquad (8.3.16)$$

where $h(\mathbf{x})$ does not involve $\boldsymbol{\theta}$. Now, a MP test rejects $H_0 : \boldsymbol{\theta} = \boldsymbol{\theta}_0$ in favor of accepting $H_1 : \boldsymbol{\theta} = \boldsymbol{\theta}_1$ for large values of $L(\mathbf{x}; \boldsymbol{\theta}_1)/L(\mathbf{x}; \boldsymbol{\theta}_0)$. But, we can write

$$L(\mathbf{x}; \boldsymbol{\theta}_1)/L(\mathbf{x}; \boldsymbol{\theta}_0) = g(\mathbf{T}(\mathbf{x}); \boldsymbol{\theta}_1)/g(\mathbf{T}(\mathbf{x}); \boldsymbol{\theta}_0)$$

which implies that a MP test rejects $H_0 : \boldsymbol{\theta} = \boldsymbol{\theta}_0$ in favor of accepting $H_1 : \boldsymbol{\theta} = \boldsymbol{\theta}_1$, for "large" values of $g(\mathbf{T}(\mathbf{x}); \boldsymbol{\theta}_1)/g(\mathbf{T}(\mathbf{x}); \boldsymbol{\theta}_0)$. So, it should not be surprising that all MP tests in the previous examples depended on a (jointly) sufficient statistic $\mathbf{T}$.

8.3.2 *Application: Parameters Are Not Involved*

In the Neyman-Pearson Lemma, we had a real valued parameter θ and a common pmf or pdf indexed by θ. We mentioned earlier that this assumption was not really crucial. As long as H_0, H_1 are simple hypotheses, the Neyman-Pearson Lemma will provide a MP level α test. An example follows.

Example 8.3.6 Suppose X_1, X_2 are iid with a common pdf $f(x)$. Consider two functions defined as follows:

$$f_0(x) = \begin{cases} \frac{3}{64}x^2 & \text{if } 0 < x < 4 \\ 0 & \text{elsewhere} \end{cases} \qquad f_1(x) = \begin{cases} \frac{3}{16}\sqrt{x} & \text{if } 0 < x < 4 \\ 0 & \text{elsewhere} \end{cases}$$

We wish to determine a MP level α test for

$$H_0 : f(x) = f_0(x) \text{ vs. } H_1 : f(x) = f_1(x) \tag{8.3.17}$$

The Neyman-Pearson Lemma applies and a MP test has the form:

$$\text{Reject } H_0 \text{ if and only if } \{f_1(x_1)f_1(x_2)\}/\{f_0(x_1)f_0(x_2)\} > k \tag{8.3.18}$$

which reduces to:

$$\text{Reject } H_0 \text{ if and only if } X_1 X_2 \text{ is ``small'' } (< k) \tag{8.3.19}$$

Let us write $F_0(x)$ for the *distribution function* (df) corresponding to the pdf $f_0(x)$:

$$F_0(x) = \begin{cases} 0 & \text{if } x \le 0 \\ \frac{x^3}{64} & \text{if } 0 < x < 4 \\ 1 & \text{if } x \ge 4 \end{cases}$$

Under H_0, we know that $-2log\{F_0(X_i)\}, i = 1, 2$ is distributed as iid χ_2^2. The test defined via Equation (8.3.19) must have size α, that is, we require:

$$\begin{aligned} \alpha &= P\{X_1 X_2 < k \text{ when } f(x) = f_0(x)\} \\ &= P\left\{\frac{X_1^3}{64}\frac{X_2^3}{64} < k \text{ when } f(x) = f_0(x)\right\} \\ &= P\left[\Sigma_{i=1}^2 \left(-2log\{F_0(X_i)\}\right) > k \text{ when } f(x) = f_0(x)\right] \\ &= P\{\chi_4^2 > k\} \end{aligned}$$

One implements a MP level α test as follows:

$$\text{Reject } H_0 \text{ if and only if } log[2^{24}/(X_1 X_2)^6] > \chi_{4,\alpha}^2 \tag{8.3.20}$$

where $\chi_{4,\alpha}^2$ is the upper $100\alpha\%$ point of a χ_4^2 distribution. ▲

8.3.3   Applications: Observations Are Non-iid

In the Neyman-Pearson Lemma, it was not crucial that $X_1, ..., X_n$ were identically distributed or independent. One merely needed to work with an explicit form of a likelihood function under simple hypotheses, H_0 and H_1. We give two examples.

Example 8.3.7 Let X_1, X_2 be independent random variables respectively distributed as $N(\mu, \sigma^2)$ and $N(\mu, 4\sigma^2)$ with unknown $\mu \in \Re$, but $\sigma^2 \in \Re^+$ is assumed known. Here, we have a situation where X_1, X_2 are independent but not identically distributed. We wish to derive a MP level α test for $H_0 : \mu = \mu_0$ vs. $H_1 : \mu = \mu_1 (> \mu_0)$. The likelihood function is

$$L(\mathbf{x}; \mu) = (4\pi\sigma^2)^{-1} exp\{-\tfrac{1}{2\sigma^2}(x_1 - \mu)^2 - \tfrac{1}{8\sigma^2}(x_2 - \mu)^2\} \qquad (8.3.21)$$

where $\mathbf{x} = (x_1, x_2)$ so that

$$L(\mathbf{x}; \mu_1)/L(\mathbf{x}; \mu_0) = exp\{\tfrac{1}{4\sigma^2}(4x_1 + x_2)(\mu_1 - \mu_0)\} exp\{\tfrac{5}{8\sigma^2}(\mu_1^2 - \mu_0^2)\}$$

The Neyman-Pearson Lemma applies, and we would reject H_0 if and only if $L(\mathbf{x}; \mu_1)/L(\mathbf{x}; \mu_0)$ is "large," that is, if $4X_1 + X_2 > k$. This will be a MP level α test if k is so chosen that $P_{\mu_0}\{4X_1 + X_2 > k\} = \alpha$. But, under H_0, the statistic $4X_1 + X_2$ is distributed as $N(5\mu_0, 20\sigma^2)$. So, we may rephrase a MP level α test as follows:

$$\text{Reject } H_0 \text{ if and only if } \{4X_1 + X_2 - 5\mu_0\}/(\sqrt{20}\sigma) > z_\alpha \qquad (8.3.22)$$

We leave some details out. ▲

Example 8.3.8 Denote $\mathbf{X}' = (X_1, X_2)$ where $\mathbf{X}$ has a $N_2(\mu, \mu, 1, 1, \tfrac{1}{\sqrt{2}})$ distribution with an unknown parameter $\mu \in \Re$. We wish to derive a MP level α test for $H_0 : \mu = \mu_0$ vs. $H_1 : \mu = \mu_1 (> \mu_0)$. The likelihood function is:

$$L(\mathbf{x}; \mu) = (\sqrt{2}\pi)^{-1} exp\{-(x_1 - \mu)^2 + \sqrt{2}(x_1 - \mu)(x_2 - \mu) - (x_2 - \mu)^2\} \quad (8.3.23)$$

so that

$$L(\mathbf{x}; \mu_1)/L(\mathbf{x}; \mu_0) = [exp\{(x_1 + x_2)(\mu_1 - \mu_0)\} exp\{(\mu_0^2 - \mu_1^2)\}]^{2-\sqrt{2}}$$

The Neyman-Pearson Lemma applies, and we would reject H_0 if and only if $L(\mathbf{x}; \mu_1)/L(\mathbf{x}; \mu_0)$ is "large," that is, if $X_1 + X_2 > k$. This will be a MP level α test if k is so chosen that $P_{\mu_0}\{X_1 + X_2 > k\} = \alpha$. Under H_0, the statistic $X_1 + X_2$ is distributed as $N(2\mu_0, 2 + \sqrt{2})$. So, we may rephrase the MP level α test as follows:

$$\text{Reject } H_0 \text{ if and only if } \{X_1 + X_2 - 2\mu_0\}/(\sqrt{2 + \sqrt{2}}) > z_\alpha \qquad (8.3.24)$$

We leave some details out. ▲

8.4 One-Sided Composite Alternative

We introduced the Neyman-Pearson methodology to test a simple null hypothesis vs. a simple alternative hypothesis. Next suppose that we have a simple null hypothesis $H_0 : \theta = \theta_0$, but the alternative hypothesis is $H_1 : \theta > \theta_0$. Here, $H_1 : \theta > \theta_0$ represents an upper-sided composite hypothesis. One could instead have $H_1 : \theta < \theta_0$, which is a lower-sided composite hypothesis. We recall that a UMP level α test of H_0 vs. H_1 will have its power function $Q(\theta)$ satisfying: (i) $Q(\theta_0) \leq \alpha$ and (ii) $Q(\theta)$ is maximized at *every* θ under H_1.

8.4.1 UMP Test via Neyman-Pearson Lemma

We explain this simple approach with the help of examples.

Example 8.4.1 (Example 8.3.1 Continued) We wish to find a UMP level α test for $H_0 : \mu = \mu_0$ vs. $H_1 : \mu > \mu_0$ where μ_0 is a fixed number. Fix $\mu_1(> \mu_0)$ and from Equation (8.3.8) recall that a MP level α test between μ_0 and (arbitrarily chosen) μ_1 will have the form:

$$\text{Reject } H_0 \text{ if and only if } \sqrt{n}(\overline{X} - \mu_0)/\sigma > z_\alpha \qquad (8.4.1)$$

This test maximizes the power at $\mu = \mu_1$, but the test itself does *not* depend on $\mu_1(> \mu_0)$. Hence, the test from Equation (8.4.1) is UMP level α for testing $H_0 : \mu = \mu_0$ vs. $H_1 : \mu > \mu_0$. ▲

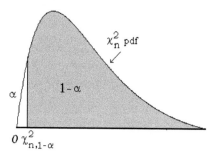

Figure 8.4.1. Chi-square lower $100\alpha\%$ or upper $100(1 - \alpha)\%$ point with the degree of freedom n.

Example 8.4.2 Suppose $X_1, ..., X_n$ are iid $N(0, \sigma^2)$ with unknown $\sigma^2 \in \mathfrak{R}^+$. We wish to find a UMP level α test for $H_0 : \sigma^2 = \sigma_0^2$ vs. $H_1 : \sigma^2 < \sigma_0^2$ where σ_0^2 is a fixed positive number. We first fix $0 < \sigma_1^2(< \sigma_0^2)$ and then check that a MP level α test between σ_0^2 and (arbitrarily chosen) σ_1^2 will have the form:

$$\text{Reject } H_0 \text{ if and only if } \Sigma_{i=1}^n X_i^2 < \sigma_0^2 \chi_{n,1-\alpha}^2 \qquad (8.4.2)$$

where $\chi_{n,1-\alpha}^2$ is the lower $100\alpha\%$ point of a χ_n^2 distribution. See Figure 8.4.1. This test maximizes the power at $\sigma^2 = \sigma_1^2$, but the test itself does *not* depend on $\sigma_1^2(< \sigma_0^2)$. Hence, the test from Equation (8.4.2) is UMP level α for testing $H_0 : \sigma^2 = \sigma_0^2$ vs. $H_1 : \sigma^2 < \sigma_0^2$. ▲

P-Value of a Test:

There is another way to report the final outcome using the *P-value*. We emphasize that the *P-value* is a data-driven quantification of the "strength" of support in favor of H_0, but its interpretation is sometimes difficult.

Consider Example 8.4.1. Suppose that we have the observed value $\overline{x}$ for the sample mean and $z_{calc} \equiv \sqrt{n}(\overline{x} - \mu_0)/\sigma$, the calculated value of the *test statistic*, $\sqrt{n}(\overline{X} - \mu_0)/\sigma$.

Now, one may ask the following question:

> What is the probability of observing data which are similar to those on hand or *more extreme*, if H_0 happened to be true? This probability will be called the *P-value*.

Since the alternative hypothesis is $\mu > \mu_0$, the phrase *more extreme* is interpreted as "$\overline{X} > \overline{x}$" and hence the *P-value* is given by:

$$P_{\mu_0}\{\sqrt{n}(\overline{X} - \mu_0)/\sigma > z_{calc}\} = P\{Z > z_{calc}\}$$

which can be obtained from a standard normal table.

If the alternative hypothesis was $\mu < \mu_0$, then the phrase *more extreme* would be interpreted as "$\overline{X} < \overline{x}$" and the *P-value* will be:

$$P_{\mu_0}\{\sqrt{n}(\overline{X} - \mu_0)/\sigma < z_{calc}\} = P\{Z < z_{calc}\}$$

A test with a "small" *P-value* indicates that a null hypothesis is less plausible than the alternative hypothesis, and in that case H_0 is rejected.

> A test would reject null hypothesis H_0 at chosen level α *if and only if* the associated *P-value* $< \alpha$.

Example 8.4.3 (Example 8.4.1 Continued) Suppose $X_1, ..., X_{15}$ are iid $N(\mu, \sigma^2)$ with unknown $\mu \in \Re$ but $\sigma^2 = 9$. We wish to test $H_0 : \mu = 3.1$ vs. $H_1 : \mu > 3.1$ at 5% level so that $z_\alpha = 1.645$. Observed data gave $\overline{x} = 5.3$ so that $z_{calc} = \sqrt{15}(5.3 - 3.1)/3 \approx 2.8402$. Hence, we will reject H_0 at 5% level since $z_{calc} > z_\alpha$. The associated *P-value* is $P\{Z > 2.8402\} \approx 0.0022543$. If we were told that the *P-value* was 0.0022543, then we would have immediately rejected H_0 at any level $\alpha > 0.0022543$. ▲

8.4.2 UMP Test via Monotone Likelihood Ratio Property

Definition 8.4.1 *A family of distributions $\{f(x; \theta): \theta \in \Theta\}$ is said to have a monotone likelihood ratio (MLR) property in a real valued statistic $T \equiv T(\mathbf{X})$ provided that the following holds: for all $\{\theta^*, \theta\} \subset \Theta$ and $\mathbf{x} \in \mathcal{X}$, we have*

$$\frac{L(\mathbf{x}; \theta^*)}{L(\mathbf{x}; \theta)} \text{ is nondecreasing in } T(\mathbf{x}) \text{ whenever } \theta^* > \theta \qquad (8.4.3)$$

Remark 8.4.1 In this definition, all we need is that $L(\mathbf{x}; \theta^*)/L(\mathbf{x}; \theta)$ should be *nonincreasing or nondecreasing in* $T(\mathbf{x})$ *whenever* $\theta^* > \theta$. If the likelihood ratio $L(\mathbf{x}; \theta^*)/L(\mathbf{x}; \theta)$ is nonincreasing instead of nondecreasing, then its primary effect would be felt in the placement of a rejection region $\mathcal{R}$. The statistic T would invariably involve sufficient statistics for the unknown parameter θ.

Example 8.4.4 Let $X_1, ..., X_n$ be iid $N(\mu, \sigma^2)$ with unknown $\mu \in \Re$, but $\sigma^2 \in \Re^+$ is known. Consider arbitrary real numbers $\mu, \mu^*(> \mu)$, and then with $T(\mathbf{x}) = \Sigma_{i=1}^n x_i$, write

$$L(\mathbf{x}; \mu^*)/L(\mathbf{x}; \mu) = exp\{[(\mu^* - \mu)T(\mathbf{x})/\sigma^2] + [n(\mu^2 - \mu^{*2})/(2\sigma^2)]\}$$

which is an increasing function in T. Then, we have a MLR (increasing) property in T. ▲

Example 8.4.5 Suppose $X_1, ..., X_n$ are iid with a common pdf

$$f(x; b) = be^{-bx}I(x > 0)$$

where $b \in \Re^+$ is unknown. Consider arbitrary real numbers $0 < b < b^*$ and with $T(\mathbf{x}) = \Sigma_{i=1}^n x_i$, write

$$L(\mathbf{x}; b^*)/L(\mathbf{x}; b) = (b^*/b)^n exp\{-(b^* - b)T(\mathbf{x})\}$$

which decreases in T. So, we have a MLR (decreasing) property in T. ▲

Let $T \equiv T(\mathbf{X})$ be a real valued sufficient statistic for θ and denote its family of pmf or pdf $\{g(t; \theta): \theta \in \Theta \subseteq \Re\}$. The domain for T is $\mathcal{T} \subseteq \Re$. Suppose that $g(t; \theta)$ belongs to a one-parameter exponential family:

$$g(t; \theta) = a(\theta)c(t)e^{tb(\theta)}, \theta \in \Theta, t \in \mathcal{T} \qquad (8.4.4)$$

where $b(\theta)$ is a nonincreasing (nondecreasing) function of θ. Then, $\{g(t; \theta): \theta \in \Theta \subseteq \Re\}$ has the MLR nonincreasing (nondecreasing) property in T. Here, $a(\theta), b(\theta)$ cannot involve t and $c(t)$ cannot involve θ. In many distributions involving a single real valued unknown parameter θ, including binomial, Poisson, normal, and gamma, the family of pmf or pdf of the associated sufficient statistic T would have a MLR property.

The following useful result is due to Karlin and Rubin (1956). We state it without giving its proof.

Theorem 8.4.1 (Karlin-Rubin Theorem) *We wish to test* $H_0 : \theta \leq \theta_0$ *vs.* $H_1 : \theta > \theta_0$. *Consider a real valued sufficient statistic* $T = T(\mathbf{X})$ *for* $\theta \in \Theta(\subseteq \Re)$. *Suppose that the family* $\{g(t; \theta) : \theta \in \Theta\}$ *of pmf or pdf induced by* T *has a MLR (nondecreasing) property. Then, a test function*

$$\psi(\mathbf{X}) = \begin{cases} 1 & \text{if } T(\mathbf{X}) > k \\ 0 & \text{if } T(\mathbf{X}) < k \end{cases} \qquad (8.4.5)$$

corresponds to a UMP level α test if k is so chosen that $E_{\theta_0}[\psi(\mathbf{X})] = \alpha$.

Example 8.4.6 (Example 8.4.1 Continued) We know that $T(\mathbf{X}) = \Sigma_{i=1}^n X_i$ is sufficient for μ and its pdf has a MLR increasing property in T. Now, we may wish to test $H_0 : \mu \le \mu_0$ vs. $H_1 : \mu > \mu_0$ with level α where μ_0 is a fixed number. By the Karlin-Rubin Theorem, a UMP level α test will be:

$$\psi(\mathbf{x}) = \begin{cases} 1 & \text{if } T(\mathbf{x}) > k \\ 0 & \text{if } T(\mathbf{x}) \le k \end{cases}$$

or equivalently

$$\psi(\mathbf{x}) = \begin{cases} 1 & \text{if } \sqrt{n}(\overline{X} - \mu_0)/\sigma > z_\alpha \\ 0 & \text{if } \sqrt{n}(\overline{X} - \mu_0)/\sigma \le z_\alpha \end{cases}$$

That is, the Type I error probability at a boundary point $\mu = \mu_0$ in the null space is exactly α. One may check directly that the same for any $\mu < \mu_0$ is smaller than α. ▲

8.5   Simple Null vs. Two-Sided Alternative

Consider testing a simple null hypothesis $H_0 : \theta = \theta_0$ vs. a two-sided alternative hypothesis $H_1 : \theta \ne \theta_0$ where θ_0 is a fixed value in the parameter space Θ. Will there exist a UMP level α test? The answer is "yes" in some situations and "no" in others. We refrain from going into general discussions of what may or may not happen. To get the important point across, we include examples. We also briefly touch upon *unbiased* tests.

8.5.1   *An Example Where the UMP Test Does Not Exist*

Suppose that $X_1, ..., X_n$ are iid $N(\mu, \sigma^2)$ with unknown $\mu \in \Re$ but known $\sigma \in \Re^+$. We wish to test a simple null hypothesis $H_0 : \mu = \mu_0$ against a two-sided alternative $H_1 : \mu \ne \mu_0$ with level α where μ_0 is a fixed number. In this problem, there exists *no* UMP level α test. For a detailed proof, one may look at Mukhopadhyay (2000, pp. 425-426).

8.5.2   *An Example Where the UMP Test Exists*

Suppose that $X_1, ..., X_n$ are iid Uniform$(0, \theta)$ with unknown $\theta \in \Re^+$. We wish to test a simple null hypothesis $H_0 : \theta = \theta_0$ against a two-sided alternative hypothesis $H_1 : \theta \ne \theta_0$ with level α where θ_0 is a fixed positive number. This is an exercise from Lehmann (1986, p. 111). In this problem, there *exists* a UMP level α test. For a detailed proof, one may look at Mukhopadhyay (2000, pp. 426-428). Incidentally, the test associated with

$$\psi(\mathbf{X}) = \begin{cases} 1 & \text{if } X_{n:n} \ge \theta_0 \text{ or } X_{n:n} \le \theta_0 \alpha^{1/n} \\ 0 & \text{otherwise} \end{cases} \tag{8.5.1}$$

is UMP level α.

8.5.3   Unbiased and UMP Unbiased Tests

If a UMP test fails to exist, one often considers a class of tests by restricting attention to *unbiased* tests, and then finds the best test within this class.

Definition 8.5.1 *Consider testing $H_0 : \theta \in \Theta_0$ vs. $H_1 : \theta \in \Theta_1$ where Θ_0, Θ_1 are subsets of Θ and Θ_0, Θ_1 are disjoint. A level α test is called unbiased if $Q(\theta)$, the power of the test (that is the probability of rejecting H_0 under θ) is at least α whenever $\theta \in \Theta_1$.*

A test with its test function $\psi(\mathbf{X})$ is unbiased level α provided that the following conditions hold:

$$
\begin{aligned}
(i) & \quad E_\theta\{\psi(\mathbf{X})\} \leq \alpha \text{ for all } \theta \in \Theta_0 \\
(ii) & \quad E_\theta\{\psi(\mathbf{X})\} \geq \alpha \text{ for all } \theta \in \Theta_1
\end{aligned}
\tag{8.5.2}
$$

The properties in Equation (8.5.2) state that an unbiased test rejects a null hypothesis more frequently when the alternative hypothesis is true than in a situation when the null hypothesis is true. This is a fairly minimal demand on any reasonable test used in practice. Now, one may set out to locate the UMP level α test within the class of level α unbiased tests. Such a test will be called a *uniformly most powerful unbiased (UMPU)* level α test. An elaborate theory of unbiased tests was given by Neyman as well as by Neyman and Pearson in a series of papers. Refer to Lehmann (1986).

From Section 8.5.1, one knows, for example, that there is no UMP level α test for a simple null against a two-sided alternative about the mean of a normal distribution with known variance. In Chapter 11, we will find a level α *likelihood ratio test* (LRT) for the same problem which would:

$$
\text{Reject } H_0 \text{ if and only if } \sqrt{n} \, | \, \overline{X} - \mu_0 \, | \, /\sigma > z_{\alpha/2}
\tag{8.5.3}
$$

This is the customary two-tailed Z-test which *is* indeed UMPU. The proof is out of scope, but the readers may be made aware of this result.

8.6   Exercises and Complements

8.2.1 Suppose that X_1, X_2, X_3, X_4 are iid from a $N(\theta, 4)$ population where $\theta (\in \Re)$ is unknown. Test $H_0 : \theta = 2$ vs. $H_1 : \theta = 5$ and consider

> Test #1: Reject H_0 if and only if $X_1 > 4.7$
>
> Test #2: Reject H_0 if and only if $\frac{1}{3}(X_1 + 2X_2) > 4.5$
>
> Test #3: Reject H_0 if and only if $\frac{1}{2}(X_1 + X_3) > 4.2$
>
> Test #4: Reject H_0 if and only if $\overline{X} > 4.1$

Find Type I and II error probabilities for each test and compare.

8.2.2 Suppose that X_1, X_2, X_3, X_4 are iid from a population having an exponential distribution with an unknown mean $\theta(\in \Re^+)$. Test $H_0 : \theta = 6$ vs. $H_1 : \theta = 2$ and consider

> Test #1: Reject H_0 if and only if $X_1 < 4$
>
> Test #2: Reject H_0 if and only if $\frac{1}{2}(X_1 + X_2) < 3.5$
>
> Test #3: Reject H_0 if and only if $\frac{1}{3}(X_1 + X_2 + X_3) < 3.4$
>
> Test #4: Reject H_0 if and only if $\overline{X} < 2.8$

Find the Type I and II error probabilities for each test and compare.

8.2.3 Suppose that X_1, X_2 are iid with a common pdf:

$$f(x; \theta) = \begin{cases} \theta x^{\theta-1} & \text{if } 0 < x < 1 \\ 0 & \text{elsewhere} \end{cases}$$

where $\theta(> 0)$ is unknown. In order to test $H_0 : \theta = 1$ vs. $H_1 : \theta = 2$, consider a critical region

$$\mathcal{R} = \{(x_1, x_2) \in (0,1) \times (0,1) : x_1 x_2 \geq \tfrac{3}{4}\}$$

(i)   Show that the level $\alpha = \frac{1}{4} + \frac{3}{4} log(\frac{3}{4})$;

(ii)   Show that the power at $\theta = 2$ is $\frac{7}{16} + \frac{9}{8} log(\frac{3}{4})$.

{*Hints*: $\alpha = P_{\theta=1}\{X_1 X_2 \geq \frac{3}{4}\}$ and power is $P_{\theta=2}(\mathcal{R}) = P_{\theta=2}\{X_1 X_2 \geq \frac{3}{4}\}$.}

8.3.1 Suppose $X_1, ..., X_n$ are iid $N(\mu, \sigma^2)$ where μ is known but σ is unknown, $\mu \in \Re, \sigma \in \Re^+$. Fix $\alpha \in (0,1)$ and two positive numbers σ_0, σ_1.

(i)   Derive the MP level α test for $H_0 : \sigma = \sigma_0$ vs. $H_1 : \sigma = \sigma_1$ ($> \sigma_0$) in the simplest implementable form;

(ii)   Derive the MP level α test for $H_0 : \sigma = \sigma_0$ vs. $H_1 : \sigma = \sigma_1$ ($< \sigma_0$) in the simplest implementable form.

8.3.2 (Example 8.3.3 Continued) With preassigned $\alpha \in (0,1)$ and two positive numbers $\theta_1 < \theta_0$, derive the MP level α test for $H_0 : \theta = \theta_0$ vs. $H_1 : \theta = \theta_1$ in the simplest implementable form. Perform power calculations.

8.3.3 (Example 8.3.5 Continued) With preassigned $\alpha \in (0,1)$, derive a randomized MP level α test for $H_0 : p = p_0$ vs. $H_1 : p = p_1$ where $p_1 < p_0$ are two numbers in $(0,1)$. Explicitly find k and γ when $n = 15, \alpha = 0.10$, and $p_0 = 0.1, 0.5, 0.7$.

8.3.4 Let $X_1, ..., X_n$ be iid Poisson(λ) where $\lambda \in \Re^+$ is unknown. With preassigned $\alpha \in (0,1)$, derive a randomized MP level α test for $H_0 : \lambda = \lambda_0$ vs. $H_1 : \lambda = \lambda_1$ where $\lambda_1 > \lambda_0$ are two positive numbers. Explicitly find k and γ numerically when $n = 10, \alpha = 0.05$ and $\lambda_0 = 0.15$.

8.3.5 Suppose that X is an observable random variable with its pdf $f(x), x \in \Re$. Consider functions defined as follows:

$$f_0(x) = \begin{cases} 1 & \text{if } 0 < x < 1 \\ 0 & \text{elsewhere} \end{cases} \qquad f_1(x) = \begin{cases} 3x^2 & \text{if } 0 < x < 1 \\ 0 & \text{elsewhere} \end{cases}$$

Determine a MP level α test for

$$H_0 : f(x) = f_0(x) \text{ vs. } H_1 : f(x) = f_1(x)$$

in the simplest implementable form. Perform power calculations. {*Hint*: Follow Example 8.3.6.}

8.3.6 Suppose that X is an observable random variable with its pdf $f(x), x \in \Re$. Consider functions defined as follows:

$$f_0(x) = \tfrac{1}{\pi}(1 + x^2)^{-1}, \;\; f_1(x) = \tfrac{1}{2} exp\{-|x|\}$$

Show that a MP level α test for

$$H_0 : f(x) = f_0(x) \text{ vs. } H_1 : f(x) = f_1(x)$$

will reject H_0 if and only if $|X| < k$. Determine k as a function of α. Perform power calculations.

8.3.7 Suppose that $X_1, ..., X_n$ are iid Uniform$(-\theta, \theta)$ with unknown $\theta(> 0)$. Test $H_0 : \theta = \theta_0$ vs. $H_1 : \theta = \theta_1(> \theta_0)$ with positive numbers θ_0, θ_1. Also, suppose that we reject H_0 if and only if $|X|_{n:n} > c$ where $|X|_{n:n} = max_{1 \le i \le n} |X_i|$ and c is a fixed positive number. Find $\alpha \in (0, 1)$ for which this test is MP level α.

8.3.8 Suppose $X_1, ..., X_n$ are iid with a common Rayleigh pdf:

$$f(x; \theta) = 2\theta^{-1} x \, exp(-x^2/\theta) I(x > 0)$$

where $\theta(> 0)$ is unknown.. With preassigned $\alpha \in (0, 1)$, derive a MP level α test for $H_0 : \theta = \theta_0$ vs. $H_1 : \theta = \theta_1(< \theta_0)$ where θ_0, θ_1 are positive numbers, in the simplest implementable form.

8.3.9 Suppose that X_1 and X_2 are independent random variables respectively distributed as $N(\mu, \sigma^2), N(3\mu, 2\sigma^2)$ where $\mu \in \Re$ is unknown and $\sigma \in \Re^+$ is assumed known. Derive a MP level α test for $H_0 : \mu = \mu_0$ vs. $H_1 : \mu = \mu_1(< \mu_0)$. Evaluate the power of the test.

8.3.10 (Example 8.3.8 Continued) Derive a MP level α test for $H_0 : \mu = \mu_0$ vs. $H_1 : \mu = \mu_1(< \mu_0)$. Evaluate the power of this test.

8.3.11 Suppose $X_1, ..., X_n$ are iid with a common pdf:

$$f(x; \mu) = [x\sigma\sqrt{2\pi}]^{-1} \; exp\{-[log(x) - \mu]^2/(2\sigma^2)\} I(x > 0)$$

and $-\infty < \mu < \infty, 0 < \sigma < \infty$. Here, μ is the only unknown parameter. Given $\alpha \in (0, 1)$, find a MP level α test in the simplest implementable form for $H_0 : \mu = \mu_0$ vs. $H_1 : \mu = \mu_1(> \mu_0)$ with μ_0, μ_1 fixed real numbers.

8.4.1 Suppose $X_1, ..., X_n$ are iid having a common Laplace pdf:

$$f(x; b) = \tfrac{1}{2}b^{-1}exp(-|x - a|/b)I(x \in \Re)$$

where $b(> 0)$ is unknown but $a(\in \Re)$ is assumed known. Show that the family has a MLR increasing property in T, a sufficient statistic for b.

8.4.2 Suppose $X_1, ..., X_n$ are iid with a common Weibull pdf:

$$f(x; a) = a^{-1}bx^{b-1}exp(-x^b/a)I(x > 0)$$

where $a(> 0)$ is unknown but $b(> 0)$ is assumed known. Show that the family has a MLR increasing property in T, a sufficient statistic for a.

8.4.3 Let $X_1, ..., X_n$ be iid Poisson(λ) with unknown $\lambda \in \Re^+$. Show that the family has a MLR increasing property in T, a sufficient statistic.

8.4.4 Suppose $X_1, ..., X_n$ are iid Uniform$(-\theta, \theta)$ with unknown $\theta(> 0)$. Show that the family has a MLR increasing property in $T = |X|_{n:n}$, a sufficient statistic for θ, where $|X|_{n:n} = \max_{1 \leq i \leq n} |X_i|$.

8.4.5 Suppose $X_1, ..., X_n$ are iid $N(\mu, \sigma^2)$ where μ is assumed known but σ^2 is unknown, $\mu \in \Re, \sigma^2 \in \Re^+$. We have fixed numbers $\alpha \in (0, 1)$ and $\sigma_0^2(> 0)$. Derive a UMP level α test for $H_0 : \sigma^2 \geq \sigma_0^2$ vs. $H_1 : \sigma^2 < \sigma_0^2$ in the simplest implementable form.

   (i)   Use the Neyman-Pearson approach;

   (ii)   Use the MLR approach.

8.4.6 Suppose $X_1, ..., X_n$ are iid $N(\mu, \sigma^2)$ where μ is unknown but σ^2 is assumed known, $\mu \in \Re, \sigma^2 \in \Re^+$. In order to test $H_0 : \mu \geq \mu_0$ vs. $H_1 : \mu < \mu_0$, suppose that we reject H_0 if and only if $\overline{X} < c$ where c is a fixed number. Determine $\alpha \in (0, 1)$ for which this test is UMP level α.

8.4.7 Suppose $X_1, ..., X_n$ are iid Geometric(p) where $p \in (0, 1)$ is unknown. With preassigned $\alpha \in (0, 1)$, derive a randomized UMP level α test for $H_0 : p \geq p_0$ vs. $H_1 : p < p_0$ where p_0 is a number between 0 and 1.

8.4.8 Suppose $X_1, ..., X_n$ are iid with a common Rayleigh pdf:

$$f(x; \theta) = 2\theta^{-1}x\,exp(-x^2/\theta)I(x > 0)$$

where $\theta(> 0)$ is unknown. With preassigned $\alpha \in (0, 1)$, derive a UMP level α test for $H_0 : \theta \geq \theta_0$ vs. $H_1 : \theta < \theta_0$ where θ_0 is a positive number, in the simplest implementable form.

   (i)   Use the Neyman-Pearson approach;

   (ii)   Use the MLR approach.

8.4.9 (Exercise 8.3.11 Continued) Given $\alpha \in (0, 1)$, derive a UMP level α test for testing $H_0 : \mu = \mu_0$ vs. $H_1 : \mu < \mu_0$ where μ_0 is a fixed number. Describe the test in its simplest implementable form.

8.4.10 Suppose $X_1, ..., X_n$ are iid with a common pdf:

$$f(x; \theta) = \theta^{-1}x^{(1-\theta)/\theta}I(0 < x < 1)$$

where $0 < \theta < \infty$ is unknown. Given $\alpha \in (0,1)$, derive a UMP level α test in its simplest implementable for $H_0 : \theta \leq \theta_0$ vs. $H_1 : \theta > \theta_0$ where θ_0 is a fixed positive number.

8.4.11 (Sample Size Determination) Let $X_1, ..., X_n$ be iid $N(\mu, \sigma^2)$ where μ is assumed unknown but σ is known, $\mu \in \Re, \sigma \in \Re^+$. Given $\alpha \in (0,1)$, a UMP level α test for $H_0 : \mu = \mu_0$ vs. $H_1 : \mu > \mu_0$

$$\text{rejects } H_0 \text{ if and only if } \sqrt{n}(\overline{X} - \mu_0)/\sigma > z_\alpha$$

The UMP test makes sure that it has *minimum* possible Type II error probability at $\mu = \mu_1 (> \mu_0)$ among all level α tests, but there is *no* guarantee that this minimum Type II error probability at $\mu = \mu_1$ will be "small" unless n is appropriately determined.

We require the UMP test to have Type II error probability $\leq \beta \in (0,1)$ when $\mu = \mu_1 (> \mu_0)$. Show that the sample size

$$n \text{ must be the smallest integer} \geq \{(z_\alpha + z_\beta)\sigma/(\mu_1 - \mu_0)\}^2$$

8.5.1 Denote the lognormal pdf:

$$f(w; \mu, \sigma) = [w\sigma\sqrt{2\pi}]^{-1} exp\left\{-[log(w) - \mu]^2/(2\sigma^2)\right\}$$

with $w > 0, -\infty < \mu < \infty$, and $0 < \sigma < \infty$. Suppose that $X_1, ..., X_m$ are iid positive random variables having a common pdf $f(x; \mu, 2)$, $Y_1, ..., Y_n$ are iid positive random variables having a common pdf $f(y; 2\mu, 3)$, and also X, Y are independent. Here, μ is unknown and $m \neq n$. Argue whether there does or does not exist a UMP level α test for $H_0 : \mu = 1$ vs. $H_1 : \mu \neq 1$. {*Hint:* Try to use the result from Section 8.5.1.}

8.5.2 Suppose X_1, X_2, X_3, X_4 are iid Uniform$(0, \theta)$ where $\theta > 0$. To test $H_0 : \theta = 1$ vs. $H_1 : \theta \neq 1$, we propose to use the critical region

$$\mathcal{R} = \{\mathbf{X} \in \Re^{+4} : X_{4:4} < \tfrac{1}{2} \text{ or } X_{4:4} > 1\}$$

Evaluate the associated level α and the power function.

9

Confidence Intervals

9.1 Introduction

As the name *confidence interval* suggests, we set out to explore methods to estimate an unknown parameter $\theta \in \Theta \subseteq \Re$ with the help of an interval. That is, we will determine two statistics $T_L(\mathbf{X}), T_U(\mathbf{X})$ based on the data $\mathbf{X}$ and propose the interval $(T_L(\mathbf{X}), T_U(\mathbf{X}))$ as a final estimator of θ. If $T_L(\mathbf{X})$ is $-\infty$, the interval $(-\infty, T_U(\mathbf{X}))$ is called an *upper* confidence interval for θ. But, if $T_U(\mathbf{X})$ is ∞, the interval $(T_L(\mathbf{X}), \infty)$ is called a *lower* confidence interval for θ.

> We prefer reporting both lower and upper end points of a confidence interval J, namely $T_L(\mathbf{X})$ and $T_U(\mathbf{X})$, that depend only on a (minimal) sufficient statistic for θ.

Definition 9.1.1 *Coverage probability associated with the confidence interval,* $J = (T_L(\mathbf{X}), T_U(\mathbf{X}))$, *for* θ *is measured by:*

$$P_\theta\{\theta \in (T_L(\mathbf{X}), T_U(\mathbf{X}))\} \qquad (9.1.1)$$

The confidence coefficient associated with J *is defined as:*

$$\min_{\theta \in \Theta} P_\theta\{\theta \in (T_L(\mathbf{X}), T_U(\mathbf{X}))\} \qquad (9.1.2)$$

However, the coverage probability, $P_\theta\{\theta \in (T_L(\mathbf{X}), T_U(\mathbf{X}))\}$, may not involve θ in many standard applications. In those situations, the confidence coefficient will coincide with the coverage probability itself. Thus, we will use these phrases interchangeably.

Before we proceed any further, we add some historical perspectives. The concepts of both "fiducial distribution" and "fiducial intervals" originated with Fisher (1930), which led to persistent and substantial philosophical arguments. Among others, Neyman came down hard on Fisher on philosophical grounds and proceeded to give the foundation of the theory of confidence intervals. This culminated in Neyman's (1935b, 1937) two path-breaking papers on this subject. After 1937, neither Neyman nor Fisher swayed from their respective philosophical stance. However, in the 1961 article, *Silver jubilee of my dispute with Fisher*, Neyman was kinder to Fisher in his exposition. It may not be out of place to note that Fisher died on July 29, 1962. Buehler (1980), Lane (1980), and Wallace (1980) gave important accounts of fiducial distributions. It looks like some researchers are recently reinventing Fisher's fiducial arguments in the name of "implicit distributions." One may refer to Mukhopadhyay (2006) for some pertinent comments.

Customarily, one first fixes a small preassigned number $\alpha \in (0,1)$ and looks for a confidence interval for θ with confidence coefficient $(1-\alpha)$. We refer to such an interval as $(1-\alpha)$ or $100(1-\alpha)\%$ *confidence interval*.

Example 9.1.1 Suppose X is distributed as $N(\mu, 1)$ where $\mu(\in \Re)$ is unknown. Let $T_L(X) = X - 1.96, T_U(X) = X + 1.96$, leading to

$$J = (X - 1.96, X + 1.96) \tag{9.1.3}$$

The coverage probability is:

$$P_\mu \{X - 1.96 < \mu < X + 1.96\}$$
$$= P\{|Z| < 1.96\}, \text{ where } Z = X_1 - \mu \text{ is distributed}$$
$$\text{as } N(0,1) \text{ if } \mu \text{ is the true population mean,}$$

which is 0.95 and it does not depend upon μ. See Figure 9.1.1. Clearly, $z_{0.025} = 1.96$. So, the confidence coefficient associated with J is 0.95. ▲

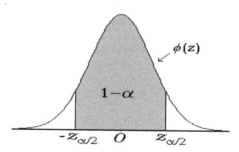

Figure 9.1.1. Standard normal pdf: the shaded area between $-z_{\alpha/2}$ and $z_{\alpha/2}$ with $z_{\alpha/2} = 1.96$ is $1 - \alpha$ where $\alpha = 0.05$.

The basic methodology involves what we call a *pivotal approach* and it is both flexible and versatile. In Section 9.2, we discuss one-sample problems. An example of simultaneous confidence intervals (Example 9.2.6) is included! We provide an interpretation of a confidence coefficient and a notion of *accuracy measure*. Section 9.3 includes some two-sample problems.

9.2   One-Sample Problems

Let $X_1, ..., X_n$ be *independent and identically distributed* (iid) real valued random variables with a common *probability density function* (pdf) $f(x; \theta)$ for $x \in \mathcal{X}$ where $\theta(\in \Theta)$ is unknown. Let $T \equiv T(\mathbf{X})$ be a real valued (minimal) sufficient statistic.

Let $\{g(t; \theta); t \in \mathcal{T}, \theta \in \Theta\}$ be the family of pdfs induced by T. In many applications, we will see that $g(t; \theta)$ will belong to a location, scale, or location-scale family of distributions which were introduced in Section 6.5.1.

> *Location Case:* With some $a(\theta)$, the distribution of $\{T - a(\theta)\}$ will not involve θ for any $\theta \in \Theta$.

> *Scale Case:* With some $b(\theta) > 0$, the distribution of $\dfrac{T}{b(\theta)}$ will not involve θ for any $\theta \in \Theta$.

> *Location-Scale Case:* With some $a(\theta), b(\theta) > 0$, the distribution of $\dfrac{T - a(\theta)}{b(\theta)}$ will not involve θ for any $\theta \in \Theta$.

Definition 9.2.1 *A pivot is a random variable* $\mathbf{U}$ *which functionally involves both (minimal) sufficient statistic* $\mathbf{T}$ *and* $\boldsymbol{\theta}$, *but the distribution of* $\mathbf{U}$ *does not involve* $\boldsymbol{\theta}$ *for any* $\boldsymbol{\theta} \in \Theta$.

In a location, scale, or a location-scale situation, when U, T, and θ are real valued, the customary pivots are often multiples of $\{T - a(\theta)\}, T/b(\theta)$ or $\{T - a(\theta)\}/b(\theta)$ respectively with suitable expressions of $a(\theta)$ and $b(\theta)$. Consider the following examples.

> One hopes that a pivotal distribution will coincide with a known distribution so that a standard statistical table may be used to find appropriate percentiles.

Example 9.2.1 Suppose X has its pdf $f(x; \theta) = \theta^{-1} exp\{-x/\theta\}I(x > 0)$ where $\theta(> 0)$ is unknown. Given $\alpha \in (0, 1)$, we wish to construct a $(1 - \alpha)$ confidence interval for θ. We know that X is minimal sufficient for θ and its pdf belongs to a scale family. Note that the pdf of $U \equiv X/\theta$ is $g(u) = e^{-u}I(u > 0)$ and find positive numbers $a < b$ such that $P(U < a) = P(U > b) = \frac{1}{2}\alpha$ so that $P(a < U < b) = 1 - \alpha$. Check that $a = -log(1 - \frac{1}{2}\alpha)$ and $b = -log(\frac{1}{2}\alpha)$. Here, U is a pivot and observe that

$$P(a < U < b) = 1 - \alpha \Rightarrow P_\theta\{\theta \in (b^{-1}X, a^{-1}X)\} = 1 - \alpha$$

so that $J = (b^{-1}X, a^{-1}X)$ is a $(1 - \alpha)$ confidence interval for θ. ▲

Example 9.2.2 Suppose $X_1, ..., X_n$ are iid Uniform$(0, \theta)$ where $\theta(> 0)$ is an unknown parameter. Given $\alpha \in (0, 1)$, we wish to construct a $(1 - \alpha)$ confidence interval for θ. We know that $T \equiv X_{n:n}$ is minimal sufficient for θ and its pdf belongs to a scale family. The pdf of $U \equiv T/\theta$ is $g(u) = nu^{n-1}I(0 < u < 1)$. We find $0 < a < b < 1$ such that $P(U < a) = P(U > b) = \frac{1}{2}\alpha$ so that $P(a < U < b) = 1 - \alpha$. Check that $a = (\frac{1}{2}\alpha)^{1/n}$ and $b = (1 - \frac{1}{2}\alpha)^{1/n}$. Here, U is a pivot and observe that

$$P(a < U < b) = 1 - \alpha \Rightarrow P_\theta\{\theta \in (b^{-1}T, a^{-1}T)\} = 1 - \alpha$$

which shows that $J = (b^{-1}X_{n:n}, a^{-1}X_{n:n})$ is a $(1 - \alpha)$ confidence interval for θ. ▲

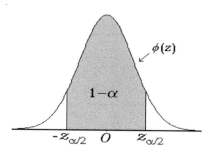

Figure 9.2.1. Standard normal pdf: the area on the right (or left) of $z_{\alpha/2}$ (or $-z_{\alpha/2}$) is $\alpha/2$.

Example 9.2.3 (Mean: Normal with Unknown μ and Known σ) Suppose $X_1, ..., X_n$ are iid $N(\mu, \sigma^2)$ with unknown $\mu \in \Re$. Assume that $\sigma^2 \in \Re^+$ is known. Given $\alpha \in (0, 1)$, we wish to construct a $(1 - \alpha)$ confidence interval for μ. The statistic $T \equiv \overline{X}$ is minimal sufficient for μ and it has a $N(\mu, \frac{1}{n}\sigma^2)$ distribution which belongs to a location family. The pivot ought to be $U \equiv \sqrt{n}(T - \mu)/\sigma$ which has a standard normal distribution for all μ. See Figure 9.2.1. We have $P\{-z_{\alpha/2} < U < z_{\alpha/2}\} = 1 - \alpha$ which implies:

$$P_\mu \left\{ T - z_{\alpha/2}n^{-1/2}\sigma < \mu < T + z_{\alpha/2}n^{-1/2}\sigma \right\} = 1 - \alpha$$

In other words,

$$\left(\overline{X} - z_{\alpha/2}n^{-1/2}\sigma, \overline{X} + z_{\alpha/2}n^{-1/2}\sigma \right) \qquad (9.2.1)$$

is a $(1 - \alpha)$ confidence interval for μ. ▲

Example 9.2.4 (Mean: Normal with Unknown μ, σ) Suppose that $X_1, ..., X_n$ are iid $N(\mu, \sigma^2)$ with unknown $\mu \in \Re$ and $\sigma^2 \in \Re^+, n \geq 2$. Given $\alpha \in (0, 1)$, we wish to construct a $(1 - \alpha)$ confidence interval for μ. Let $\overline{X}, S^2$ respectively be the sample mean and variance. The statistic $T \equiv (\overline{X}, S^2)$ is minimal sufficient for (μ, σ^2). Here, the distribution of $\overline{X}$ belongs to a location-scale family. The pivot ought to be $U \equiv \sqrt{n}(\overline{X} - \mu)/S$ which has a t_{n-1} distribution for all μ, σ^2. So, we have $P\{-t_{n-1,\alpha/2} < U < t_{n-1,\alpha/2}\} = 1 - \alpha$ where $t_{n-1,\alpha/2}$ is the upper $50\alpha\%$ point of Student's t_{n-1} distribution. See Figure 9.2.2

Thus, we have:

$$P_{\mu,\sigma^2} \left\{ \overline{X} - t_{n-1,\alpha/2}n^{-1/2}S < \mu < \overline{X} + t_{n-1,\alpha/2}n^{-1/2}S \right\} = 1 - \alpha$$

so that

$$\left(\overline{X} - t_{n-1,\alpha/2}n^{-1/2}S, \overline{X} + t_{n-1,\alpha/2}n^{-1/2}S \right) \qquad (9.2.2)$$

is a $(1 - \alpha)$ confidence interval for μ. ▲

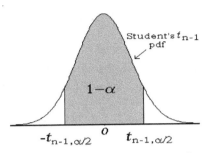

Figure 9.2.2. The area on the right (or left) of $t_{n-1,\alpha/2}$
(or $-t_{n-1,\alpha/2}$) is $\alpha/2$.

Example 9.2.5 (Variance: Normal with Unknown μ,σ) Suppose
$X_1, ..., X_n$ are iid $N(\mu, \sigma^2)$ with unknown $\mu \in \Re$ and $\sigma^2 \in \Re^+, n \geq 2$.
Given $\alpha \in (0,1)$, we wish to construct a $(1 - \alpha)$ confidence interval for σ^2.
Again, $T \equiv (\overline{X}, S^2)$ is minimal sufficient for (μ, σ^2). The distribution of S^2
belongs to a scale family. The pivot ought to be $U \equiv (n - 1)S^2/\sigma^2$ which
has a χ^2_{n-1} distribution for all μ, σ^2. Recall that $\chi^2_{\nu,\gamma}$ is the upper $100\gamma\%$
point of χ^2_ν distribution.

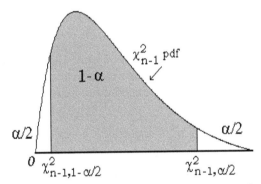

Figure 9.2.3. The area on the right (or left) of $\chi^2_{n-1,\alpha/2}$
(or $\chi^2_{n-1,1-\alpha/2}$) is $\alpha/2$.

See Figure 9.2.3. We have $P\{\chi^2_{n-1,1-\alpha/2} < U < \chi^2_{n-1,\alpha/2}\} = 1 - \alpha$ so
that we can immediately write:

$$P_{\mu,\sigma^2}\left\{(\chi^2_{n-1,\alpha/2})^{-1}(n-1)S^2 < \sigma^2 < (\chi^2_{n-1,1-\alpha/2})^{-1}(n-1)S^2\right\}$$
$$= 1 - \alpha$$

Thus,

$$\left((\chi^2_{n-1,\alpha/2})^{-1}(n-1)S^2, (\chi^2_{n-1,1-\alpha/2})^{-1}(n-1)S^2\right) \qquad (9.2.3)$$

is a $(1 - \alpha)$ confidence interval for σ^2. Obviously,

$$\left(\sqrt{(\chi^2_{n-1,\alpha/2})^{-1}(n-1)S^2}, \sqrt{(\chi^2_{n-1,1-\alpha/2})^{-1}(n-1)S^2} \right)$$

will be a $(1 - \alpha)$ confidence interval for σ. ▲

Example 9.2.6 (Joint Confidence Intervals: Normal with Unknown μ,σ) Suppose $X_1, ..., X_n$ are iid $N(\mu, \sigma^2)$ with unknown $\mu \in \Re$ and $\sigma^2 \in \Re^+, n \geq 2$. Given $\alpha \in (0, 1)$, we wish to construct $(1 - \alpha)$ joint confidence intervals for both μ and σ. From Example 9.2.4, we have

$$J_1 = \left(\overline{X} - t_{n-1,\gamma/2}n^{-1/2}S, \overline{X} + t_{n-1,\gamma/2}n^{-1/2}S \right) \qquad (9.2.4)$$

is a $(1 - \gamma)$ confidence interval for μ with fixed $\gamma \in (0, 1)$. Also, from Example 9.2.5, we have

$$J_2 = \left(\sqrt{(\chi^2_{n-1,\delta/2})^{-1}(n-1)S^2}, \sqrt{(\chi^2_{n-1,1-\delta/2})^{-1}(n-1)S^2} \right) \qquad (9.2.5)$$

is a $(1 - \delta)$ confidence interval for σ with fixed $\delta \in (0, 1)$. Now, we write:

$$\begin{aligned} P_{\mu,\sigma^2}\{\mu \in J_1 \cap \sigma \in J_2\} \\ \geq P_{\mu,\sigma^2}\{\mu \in J_1\} + P_{\mu,\sigma}\{\sigma \in J_2\} - 1, \text{ using} \\ \text{Bonferroni Inequality (3.8.7)} \\ = 1 - \gamma - \delta. \end{aligned} \qquad (9.2.6)$$

If we choose $0 < \gamma, \delta < 1$ so that $\gamma + \delta = \alpha$, then we can think of J_1, J_2 as the *joint* confidence intervals for unknown μ, σ with a *joint* confidence coefficient at least $(1 - \alpha)$. Customarily, one picks $\gamma = \delta = \frac{1}{2}\alpha$. ▲

9.2.1 Interpretation of Confidence Coefficient

Let us now *interpret* a confidence coefficient or the coverage probability. Consider a proposed confidence interval J for θ. Once we observe the data $\mathbf{X} = \mathbf{x}$, a two-sided confidence interval *estimate* of θ will be $(T_L(\mathbf{x}), T_U(\mathbf{x}))$, a *fixed* subinterval of $\Re$. Note that there is nothing random about this observed interval estimate. Also, recall that while θ is unknown $(\in \Theta)$, it is a *fixed* entity. The interpretation of the phrase "$(1 - \alpha)$ confidence" simply means: hypothetically, observe data $\mathbf{X} = \mathbf{x}_1, \mathbf{x}_2, \mathbf{x}_3, ...$ and sequentially construct corresponding observed confidence interval estimates,

$$(T_L(\mathbf{x}_1), T_U(\mathbf{x}_1)), (T_L(\mathbf{x}_2), T_U(\mathbf{x}_2)), (T_L(\mathbf{x}_3), T_U(\mathbf{x}_3)), ...$$

In a long haul, out of these conceptual interval estimates found, approximately $100(1 - \alpha)\%$ would include the unknown value of θ. This interpretation goes hand in hand with the relative frequency idea explained in Chapter 1.

In a *frequentist paradigm*, one may *not* talk about the *probability* of a fixed interval estimate covering (or not covering) some unknown but fixed θ. However, we can claim:

$$\lim_{k \to \infty} k^{-1} \sum_{i=1}^{k} I\left(\theta \in (T_L(\mathbf{x}_i), T_U(\mathbf{x}_i))\right) = 1 - \alpha \qquad (9.2.7)$$

9.2.2 Accuracy Measure

In Examples 9.2.3 and 9.2.4, we used equal tail percentage points of standard normal and Student's t distributions. Both pivotal distributions were symmetric about the origin. Both standard normal and Student's t_{n-1} pdfs obviously integrated to $(1 - \alpha)$ respectively on the intervals $(-z_{\alpha/2}, z_{\alpha/2})$ and $(-t_{n-1,\alpha/2}, t_{n-1,\alpha/2})$. But, why did we place $(1-\alpha)$ probability around the center of a distribution? The following result will provide some necessary insight.

Theorem 9.2.1 *Suppose that X is a continuous random variable having a unimodal pdf $f(x)$ with its support $\Re$. Assume that $f(x)$ is symmetric around $x = 0$, that is, $f(-x) = f(x)$ for all $x > 0$. Let $P(-a < X < a) = 1 - \alpha$ for some $0 < \alpha < \frac{1}{2}$. Let positive numbers g, h be such that $P(-a - g < X < a - h) = 1 - \alpha$. Then, the interval $(-a - g, a - h)$ must be wider than the interval $(-a, a)$.*

Theorem 9.2.1 proves that among all $(1 - \alpha)$ intervals going to the left or right of $(-a, a)$, the interval $(-a, a)$ in the center is the shortest one. Since Examples 9.2.3 and 9.2.4 handled the location parameter case, we can claim this *optimality* (*shortest width*) property associated with the proposed $(1 - \alpha)$ confidence intervals. In nonsymmetric cases, the problem becomes more involved, but practitioners often follow the convention below.

> *Convention:* For standard pivotal distributions such as Normal, Student's t, Chi-square, and F, we customarily assign the tail area probability $\frac{1}{2}\alpha$ on both sides in order to construct a $100(1 - \alpha)\%$ two-sided confidence interval.

9.3 Two-Sample Problems

Here again the fundamental approach of using a pivot remains in effect. That is, we continue to emphasize working with some appropriate pivot involving a minimal sufficient statistic $\mathbf{T}$ and $\boldsymbol{\theta}$.

9.3.1 Comparing Location Parameters

Examples include estimation of the difference of (i) means of independent normal populations, (ii) location parameters of independent negative exponential populations, and (iii) means of a bivariate normal population.

Example 9.3.1 (Means: Independent Normal Common Unknown Variance) Let $X_{i1}, ..., X_{in_i}$ be iid $N(\mu_i, \sigma^2), i = 1, 2$, and X_1, X_2 be independent. Let all three parameters be unknown and denote $\boldsymbol{\theta} = (\mu_1, \mu_2, \sigma^2) \in \Re \times \Re \times \Re^+$. Given $\alpha \in (0, 1)$, we wish to construct a $(1 - \alpha)$ confidence interval for $\mu_1 - \mu_2 (\equiv \Delta)$ based on minimal sufficient statistics for $\boldsymbol{\theta}$. With $n_i \geq 2$, denote:

$$\overline{X}_i = n_i^{-1}\Sigma_{j=1}^{n_i}X_{ij}, \ S_i^2 = (n_i - 1)^{-1}\Sigma_{j=1}^{n_i}(X_{ij} - \overline{X}_i)^2$$
$$\text{and } S_P^2 = (n_1 + n_2 - 2)^{-1}\left\{(n_1 - 1)S_1^2 + (n_2 - 1)S_2^2\right\} \tag{9.3.1}$$

for $i = 1, 2$. S_P^2 is the pooled sample variance.

Using Example 4.5.2, we construct the pivot

$$U \equiv \{n_1^{-1} + n_2^{-1}\}^{-\frac{1}{2}}[(\overline{X}_1 - \overline{X}_2) - \Delta]S_P^{-1} \tag{9.3.2}$$

which has a t_ν distribution with $\nu = (n_1 + n_2 - 2)$. Now, we have $P\{-t_{\nu,\alpha/2} < U < t_{\nu,\alpha/2}\} = 1 - \alpha$ so that we claim:

$$P_{\boldsymbol{\theta}}\left[(\overline{X}_1 - \overline{X}_2) - t_{\nu,\alpha/2}S_P\{n_1^{-1} + n_2^{-1}\}^{1/2} < \Delta\right.$$
$$\left. < (\overline{X}_1 - \overline{X}_2) + t_{\nu,\alpha/2}S_P\{n_1^{-1} + n_2^{-1}\}^{1/2}\right] = 1 - \alpha$$

Now, writing $W = \overline{X}_1 - \overline{X}_2$, we have

$$\left(W - t_{\nu,\alpha/2}\{n_1^{-1} + n_2^{-1}\}^{1/2}S_P, W + t_{\nu,\alpha/2}\{n_1^{-1} + n_2^{-1}\}^{1/2}S_P\right) \tag{9.3.3}$$

as a $(1 - \alpha)$ confidence interval for $\mu_1 - \mu_2 (\equiv \Delta)$. ▲

> What is one supposed to do when variances are unknown and unequal? It is a complicated scenario that goes by the name, *Behrens-Fisher problem*. See a simple-minded suggestion in Exercise 9.3.6.

Example 9.3.2 (Locations: Independent Negative Exponential with Common Unknown Scale) Suppose $X_{i1}, ..., X_{in}$ are iid having a common pdf $f(x; \mu_i, \sigma), i = 1, 2$, where $f(x; \mu, \sigma) = \sigma^{-1}exp\{-(x - \mu)/\sigma\}I(x > \mu)$. Also, X_1, X_2 are independent. All three parameters are unknown and $\boldsymbol{\theta} = (\mu_1, \mu_2, \sigma) \in \Re \times \Re \times \Re^+$. Given $\alpha \in (0, 1)$, we wish to construct a $(1 - \alpha)$ confidence interval for $\mu_1 - \mu_2 (\equiv \Delta)$ based on minimal sufficient statistics for $\boldsymbol{\theta}$. With $n \geq 2$, denote:

$$X_i^{(1)} = \min_{1 \leq j \leq n} X_{ij}, \ W_i = (n - 1)^{-1}\Sigma_{j=1}^n(X_{ij} - X_i^{(1)})$$
$$\text{and } W_P = \tfrac{1}{2}\{W_1 + W_2\}, \text{ for } i = 1, 2 \tag{9.3.4}$$

Here, W_P is a *pooled* estimator of σ. It is easy to verify that W_P estimates σ unbiasedly and

$$V[W_1] = V[W_2] = (n - 1)^{-1}\sigma^2 \text{ and } V[W_P] = \tfrac{1}{2}(n - 1)^{-1}\sigma^2$$

so that the pooled estimator W_P is indeed better than either $W_i, i = 1, 2$.

One may check that $2(n-1)W_i/\sigma$ is distributed as a $\chi^2_{2n-2}, i = 1, 2$, and these are independent. Using the reproductive property of independent Chi-squares, we can claim that $4(n-1)W_P\sigma^{-1}$ has a χ^2_{4n-4} distribution. Also, $(X_1^{(1)}, X_2^{(1)}), W_P$ are independent, and $(X_1^{(1)}, X_2^{(1)}, W_P)$ is minimal sufficient for $\boldsymbol{\theta}$. So, we may use the pivot:

$$U \equiv n[(X_1^{(1)} - X_2^{(1)}) - (\mu_1 - \mu_2)]W_P^{-1} \qquad (9.3.5)$$

One may verify that the pivotal distribution of $|U|$ is $F_{2,4n-4}$. This is a two-sample version of Exercise 4.4.14. Let $F_{2,4n-4,\alpha}$ be the upper $100\alpha\%$ point of $F_{2,4n-4}$ distribution. See Figure 9.3.1. We can say that $P\{-F_{2,4n-4,\alpha} < U < F_{2,4n-4,\alpha}\} = 1 - \alpha$ and claim:

$$P_\theta \left\{ \Delta \in [(X_1^{(1)} - X_2^{(1)}) \pm F_{2,4n-4,\alpha} n^{-1} W_P] \right\} = 1 - \alpha$$

In other words,

$$\left((X_1^{(1)} - X_2^{(1)}) - F_{2,4n-4,\alpha} n^{-1} W_P, (X_1^{(1)} - X_2^{(1)}) + F_{2,4n-4,\alpha} n^{-1} W_P \right)$$

$$(9.3.6)$$

is a $(1 - \alpha)$ confidence interval for $\mu_1 - \mu_2 (\equiv \Delta)$. ▲

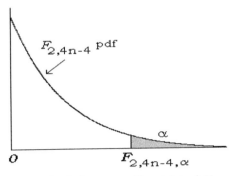

Figure 9.3.1. The shaded area on the right of $F_{2,4n-4,\alpha}$ is α.

Example 9.3.3 (Paired t Method: Bivariate Normal) Sometimes two populations may be assumed normal, but they may be dependent. In a large establishment, for example, suppose that X_{1j}, X_{2j} respectively denote "job performance" scores for the jth employee, before and after going through a week-long job enhancement program, $j = 1, ..., n(\geq 2)$. We assume that these employees are selected randomly and independently of each other and wish to compare average job performance scores before and after the training program. Observe that X_{1j}, X_{2j} are dependent random variables and the methodology from Example 9.3.1 will not apply here. This refers to customary "paired" data.

Let (X_{1j}, X_{2j}) be iid $N_2(\mu_1, \mu_2, \sigma_1^2, \sigma_2^2, \rho), j = 1, ..., n(\geq 2)$ with all parameters unknown, $(\mu_i, \sigma_i^2) \in \Re \times \Re^+, i = 1, 2$ and $-1 < \rho < 1$. Given $\alpha \in (0, 1)$, we wish to construct a $(1-\alpha)$ confidence interval for $\mu_1 - \mu_2 (\equiv \Delta)$ based on minimal sufficient statistics for $\boldsymbol{\theta}(= (\mu_1, \mu_2, \sigma_1, \sigma_2, \rho))$. Denote:

$$Y_j = X_{1j} - X_{2j}, j = 1, ..., n, \quad \overline{Y} = n^{-1}\Sigma_{j=1}^n Y_j, \text{ and}$$
$$S^2 = (n-1)^{-1}\Sigma_{j=1}^n(Y_j - \overline{Y})^2 \tag{9.3.7}$$

In Equation (9.3.7), observe that $Y_1, ..., Y_n$ are iid $N(\Delta, \sigma^2)$ where $\sigma^2 = \sigma_1^2 + \sigma_2^2 - 2\rho\sigma_1\sigma_2$, but Δ and σ^2 are unknown. Now, this matches with the scenario that we had tackled in Example 9.3.1. One ought to proceed with the pivot: $U \equiv \sqrt{n}(\overline{Y} - \Delta)/S$. Further details are left out. ▲

9.3.2   Comparing Variances

Example 9.3.4 (Ratio of Variances: Independent Normal with Unknown Means and Variances) Suppose $X_{i1}, ..., X_{in_i}$ are iid $N(\mu_i, \sigma_i^2), i = 1, 2$, and X_1, X_2 are independent. Let all four parameters be unknown and denote $\boldsymbol{\theta} = (\mu_1, \mu_2, \sigma_1^2, \sigma_2^2) \in \Re \times \Re \times \Re^+ \times \Re^+$. Given $\alpha \in (0, 1)$, we wish to construct a $(1 - \alpha)$ confidence interval for $\sigma_1^2/\sigma_2^2 (\equiv \Lambda)$ based on minimal sufficient statistics for $\boldsymbol{\theta}$. With $n_i \geq 2$, denote:

$$\overline{X}_i = n_i^{-1}\Sigma_{j=1}^{n_i} X_{ij}, \quad S_i^2 = (n_i - 1)^{-1}\Sigma_{j=1}^{n_i}(X_{ij} - \overline{X}_i)^2, i = 1, 2 \tag{9.3.8}$$

Consider the pivot:

$$U \equiv [S_1^2/\sigma_1^2] \div [S_2^2/\sigma_2^2] = \Lambda^{-1}S_1^2 S_2^{-2} \tag{9.3.9}$$

It should be clear that U is distributed as F_{n_1-1, n_2-1} and denote its upper $50\alpha\%$ point $F_{n_1-1, n_2-1, \alpha/2}$. See Figure 9.3.2.

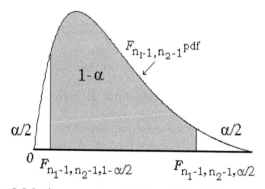

Figure 9.3.2. Area on the right (or left) of $F_{n_1-1, n_2-1, \alpha/2}$ (or $F_{n_1-1, n_2-1, 1-\alpha/2}$) is $\alpha/2$.

Thus, $P\left\{F_{n_1-1, n_2-1, 1-\alpha/2} < U < F_{n_1-1, n_2-1, \alpha/2}\right\} = 1 - \alpha$ and claim:

$$P_\theta \left\{ F^{-1}_{n_1-1,n_2-1,\alpha/2} S_1^2/S_2^2 < \Lambda < F^{-1}_{n_1-1,n_2-1,1-\alpha/2} S_1^2/S_2^2 \right\}$$

$$= 1 - \alpha \text{ for all } \theta$$

so that

$$\left(F^{-1}_{n_1-1,n_2-1,\alpha/2} S_1^2/S_2^2, F^{-1}_{n_1-1,n_2-1,1-\alpha/2} S_1^2/S_2^2 \right) \qquad (9.3.9)$$

is a $(1 - \alpha)$ confidence interval for the variance ratio, σ_1^2/σ_2^2. ▲

9.4   Exercises and Complements

9.1.1 Let X have a Laplace pdf $\frac{1}{2} exp\{-|x - \theta|\} I(x \in \Re)$ where $\theta (\in \Re)$ is unknown. Given $\alpha \in (0,1)$, find a $(1 - \alpha)$ confidence interval for θ.

9.1.2 Let X have a Cauchy pdf $\frac{1}{\pi} \{1 + (x - \theta)^2\}^{-1} I(x \in \Re)$ where $\theta (\in \Re)$ is unknown. Given $\alpha \in (0,1)$, find a $(1 - \alpha)$ confidence interval for θ.

9.1.3 Let X have a Laplace pdf $\frac{1}{2\theta} exp\{-|x|/\theta\} I(x \in \Re)$ where $\theta (\in \Re^+)$ is unknown. Given $\alpha \in (0,1)$, find a $(1 - \alpha)$ confidence interval for θ.

9.2.1 Suppose that X has $N(0, \sigma^2)$ distribution where $\sigma (\in \Re^+)$ is unknown. Consider the confidence interval $J = (|X|, 10|X|)$ for σ.

(i)   Find the confidence coefficient associated with J;

(ii)   What is the expected length of the interval J?

9.2.2 Suppose that X has its pdf:

$$2(\theta - x)\theta^{-2} I(0 < x < \theta)$$

where $\theta (\in \Re^+)$ is unknown. Given $\alpha \in (0,1)$, consider the pivot $U \equiv X/\theta$ and find a $(1 - \alpha)$ confidence interval for θ.

9.2.3 Let $X_1, ..., X_n$ be iid Gamma(a, b) where $a(> 0)$ is known but $b(> 0)$ is unknown. Given $\alpha \in (0,1)$, find an appropriate pivot based on a minimal sufficient statistic, and find a $(1 - \alpha)$ confidence interval for b.

9.2.4 Suppose that $X_1, ..., X_n$ are iid with a common pdf:

$$\sigma^{-1} exp\{-(x - \theta)/\sigma\} I(x > \theta).$$

Also, suppose that $\theta (\in \Re)$ is known but $\sigma (\in \Re^+)$ is unknown. Given $\alpha \in (0,1)$, find a $(1 - \alpha)$ confidence interval for σ. {$Hint$: $\Sigma_{i=1}^n (X_i - \theta)$ is a minimal sufficient statistic for σ.}

9.2.5 Suppose that $X_1, ..., X_n$ are iid with a common pdf:

$$\sigma^{-1} exp\{-(x - \theta)/\sigma\} I(x > \theta)$$

with both $\theta(\in \Re)$ and $\sigma(\in \Re^+)$ unknown. Given $\alpha \in (0,1)$, find a $(1-\alpha)$ confidence interval for σ by using an appropriate pivot. {*Hint:* Can $\Sigma_{i=1}^n(X_i - X_{n:1})$ be a starter?}

9.2.6 Suppose that $X_1, ..., X_n$ are iid Uniform$(-\theta, \theta)$ where $\theta(\in \Re^+)$ is unknown. Given $\alpha \in (0,1)$, find a $(1-\alpha)$ confidence interval for θ. {*Hint:* Can $\max_{1 \le i \le n} |X_i|$ be a starter?}

9.2.7 (Example 9.2.6 Continued) Given $\alpha \in (0,1)$, find a confidence interval for each parametric function given below with a confidence coefficient at least $(1-\alpha)$.

(i) $\mu + \sigma$ (ii) $\mu + \sigma^2$ (iii) μ/σ (iv) μ/σ^2

9.3.1 Suppose $X_{i1}, ..., X_{in_i}$ are iid $N(\mu_i, \sigma_i^2), i = 1, 2$, and X_1, X_2 are independent, with μ_1, μ_2 unknown but σ_1^2, σ_2^2 known, $(\mu_i, \sigma_i^2) \in \Re \times \Re^+, i = 1, 2$. Given $\alpha \in (0,1)$, construct a $(1-\alpha)$ confidence interval for $\mu_1 - \mu_2$ based on the minimal sufficient statistics for (μ_1, μ_2).

9.3.2 (Example 9.3.1 Continued) Let $X_{i1}, ..., X_{in_i}$ be iid $N(\mu_i, k_i\sigma^2), n_i \ge 2, i = 1, 2$, and X_1, X_2 be independent. Assume that all three parameters μ_1, μ_2, σ^2 are unknown but k_1, k_2 are positive and known, $(\mu_1, \mu_2, \sigma^2) \in \Re \times \Re \times \Re^+$. Given $\alpha \in (0,1)$, construct a $(1-\alpha)$ confidence interval for $\mu_1 - \mu_2$ based on the minimal sufficient statistics for (μ_1, μ_2, σ^2).

9.3.3 (Example 9.3.1 Continued) Given $\alpha \in (0,1)$, construct a $(1-\alpha)$ confidence interval for $(\mu_1 - \mu_2)/\sigma$ based on sufficient statistics for (μ_1, μ_2, σ^2). {*Hint:* Combine estimation problems for $(\mu_1 - \mu_2), \sigma$ via the Bonferroni Inequality.}

9.3.4 Suppose $X_{i1}, ..., X_{in_i}$ are iid with a common pdf $f(x; \theta_i), i = 1, 2$, where $f(x; a) = a^{-1}\exp\{-x/a\}I(x > 0), a \in \Re^+$, and X_1, X_2 are independent. Let θ_1, θ_2 be unknown, $(\theta_1, \theta_2) \in \Re^+ \times \Re^+$. Given $\alpha \in (0,1)$, find a $100(1-\alpha)\%$ confidence interval for θ_2/θ_1 based on minimal sufficient statistics for (θ_1, θ_2).

9.3.5 Suppose $X_{i1}, ..., X_{in}$ are iid Uniform$(0, \theta_i), i = 1, 2$, and X_1, X_2 are independent. Let θ_1, θ_2 be unknown, $(\theta_1, \theta_2) \in \Re^+ \times \Re^+$. Given $\alpha \in (0,1)$, derive a $(1-\alpha)$ confidence interval for $\theta_1/(\theta_1 + \theta_2)$ based on sufficient statistics for (θ_1, θ_2).

9.3.6 (Means: Independent Normal with Unknown Means and Variances) We have $X_{i1}, ..., X_{in_i}$ iid $N(\mu_i, \sigma_i^2), i = 1, 2$, and X_1, X_2 are independent. Let all parameters be unknown and denote $\boldsymbol{\theta} = (\mu_1, \mu_2, \sigma_1^2, \sigma_2^2) \in \Re \times \Re \times \Re^+ \times \Re^+$. Given $\alpha \in (0,1)$, we wish to construct a confidence interval for $\mu_1 - \mu_2 (\equiv \Delta)$ based on the minimal sufficient statistics for $\boldsymbol{\theta}$. With $n_i \ge 2$, denote:

$$\overline{X}_i = n_i^{-1}\Sigma_{j=1}^{n_i}X_{ij}, \quad S_i^2 = (n_i - 1)^{-1}\Sigma_{j=1}^{n_i}(X_{ij} - \overline{X}_i)^2$$

$$J_i \equiv \left(\overline{X}_i - t_{n_i-1,\alpha/4}n_i^{-1/2}S_i, \overline{X}_i + t_{n_i-1,\alpha/4}n_i^{-1/2}S_i\right), i = 1, 2$$

$$U = t_{n_1-1,\alpha/4}n_1^{-1/2}S_1 + t_{n_2-1,\alpha/4}n_2^{-1/2}S_2$$

Using Equation (9.2.2), one can claim that J_i is a $100(1 - \frac{1}{2}\alpha)\%$ confidence interval for $\mu_i, i = 1, 2$. Given $0 < \alpha < 1$, show that

(i) $P_\theta \{\mu_1 \in J_1 \cap \mu_2 \in J_2\} \geq 1 - \alpha$;

(ii) $P_\theta \{\Delta \in (\overline{X}_1 - \overline{X}_2) \pm U\} \geq 1 - \alpha$.

Now, one has a confidence interval

$$\left((\overline{X}_1 - \overline{X}_2) - U, (\overline{X}_1 - \overline{X}_2) + U\right)$$

for Δ with a confidence coefficient at least $1 - \alpha$ in the *Behrens-Fisher* situation.

9.3.7 Let $X_{i1}, ..., X_{in_i}$ be iid $N(\mu_i, k_i\sigma^2), n_i \geq 2, i = 1, 2$, and X_1, X_2 be independent. Assume that all three parameters μ_1, μ_2, σ^2 are unknown but k_1, k_2 are positive and known, $(\mu_1, \mu_2, \sigma^2) \in \Re \times \Re \times \Re^+$. Given $\alpha \in (0, 1)$, construct a $(1 - \alpha)$ confidence interval for μ_1/σ based on the minimal sufficient statistics for (μ_1, μ_2, σ^2).

9.3.8 (Example 9.3.1 Continued) Given $\alpha \in (0, 1)$, construct a $(1 - \alpha)$ confidence interval for μ_1^2/σ^2 based on sufficient statistics for (μ_1, μ_2, σ^2). {*Hint*: Combine estimation problems for μ_1^2 and σ^2 via the Bonferroni Inequality.}

9.3.9 (Exercise 9.3.4 Continued) Suppose $X_{i1}, ..., X_{in_i}$ are iid with a common pdf $f(x; \theta_i), i = 1, 2$, where $f(x; a) = a^{-1} exp\{-x/a\} I(x > 0), a \in \Re^+$, and X_1, X_2 are independent. Let θ_1, θ_2 be unknown, $(\theta_1, \theta_2) \in \Re^+ \times \Re^+$. Given $\alpha \in (0, 1)$, find a $100(1-\alpha)\%$ confidence interval for θ_1^2/θ_2^2 based on minimal sufficient statistics for (θ_1, θ_2).

9.3.10 (Exercise 9.3.5 Continued) Suppose $X_{i1}, ..., X_{in}$ are iid with a common Uniform$(0, \theta_i)$ distribution, $i = 1, 2$. Also, suppose X_1, X_2 are independent. Let θ_1, θ_2 be unknown, $(\theta_1, \theta_2) \in \Re^+ \times \Re^+$. Given $\alpha \in (0, 1)$, derive a $(1 - \alpha)$ confidence interval for θ_1^3/θ_2^3 based on sufficient statistics for (θ_1, θ_2).

9.3.11 (Exercise 9.3.2 Continued) Let $X_{i1}, ..., X_{in_i}$ be iid $N(\mu_i, k_i\sigma^2), n_i \geq 2, i = 1, 2$, and X_1, X_2 be independent. Assume that all three parameters μ_1, μ_2, σ^2 are unknown but k_1, k_2 are positive and known, $(\mu_1, \mu_2, \sigma^2) \in \Re \times \Re \times \Re^+$. Given $\alpha \in (0, 1)$, construct a $(1 - \alpha)$ confidence interval for $(\mu_1 - \mu_2)/\sigma$ based on the minimal sufficient statistics for (μ_1, μ_2, σ^2).

9.3.12 (Example 9.3.1 Continued) Given $\alpha \in (0, 1)$, construct a $(1 - \alpha)$ confidence interval for $(\mu_1 - \mu_2)^2/\sigma^2$ based on sufficient statistics for (μ_1, μ_2, σ^2). {*Hint*: Combine estimation problems for $(\mu_1 - \mu_2)^2, \sigma^2$ via the Bonferroni Inequality.}

9.3.13 (Exercise 9.3.4 Continued) Suppose $X_{i1}, ..., X_{in_i}$ are iid with a common pdf $f(x; \theta_i), i = 1, 2$, where $f(x; a) = a^{-1} exp\{-x/a\} I(x > 0), a \in \Re^+$, and X_1, X_2 are independent. Let θ_1, θ_2 be unknown, $(\theta_1, \theta_2) \in \Re^+ \times \Re^+$. Given $\alpha \in (0, 1)$, find a $100(1-\alpha)\%$ confidence interval for $\theta_2/(\theta_1+2\theta_2)$ based on minimal sufficient statistics for (θ_1, θ_2).

9.3.14 (Exercise 9.3.5 Continued) Suppose $X_{i1}, ..., X_{in}$ are iid with a common Uniform$(0, \theta_i)$ distribution, $i = 1, 2$. Also, suppose X_1, X_2 are independent. Let θ_1, θ_2 be unknown, $(\theta_1, \theta_2) \in \Re^+ \times \Re^+$. Given $\alpha \in (0, 1)$, derive a $(1 - \alpha)$ confidence interval for $\theta_1^3/(2\theta_1 + 3\theta_2)^3$ based on sufficient statistics for (θ_1, θ_2).

10

Bayesian Methods

10.1 Introduction

The nature of what we are about to discuss is conceptually very different from anything we had included in previous chapters. So far, we developed methodologies from the point of view of a frequentist. So far, we have started with a random sample $X_1, ..., X_n$ from a population having the *probability mass function* (pmf) or *probability density function* (pdf) $f(x; \vartheta)$ with $x \in \mathcal{X}$ and $\vartheta \in \Theta$ where we assumed that the unknown parameter ϑ was *fixed*. Thus, all inference procedures relied upon the likelihood function, $L(\vartheta) = \Pi_{i=1}^{n} f(x_i; \vartheta)$.

Under the Bayesian approach, an experimenter believes that the unknown parameter ϑ *is* a random variable having its own probability distribution on Θ. Now that ϑ is assumed *random*, a likelihood function will be the same as $L(\theta)$ *given that* $\vartheta = \theta$. Let $h(\theta)$ denote a pmf or pdf of ϑ at $\vartheta = \theta$ which is called a *prior* distribution of ϑ. A prior, $h(\theta)$, reflects an experimenter's *subjective* belief regarding ϑ values that are more (or less) likely. Ideally, $h(\theta)$ is specified before collecting data.

> $f(x; \theta)$ denotes the *conditional* pmf or pdf of X given $\vartheta = \theta$.

A Bayesian paradigm requires one to perform all statistical inferences after combining the information about ϑ supplied by collected data (evidenced by the likelihood function $L(\theta)$ given that $\vartheta = \theta$), and that from a prior $h(\theta)$. One combines these evidences by means of Bayes' Theorem 1.4.3 and comes up with what is called a *posterior* distribution. All inferences are then dictated by the posterior distribution. This approach was due to Rev. Thomas Bayes (1783). A strong theoretical foundation evolved from fundamental contributions of de Finetti (1937), Savage (1954), and Jeffreys (1957), among others. The contributions of L. J. Savage were beautifully synthesized by Lindley (1980).

R. A. Fisher was vehemently opposed to anything remotely Bayesian. Illuminating accounts of his philosophy and interactions with some of the key Bayesian researchers of his time can be found in the biography written by his daughter, Joan Fisher Box (1978). Some interesting exchanges between Fisher and H. Jeffreys as well as L. J. Savage are included in the edited volume of Bennett (1990). Also, Buehler (1980), Lane (1980), and Wallace (1980) gave important perspectives on possible connections between Fisher's fiducial inference and the Bayesian doctrine. It appears that researchers are recently reinventing fiducial arguments in the name of "implicit distributions." One may refer to Mukhopadhyay (2006) for some pertinent comments.

In Section 10.2, we discuss prior and posterior distributions. Section 10.3 introduces *conjugate priors*. In Section 10.4, we discuss point estimation problems under a *squared error loss* function and introduce *Bayes estimator*. Section 10.5 includes examples involving nonconjugate priors. Unfortunately, we cannot present a full-blown Bayes theory. It is hoped that readers will get a taste of underlying principles from this brief exposure.

10.2   Prior and Posterior Distributions

For simplicity, let the parameter ϑ represent a continuous real valued random variable defined on Θ which will customarily be a subinterval of $\Re$. The unknown ϑ has its pdf $h(\theta)$ on Θ. We call $h(\theta)$ a *prior* distribution.

The evidence about ϑ derived from the prior is combined with that from the conditional likelihood function by means of Bayes' Theorem 1.4.3. A likelihood function is a conditional joint pmf or pdf of $\mathbf{X} = (X_1, ..., X_n)$ given $\vartheta = \theta$. Let T be a (minimal) sufficient statistic for θ given that $\vartheta = \theta$. We let T be real valued and we work with its pmf or pdf $g(t; \theta)$, given $\vartheta = \theta$, for $t \in \mathcal{T} \subseteq \Re$.

For uniformity of notation, however, we treat T as a continuous variable and the associated probabilities and expectations will be expressed as integrals over $\mathcal{T}$. Such integrals will be replaced by appropriate sums when T is discrete.

A *joint pdf* of T and ϑ is:

$$g(t; \theta)h(\theta) \text{ for all } t \in \mathcal{T} \text{ and } \theta \in \Theta \qquad (10.2.1)$$

The *marginal pdf* of T is:

$$m(t) = \int_{\theta \in \Theta} g(t; \theta)h(\theta)d\theta \text{ for all } t \in \mathcal{T} \qquad (10.2.2)$$

so that a conditional pdf of ϑ given $T = t$ is:

$$k(\theta; t) \equiv k(\theta \mid T = t) = g(t; \theta)h(\theta)/m(t) \text{ for all } t \in \mathcal{T}$$
$$\text{and } \theta \in \Theta \text{ such that } m(t) > 0 \qquad (10.2.3)$$

Note that the expression for $k(\theta; t)$ follows directly from Bayes' Theorem 1.4.3 by simply replacing $P(A_j \mid B), P(A_j), P(B \mid A_i)$ and $\Sigma_{i=1}^k$ by $k(\theta; t), h(\theta), g(t; \theta)$, and $\int_{\theta \in \Theta}$, respectively.

Definition 10.2.1 *The conditional pdf $k(\theta; t)$ of ϑ given that $T = t$ is called a posterior distribution of ϑ.*

Under a Bayesian paradigm, a posterior pdf $k(\theta; t)$ epitomizes how one combines the information about ϑ obtained from two separate sources, the prior and collected data. The tractability of $k(\theta; t)$ largely depends on how easy it is to obtain an expression of $m(t)$. In some cases, a marginal and a posterior distribution can be evaluated only numerically.

Example 10.2.1 Suppose $X_1, ..., X_n$ are *independent and identically distributed* (iid) Bernoulli(θ) given $\vartheta = \theta$ where ϑ is an unknown probability of success, $0 < \vartheta < 1$. Given $\vartheta = \theta$, $T = \Sigma_{i=1}^{n} X_i$ is minimal sufficient for θ, and

$$g(t; \theta) = \binom{n}{t}\theta^t(1-\theta)^{n-t} \text{ for } t \in \mathcal{T} \tag{10.2.4}$$

where $\mathcal{T} = \{0, 1, ..., n\}$. Suppose that a prior distribution of ϑ on $\Theta = (0, 1)$ is Uniform$(0, 1)$, that is $h(\theta) = I(0 < \theta < 1)$. From Equation (10.2.2), for $t \in \mathcal{T}$, we express $m(t)$ as:

$$\int_{\theta=0}^{1} \binom{n}{t}\theta^t(1-\theta)^{n-t}d\theta$$
$$= \binom{n}{t}\int_{\theta=0}^{1}\theta^t(1-\theta)^{n-t}d\theta \tag{10.2.5}$$
$$= \binom{n}{t}b(t+1, n-t+1) \text{ where } b(\alpha, \beta) \text{ is}$$
$$\text{a Beta function from Equation (1.6.25)}$$

Thus, for any fixed value $t \in \mathcal{T}$, a posterior pdf of ϑ given $T = t$ becomes:

$$k(\theta; t)$$
$$= \left\{\binom{n}{t}\theta^t(1-\theta)^{n-t}h(\theta)\right\} / \left\{\binom{n}{t}b(t+1, n-t+1)\right\} \tag{10.2.6}$$
$$= [b(t+1, n-t+1)]^{-1}\theta^t(1-\theta)^{n-t}, \text{ for all } \theta \in (0, 1)$$

That is, the posterior distribution is Beta$(t+1, n-t+1)$. ▲

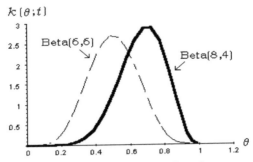

Figure 10.2.1. Posterior pdfs of ϑ under two prior distributions of ϑ when $t = 7$.

Example 10.2.2 (Example 10.2.1 Continued) Let $X_1, ..., X_{10}$ be iid Bernoulli(θ) given $\vartheta = \theta$ where ϑ is an unknown probability of success, $0 < \vartheta < 1$. The statistic $T = \Sigma_{i=1}^{10} X_i$ is minimal sufficient for θ given that $\vartheta = \theta$. Assume a Uniform$(0, 1)$ prior. Suppose that we have observed $T = 7$, the number of successes out of ten trials. From Equation (10.2.6) we know that the posterior distribution of ϑ is Beta$(8,4)$. The posterior density has been plotted as a solid curve in Figure 10.2.1. This curve is skewed to the left. Instead, if we had observed $T = 5$, then the posterior distribution of ϑ will be Beta$(6,6)$. The corresponding pdf has been plotted as a dashed curve in Figure 10.2.1. This curve is symmetric about $\theta = 0.5$. That is, under the same *uniform* prior, the shape of a posterior distribution may change one's perception about ϑ. ▲

10.3   Conjugate Prior

Definition 10.3.1 *A prior pdf $h(\theta)$ belonging to a family of distributions, $\mathcal{P}$, is called a conjugate prior for ϑ if and only if the posterior pdf $k(\theta; t)$ also belongs to the same family, $\mathcal{P}$.*

Example 10.3.1 (Example 10.2.1 Continued) In the expression for the pdf $g(t; \theta)$, look carefully at the part that involves θ, namely $\theta^t(1-\theta)^{n-t}$. It resembles a beta pdf without a normalizing constant. Hence, we may suppose that a prior distribution of ϑ on Θ is Beta(α, β) where $\alpha(> 0)$ and $\beta(> 0)$ are known numbers. From Equation (10.2.2), for $t \in \mathcal{T}$, we obtain a marginal pmf of T as follows:

$$m(t) = \binom{n}{t} \frac{b(t + \alpha, n + \beta - t)}{b(\alpha, \beta)} \text{ where } b(\alpha, \beta)$$

is the Beta function from Equation (1.6.25)

$$(10.3.1)$$

Using Equations (10.2.3) and (10.3.1), the posterior pdf of ϑ reduces to:

$$k(\theta; t) = [b(t + \alpha, n - t + \beta)]^{-1} \theta^{t+\alpha-1}(1-\theta)^{n+\beta-t-1} \qquad (10.3.2)$$

for all $\theta \in (0, 1), t \in \mathcal{T}$. That is, the posterior pdf of ϑ would correspond to a Beta$(t + \alpha, n - t + \beta)$ distribution.

> We started with a beta prior and ended up with a beta posterior.
> Implies: The beta pdf for ϑ is a conjugate prior.

Note that in Example 10.2.1, Uniform$(0, 1)$ prior was same as a Beta$(1, 1)$ distribution. ▲

Example 10.3.2 Let $X_1, ..., X_n$ be iid Poisson(θ) given $\vartheta = \theta$ where $\vartheta(> 0)$ is an unknown parameter. Given $\vartheta = \theta$, $T = \Sigma_{i=1}^n X_i$ is minimal sufficient for θ and $g(t; \theta) = e^{-n\theta}(n\theta)^t/t!$ for $t \in \mathcal{T} = \{0, 1, 2, ...\}$.

In the expression of $g(t; \theta)$, look at the part which involves θ, namely $e^{-n\theta}\theta^t$. It resembles a gamma pdf without a normalizing constant. Hence, we may suppose that a prior distribution of ϑ on $(0, \infty)$ is Gamma(α, β) with $\alpha(> 0)$ and $\beta(> 0)$ known. From Equation (10.2.2), for $t \in \mathcal{T}$, we obtain a marginal pmf of T as follows:

$$m(t) = \frac{n^t}{t!\beta^\alpha\Gamma(\alpha)} \int_{\theta=0}^{\infty} e^{-\theta(n\beta+1)/\beta}\theta^{\alpha+t-1}d\theta = \frac{n^t\beta^t\Gamma(\alpha+t)}{t!(n\beta+1)^{\alpha+t}\Gamma(\alpha)}$$

(10.3.3)

Thus, the posterior pdf of ϑ given $T = t$ simplifies to:

$$k(\theta;t) = \frac{\{(n\beta+1)/\beta\}^{\alpha+t}}{\Gamma(\alpha+t)}e^{-\theta(n\beta+1)/\beta}\theta^{\alpha+t-1}$$

(10.3.4)

for all $\theta \in (0,\infty), t \in T$. That is, the posterior pdf of ϑ would correspond to a Gamma$(t+\alpha, \beta(n\beta+1)^{-1})$ distribution.

> We plugged a gamma prior and ended with a gamma posterior.
> Implies: The gamma pdf for ϑ is a conjugate prior.

Observe that a conjugate prior is a gamma pdf for ϑ. ▲

Example 10.3.3 Suppose $X_1,...,X_n$ are iid $N(\theta,\sigma^2)$ given $\vartheta = \theta$, where $\vartheta(\in \Re)$ is unknown and $\sigma^2(> 0)$ is assumed known. Consider $T = \Sigma_{i=1}^n X_i$ which is minimal sufficient for θ given that $\vartheta = \theta$. Let the prior distribution be $N(\tau,\delta^2)$ on Θ where $\tau(\in \Re)$ and $\delta(\in \Re^+)$ are known numbers. The joint pdf of (ϑ, T) is *proportional* to

$$exp[\theta\{(t/\sigma^2)+(\tau/\delta^2)\}]\ exp[-\theta^2\{(n/\sigma^2)+(1/\delta^2)\}/2]$$
$$\propto exp\left\{-\tfrac{1}{2}\left(\tfrac{n}{\sigma^2}+\tfrac{1}{\delta^2}\right)\left[\theta-\left(\tfrac{t}{\sigma^2}+\tfrac{\tau}{\delta^2}\right)\left(\tfrac{n}{\sigma^2}+\tfrac{1}{\delta^2}\right)^{-1}\right]\right\}$$

(10.3.5)

for $\theta \in \Re, t \in \Re$. The expression in Equation (10.3.5) resembles a normal density without a normalizing constant.

> We plugged a normal prior and ended with a normal posterior.
> Implies: A normal pdf for ϑ is a conjugate prior.

With

$$\mu = \left(\frac{t}{\sigma^2}+\frac{\tau}{\delta^2}\right)\left(\frac{n}{\sigma^2}+\frac{1}{\delta^2}\right)^{-1} \text{ and } \sigma_0^2 = \left(\frac{n}{\sigma^2}+\frac{1}{\delta^2}\right)^{-1}$$

the posterior distribution of ϑ is $N(\mu,\sigma_0^2)$. ▲

> A conjugate prior may not be reasonable in every problem.
> Look at Examples 10.5.1 and 10.5.2.

10.4 Point Estimation

Now, we explore briefly how we may approach point estimation problems under a particular *loss function*. The data consist of a random sample

$\mathbf{X} = (X_1, ..., X_n)$ given $\vartheta = \theta$. Suppose that a real valued statistic T is (minimal) sufficient for θ given that $\vartheta = \theta$. Let $h(\theta)$ be a *prior* for ϑ.

An arbitrary point estimator of ϑ is denoted by $\delta \equiv \delta(T)$, which takes a value $\delta(t)$ when one observes $T = t, t \in \mathcal{T}$. Suppose that the discrepancy in estimating ϑ by $\delta(T)$ is measured by:

$$L^*(\vartheta, \delta) \equiv L^*(\vartheta, \delta(T)) = [\delta(T) - \vartheta]^2 \qquad (10.4.1)$$

which is referred to as the *squared error loss*.

The mean squared error (MSE) discussed in Section 7.3.1 will be the weighted average of loss function from Equation (10.4.1) with respect to weights assigned by $g(t; \theta)$. In other words, this average is actually a conditional average given that $\vartheta = \theta$. Given $\vartheta = \theta$, the *risk function* associated with δ is:

$$R^*(\theta, \delta) \equiv E_{T|\vartheta=\theta}[L^*(\theta, \delta)] = \int_{\mathcal{T}} L^*(\theta, \delta(t))g(t; \theta)dt \qquad (10.4.2)$$

This is the *frequentist risk* which would be referred to as MSE_δ in the context of Section 7.3. Now, it may be possible to have δ_1 and δ_2 with risk functions $R^*(\theta, \delta_i), i = 1, 2$ where $R^*(\theta, \delta_1) > R^*(\theta, \delta_2)$ for some θ values, whereas $R^*(\theta, \delta_1) \leq R^*(\theta, \delta_2)$ for other θ values. So, by comparing the frequentist risks of δ_1 and δ_2, it may be hard to judge which estimator is decisively superior! At this point, one could employ more advanced decision-theoretic principles such as *minimaxity* or *admissibility* in order to arrive at some resolution.

A Bayesian would follow a different route. The prior, $h(\theta)$, sets a sense of preference ordering on Θ. So, while comparing δ_1 and δ_2, one may consider averaging the frequentist risks $R^*(\theta, \delta_i), i = 1, 2$, with respect to the prior, $h(\theta)$, and then check to see which weighted average frequentist risk is smaller. Obviously, the estimator with a smaller average risk should be preferred.

Define the *Bayes risk* (as opposed to frequentist risk):

$$r^*(\vartheta, \delta) \equiv E_\vartheta[R^*(\vartheta, \delta)] = \int_\Theta R^*(\theta, \delta)h(\theta)d\theta \qquad (10.4.3)$$

Now, let $\mathcal{D}$ be a class of all estimators of ϑ whose Bayes risks are finite. Then, the *best* estimator under the Bayesian paradigm will be δ^* from $\mathcal{D}$ such that

$$r^*(\vartheta, \delta^*) = \min_{\delta \in \mathcal{D}} r^*(\vartheta, \delta) \qquad (10.4.4)$$

Such an estimator is called the *Bayes* estimator of ϑ. In many problems, the Bayes estimator δ^* is unique.

Let us suppose that we consider only those estimators δ and prior $h(\theta)$ so that both $R^*(\theta, \delta)$ and $r^*(\vartheta, \delta)$ are finite.

Theorem 10.4.1 *A Bayes estimator $\delta^* \equiv \delta^*(T)$ is determined in such a way that the posterior risk of $\delta^*(t)$ is the least possible, that is,*

$$\int_\Theta L^*(\theta, \delta^*(t))k(\theta; t)d\theta = \min_{\delta \in \mathcal{D}} \int_\Theta L^*(\theta, \delta(t))k(\theta; t)d\theta \qquad (10.4.5)$$

for all observed data $t \in \mathcal{T}$.

Proof Let $m(t) > 0$, and express the Bayes risk as follows:

$$
\begin{aligned}
r^*(\vartheta, \delta) &= \int_\Theta \int_{\mathcal{T}} L^*(\theta, \delta(t)) g(t; \theta) h(\theta) dt d\theta \\
&= \int_{\mathcal{T}} \left[\int_\Theta L^*(\theta, \delta(t)) k(t; \theta) d\theta \right] m(t) dt
\end{aligned}
\tag{10.4.6}
$$

In the last step, we first used the relationship $g(t; \theta) h(\theta) = k(t; \theta) m(t)$. Then, we used the fact that the order of the double integral $\int_\Theta \int_{\mathcal{T}}$ may be changed to $\int_{\mathcal{T}} \int_\Theta$ because the integrands are nonnegative. This interchanging of the order of integration is justified in view of a result known as Fubini's Theorem (Exercise 10.4.5).

Now, having observed the data $T = t$, the Bayes estimate $\delta^*(t)$ must be the one associated with the $\min_{\delta \in \mathcal{D}} \int_\Theta L^*(\theta, \delta(t)) k(\theta; t) d\theta$, that is, the smallest possible posterior risk. ∎

An attractive feature of a Bayes estimator δ^* is this: having observed $T = t$, one can explicitly determine $\delta^*(t)$ by minimizing the posterior risk as stated in Equation (10.4.5). In the case of a squared error loss function, the determination of a Bayes estimator happens to be very simple.

Theorem 10.4.2 *In the case of a squared error loss function from Equation (10.4.1), a Bayes estimate $\delta^* \equiv \delta^*(t)$ is the mean of the posterior distribution $k(\theta; t)$, that is,*

$$
\delta^*(t) = \int_\Theta \theta k(\theta; t) d\theta \equiv E_{\theta|T=t}[\vartheta]
\tag{10.4.7}
$$

for all observed data $t \in \mathcal{T}$.

Proof To determine $\delta^*(T)$, we must minimize $\int_\Theta L^*(\theta, \delta(t)) k(\theta; t) d\theta$ with respect to δ, for every *fixed* $t \in \mathcal{T}$. Rewrite

$$
\begin{aligned}
\int_\Theta L^*(\theta, \delta(t)) k(\theta; t) d\theta &= \int_\Theta [\theta^2 - 2\theta\delta + \delta^2] k(\theta; t) d\theta \\
&= a - 2\delta E_{\theta|T=t}[\vartheta] + \delta^2
\end{aligned}
\tag{10.4.8}
$$

where we denote $a \equiv a(t) = \int_\Theta \theta^2 k(\theta; t) d\theta$ and use the fact that $\int_\Theta k(\theta; t) d\theta = 1$. Now, we look at the expression $a - 2\delta E_{\theta|T=t}[\vartheta] + \delta^2$ involving $\delta \equiv \delta(t)$ and wish to minimize this with respect to δ. One can accomplish this task easily. We leave out the details as an exercise. ∎

Example 10.4.1 (Example 10.3.1 Continued) From Equation (10.3.2), the posterior distribution of ϑ given $T = t$ was Beta$(t + \alpha, n - t + \beta)$ for $t \in \mathcal{T} = \{0, 1, ..., n\}$. Under the squared error loss, the Bayes estimate of ϑ will be the mean of this posterior distribution. One can check easily that the mean simplifies to $(t + \alpha)/(\alpha + \beta + n)$ so that:

$$
\text{Bayes estimator of } \vartheta \text{ is } \widehat{\vartheta}_B \equiv \frac{(T + \alpha)}{(\alpha + \beta + n)}
\tag{10.4.9}
$$

We may rewrite this estimator as follows:

$$\hat{\vartheta}_B = \frac{n(T/n) + (\alpha + \beta)[\alpha/(\alpha + \beta)]}{(\alpha + \beta + n)} = \frac{n\overline{X} + (\alpha + \beta)[\alpha/(\alpha + \beta)]}{n + \alpha + \beta}$$

(10.4.10)

Given $\vartheta = \theta$, note that the *maximum likelihood estimator* (MLE) or *uniformly minimum variance unbiased estimator* (UMVUE) of θ would be $\overline{X}$, whereas the mean of the prior distribution is $\alpha/(\alpha + \beta)$. The Bayes estimator is a weighted average of $\overline{X}$ and $\alpha/(\alpha + \beta)$. If n is large, the classical estimator $\overline{X}$ is weighted more than the mean of the prior belief, $\alpha/(\alpha + \beta)$. For small n, the sample mean $\overline{X}$ is weighted less. ▲

Example 10.4.2 (Example 10.3.3 Continued) Posterior distribution of ϑ was $N(\mu, \sigma_0^2)$ where

$$\mu = \left(\frac{t}{\sigma^2} + \frac{\tau}{\delta^2}\right)\left(\frac{n}{\sigma^2} + \frac{1}{\delta^2}\right)^{-1} \text{ and } \sigma_0^2 = \left(\frac{n}{\sigma^2} + \frac{1}{\delta^2}\right)^{-1}$$

Under the squared error loss, a Bayes estimator of ϑ will be the mean of this posterior distribution, namely, μ. In other words,

$$\text{Bayes estimator of } \vartheta \text{ is } \left(\frac{t}{\sigma^2} + \frac{\tau}{\delta^2}\right)\left(\frac{n}{\sigma^2} + \frac{1}{\delta^2}\right)^{-1} \qquad (10.4.11)$$

From the likelihood function, given $\vartheta = \theta$, note that the MLE or UMVUE of θ would be $\overline{X}$, whereas the mean of the prior distribution is τ. The Bayes estimator is a weighted average of $\overline{X}$ and τ. If $\frac{n}{\sigma^2}$ is larger than $\frac{1}{\delta^2}$, then $\overline{X}$ is weighted more than the prior belief. When $\frac{n}{\sigma^2}$ is smaller than $\frac{1}{\delta^2}$, the prior mean of ϑ is trusted more than the sample mean $\overline{X}$. ▲

10.5   Examples with a Nonconjugate Prior

The following examples use nonconjugate priors. These examples emphasize the point that even though a prior is nonconjugate, in some cases, one may be able to derive *analytical* expressions of the posterior distribution and the Bayes estimator.

Example 10.5.1 Let X be $N(\theta, 1)$ given $\vartheta = \theta$ where ϑ is unknown so that X is minimal sufficient for θ given $\vartheta = \theta$. We are told that ϑ is *positive* and so there is no point in assuming a normal prior for ϑ. For simplicity, let the prior ϑ be exponential with a known mean $\alpha^{-1}(> 0)$. We proceed to find the marginal pdf $m(x)$. The joint distribution of (ϑ, X) is:

$$\alpha(\sqrt{2\pi})^{-1}exp\{-\tfrac{1}{2}(x^2 - 2\theta x + \theta^2) - \alpha\theta\}$$
$$= \alpha(\sqrt{2\pi})^{-1}exp\{\tfrac{1}{2}\alpha^2 - \alpha x\}\times$$
$$exp\{-\tfrac{1}{2}[\theta - (x - \alpha)]^2\} \text{ for } x \in \Re, \theta \in \Re^+$$

(10.5.1)

Denote a standard normal pdf $\phi(y) = (\sqrt{2\pi})^{-1} exp\{-\frac{1}{2}y^2\}$ and df $\Phi(y) = \int_{-\infty}^{y} \phi(u)du$, $y \in \Re$. Thus, for all $x \in \Re$, the marginal pdf of X is:

$$
\begin{aligned}
m(x) \\
&= \alpha exp\{\tfrac{1}{2}\alpha^2 - \alpha x\} \int_{\theta=0}^{\infty} \phi(\theta - [x-\alpha])d\theta \\
&= \alpha exp\{\tfrac{1}{2}\alpha^2 - \alpha x\}\{1 - \Phi(-[x-\alpha])\} \\
&= \alpha exp\{\tfrac{1}{2}\alpha^2 - \alpha x\}\Phi([x-\alpha])
\end{aligned}
\tag{10.5.2}
$$

Combining Equations (10.5.1) and (10.5.2), we obtain the posterior pdf of ϑ as follows: For all $x \in \Re, \theta \in \Re^+$,

$$
\begin{aligned}
k(\theta; x) \\
&= \{\Phi([x-\alpha])\}^{-1}(\sqrt{2\pi})^{-1} exp\{-\tfrac{1}{2}[\theta - (x-\alpha)]^2\} \\
&= \{\Phi([x-\alpha])\}^{-1}\phi(\theta - [x-\alpha])
\end{aligned}
\tag{10.5.3}
$$

The prior is *not* conjugate and yet we have an expression of the posterior pdf in a closed form. ▲

Example 10.5.2 (Example 10.5.1 Continued) The posterior distribution of ϑ was given by $k(\theta; x) = \{\Phi([x-\alpha])\}^{-1}\phi(\theta - [x-\alpha])$ for $\theta > 0, -\infty < x < \infty$.

In view of Theorem 10.4.2, under the squared error loss, a Bayes estimate of ϑ is the posterior mean. In other words, the Bayes estimate $\widehat{\vartheta}_B$ is:

$$
\begin{aligned}
\int_{\theta=0}^{\infty} \theta k(\theta; x)d\theta \\
&= \{\Phi([x-\alpha])\}^{-1} \int_{\theta=0}^{\infty} \theta\phi(\theta - [x-\alpha])d\theta \\
&= \{\Phi([x-\alpha])\}^{-1} \int_{\theta=0}^{\infty} [\theta - (x-\alpha)]\phi(\theta - [x-\alpha])d\theta \\
&\quad + \{\Phi([x-\alpha])\}^{-1}(x-\alpha) \int_{\theta=0}^{\infty} \phi(\theta - [x-\alpha])d\theta \\
&= I_1 + I_2, \text{ say.}
\end{aligned}
\tag{10.5.4}
$$

It is easy to check that $I_2 = x - \alpha$. Next, rewrite I_1 as

$$
\begin{aligned}
\{\Phi([x-\alpha])\}^{-1} \int_{y=-(x-\alpha)}^{\infty} y\phi(y)dy \\
&= (\sqrt{2\pi})^{-1}\{\Phi([x-\alpha])\}^{-1} \int_{u=(x-\alpha)^2/2}^{\infty} e^{-u}du \\
&= \phi(x-\alpha)/\Phi(x-\alpha)
\end{aligned}
\tag{10.5.5}
$$

Combining Equations (10.5.4) and (10.5.5), we write:

$$
\widehat{\vartheta}_B \equiv (x-\alpha) + \frac{\phi(X-\alpha)}{\Phi(X-\alpha)}
\tag{10.5.6}
$$

for the Bayes estimator of ϑ. ▲

10.6   Exercises and Complements

10.2.1 Suppose that $X_1, ..., X_n$ are iid with a common pdf:

$$f(x; \theta) = \theta e^{-\theta x} I(x > 0)$$

given $\vartheta = \theta$ where $\vartheta(> 0)$ is unknown. Assume a prior density:

$$h(\theta) = \alpha e^{-\alpha \theta} I(\theta > 0)$$

with $\alpha(> 0)$ is known. Derive the posterior pdf of ϑ where $T \equiv \Sigma_{i=1}^{n} X_i$ takes the value $t > 0$.

10.3.1 Suppose that X_1, X_2 are independent, X_1 is $N(\theta, 1)$, and X_2 is $N(2\theta, 3)$ given $\vartheta = \theta$ where $\vartheta(\in \Re)$ is unknown. Consider the minimal sufficient statistic T for θ given $\vartheta = \theta$. Suppose that a prior for ϑ is $N(5, \tau^2)$ with $\tau(> 0)$ known. Derive the posterior pdf of ϑ. {*Hint*: Given $\vartheta = \theta$, is the statistic T normally distributed?}

10.3.2 Suppose that $X_1, ..., X_n$ are iid Exponential(θ) given $\vartheta = \theta$ where $\vartheta(> 0)$ is unknown. Let ϑ have an *inverted gamma prior*, denoted by IGamma(α, β), that is:

$$h(\theta) = \{\beta^\alpha \Gamma(\alpha) \theta^{\alpha+1}\}^{-1} exp\{-1/(\theta\beta)\} I(\theta > 0)$$

where α, β are known positive numbers. Denote the minimal sufficient statistic $T \equiv \Sigma_{i=1}^{n} X_i$ given $\vartheta = \theta$.

(i)   Show that the prior distribution of ϑ is IGamma(α, β) if and only if ϑ^{-1} has a Gamma(α, β) distribution;

(ii)   Show that the posterior distribution of ϑ is IGamma(α^*, β^*) with $\alpha^* = n + \alpha, \beta^* = \{t + \beta^{-1}\}^{-1}$.

10.3.3 Suppose $X_1, ..., X_n$ are iid $N(0, \theta^2)$ given $\vartheta = \theta$ where $\vartheta(> 0)$ is unknown. Assume that ϑ^2 has an inverted gamma prior IGamma(α, β) where α, β are known positive numbers. Denote the minimal sufficient statistic $T \equiv \Sigma_{i=1}^{n} X_i^2$ given $\vartheta = \theta$. Show that the posterior distribution of ϑ^2 is an inverted gamma distribution.

10.3.4 (Pareto Prior) Suppose that $X_1, ..., X_n$ are iid Uniform$(0, \theta)$ given $\vartheta = \theta$ where $\vartheta(> 0)$ is unknown. Let ϑ have a *Pareto* prior, denoted by Pareto(α, β), that is:

$$h(\theta) = \beta \alpha^\beta \theta^{-(\beta+1)} I(\alpha < \theta < \infty)$$

where α, β are known positive numbers. Denote the minimal sufficient statistic $T \equiv X_{n:n}$ given $\vartheta = \theta$. Show that the posterior distribution of ϑ is Pareto$(\max(t, \alpha), n + \beta)$.

10.3.5 Suppose that $X_1, ..., X_n$ are iid Uniform$(-\theta, \theta)$ given $\vartheta = \theta$ where $\vartheta(> 0)$ is unknown. Let ϑ have a Pareto(α, β) prior where α, β are known positive numbers. Denote the minimal sufficient statistic

$$T \equiv |X|_{n:n}, \text{ the largest among } |X_1|, ..., |X_n|, \text{ given } \vartheta = \theta$$

Show that the posterior distribution of ϑ is a Pareto distribution.

10.4.1 (Example 10.3.2 Continued) Under a squared error loss function, find the Bayes estimator for ϑ.

10.4.2 (Exercise 10.3.1 Continued) Under a squared error loss function, find the Bayes estimator for ϑ.

10.4.3 (Exercise 10.3.3 Continued) Under a squared error loss function, find the Bayes estimator for ϑ.

10.4.4 (Exercise 10.3.4 Continued) Under a squared error loss function, find the Bayes estimator for ϑ.

10.4.5 (Fubini's Theorem) Suppose that a function of two real variables $g(x_1, x_2)$ is either nonnegative or integrable on a space $\mathcal{X} = \mathcal{X}_1 \times \mathcal{X}_2 (\subseteq \Re \times \Re)$. That is, for all $(x_1, x_2) \in \mathcal{X}$, the function $g(x_1, x_2)$ is either nonnegative or integrable. Show that the order of (two-dimensional) integrals can be interchanged, that is:

$$\int_{\mathcal{X}_2} \int_{\mathcal{X}_1} g(x_1, x_2) dx_1 dx_2 = \int_{\mathcal{X}_2} \left\{ \int_{\mathcal{X}_1} g(x_1, x_2) dx_1 \right\} dx_2$$
$$= \int_{\mathcal{X}_1} \left\{ \int_{\mathcal{X}_2} g(x_1, x_2) dx_2 \right\} dx_1$$

{*Note*: This result is included here for reference purposes and completeness. It was needed in the proof of Theorem 10.4.1.}

10.5.1 (Example 10.5.2 Continued) Check directly that $\widehat{\vartheta}_B$ is always positive. {*Hint*: It will suffice to prove directly that

$$p(x) = x\Phi(x) + \phi(x) > 0$$

for all $x \in \Re$.}

10.5.2 Suppose that $X_1, ..., X_n$ are iid $N(\theta, 4)$ given $\vartheta = \theta$ where $\vartheta (\in \Re^+)$ is unknown. Let a prior pdf be:

$$h(\theta) = 2\alpha e^{-2\alpha\theta} I(\theta > 0) \text{ with } \alpha(> 0) \text{ known}$$

Consider $T \equiv \overline{X}$, the sample mean, which is minimal sufficient for θ given $\vartheta = \theta$.

(i)   Derive the marginal pdf $m(t)$ of T for $t \in \Re$;
(ii)   Derive the posterior pdf $k(\theta; t)$ for ϑ given $T = t$, for $\theta > 0$ and $-\infty < t < \infty$;
(iii) Under the squared error loss, find $\widehat{\vartheta}_B$ and show that $\widehat{\vartheta}_B$ can take only positive values.

10.5.3 Suppose that an observation X is distributed exponentially with mean θ^{-1} given $\vartheta = \theta$ where $\vartheta(\in \Re^+)$ is unknown. Let a prior pdf be:

$$h(\theta) = \theta\exp(-\theta^2/2)I(\theta > 0)$$

Consider $T = X$, which is minimal sufficient for θ given $\vartheta = \theta$.

(i)   Derive the marginal pdf $m(t)$ of T for $t \in \Re^+$;

(ii)   Derive the posterior pdf $k(\theta; t)$ for ϑ given $T = t$, for $\theta > 0$ and $0 < t < \infty$;

(iii)   Under the squared error loss, find $\widehat{\vartheta}_B$.

11

Likelihood Ratio and Other Tests

11.1 Introduction

In Chapter 8, the Neyman-Pearson theory of *uniformly most powerful* (UMP) level α tests was developed. We recall that even in a one-parameter situation, sometimes a UMP level α test does not exist for deciding between a simple null and a two-sided alternative hypothesis. In Section 8.5.1, we cited a testing problem for the mean of a normal distribution (with known variance) when the alternative hypothesis was two-sided! In many situations where a UMP test cannot be found, likelihood ratio tests often save the day by providing an indispensable statistical tool. A general approach for testing composite null and alternative hypotheses was developed by Neyman and Pearson (1928a,b, 1933a,b).

We begin with *independent and identically distributed* (iid) real valued observations $X_1, ..., X_n$ having a common *probability density function* (pdf) $f(x; \boldsymbol{\theta})$ where the unknown parameter $\boldsymbol{\theta} \equiv (\theta_1, ..., \theta_p)$ consists of $p(\geq 1)$ components, $\boldsymbol{\theta} \in \Theta(\subseteq \Re^p)$. We wish to test:

a null hypothesis   an alternative hypothesis
$H_0 : \boldsymbol{\theta} \in \Theta_0$ vs. $H_1 : \boldsymbol{\theta} \in \Theta_1$

with given level α where $\Theta_0 \subset \Theta, \Theta_1 = \Theta - \Theta_0, 0 < \alpha < 1$. We write down the likelihood function:

$$L(\boldsymbol{\theta}) = \Pi_{i=1}^n f(x_i; \boldsymbol{\theta}), \boldsymbol{\theta} \in \Theta \qquad (11.1.1)$$

It is customary to interpret $\max_{\boldsymbol{\theta} \in \Theta_0} L(\boldsymbol{\theta})$ as the best evidence in favor of H_0 and interpret $\max_{\boldsymbol{\theta} \in \Theta} L(\boldsymbol{\theta})$ as the overall best evidence. Now, a *likelihood ratio* (LR) test statistic is defined as:

$$\Lambda = \max_{\boldsymbol{\theta} \in \Theta_0} L(\boldsymbol{\theta}) / \max_{\boldsymbol{\theta} \in \Theta} L(\boldsymbol{\theta}) \qquad (11.1.2)$$

whereas the *LR test* (LRT) is implemented as follows:

Reject H_0 if and only if Λ is "small" $(< k)$ (11.1.3)

Note that "small" values of Λ are associated with "small" values of $\max_{\boldsymbol{\theta} \in \Theta_0} L(\boldsymbol{\theta})$ relative to $\max_{\boldsymbol{\theta} \in \Theta} L(\boldsymbol{\theta})$. The rationale is that if the best evidence in favor of H_0 appears weak compared with the overall evidence, then H_0 ought to be rejected!

It is easy to see that $0 < \Lambda < 1$ because in the definition of Λ, the maximum in the numerator (denominator) is taken over a smaller (larger) set $\Theta_0(\Theta)$. The cut-off number $k \in (0,1)$ has to be so chosen that the LRT from Equation (11.1.3) has level α, that is:

$$\max_{\boldsymbol{\theta} \in \Theta_0} P_{\boldsymbol{\theta}}\{\Lambda < k\} = \alpha \tag{11.1.4}$$

In order to maintain the level of this text, we handle only a special kind of null hypothesis. We test:

$$H_0 : \theta_1 = \theta_1^* \text{ vs. } H_1 : \theta_1 \neq \theta_1^* \tag{11.1.5}$$

where θ_1^* is a specified value for the (sub)parameter θ_1. One may note that H_0 may not be simple in general because even though it specifies a fixed value for θ_1, other components $\theta_2, ..., \theta_p$ remain unknown and arbitrary!

How should one evaluate $\max_{\boldsymbol{\theta} \in \Theta_0} L(\boldsymbol{\theta})$? First, in the expression of $L(\boldsymbol{\theta})$, one will replace θ_1 by θ_1^* and then maximize $L(\theta_1^*, \theta_2, ..., \theta_p)$ with respect to $\theta_2, ..., \theta_p$ by substituting their respective *maximum likelihood estimators* (MLEs) when $\theta_1 = \theta_1^*$. On the other hand, $\max_{\boldsymbol{\theta} \in \Theta} L(\boldsymbol{\theta})$ is found by replacing all parameters $\theta_1, ..., \theta_p$ with their respective MLEs in $L(\boldsymbol{\theta})$.

Section 11.2 introduces LR tests for the mean and variance of a normal population. In Section 11.3, we discuss LR tests for comparing the means and variances of two independent normal populations. Section 11.4 summarizes test procedures for the population correlation coefficient ρ and for comparing the means as well as the variances under bivariate normality.

11.2   One-Sample LR Tests: Normal

In this section, LR tests are developed for the mean and variance in a *normal* population that match with the two-sided tests that are customarily employed in practice. Suppose that $X_1, ..., X_n$ are iid $N(\mu, \sigma^2)$ where $\mu \in \Re, \sigma \in \Re^+$. As usual, denote:

$$\overline{X} = n^{-1}\Sigma_{i=1}^n X_i \text{ and } S^2 = (n-1)^{-1}\Sigma_{i=1}^n (X_i - \overline{X})^2$$

11.2.1   Mean: Unknown Mean and Known Variance

Assume that only μ is unknown, but σ^2 is known. Given $\alpha \in (0,1)$, we test $H_0 : \mu = \mu_0$ vs. $H_1 : \mu \neq \mu_0$ with level α. Here, μ_0 is a fixed real number. Since σ is known, we have $\theta = \mu, \Theta_0 = \{\mu_0\}, \Theta = \Re$, and we do not require S^2. The likelihood function is:

$$L(\mu) = \{\sigma\sqrt{2\pi}\}^{-n} exp\{-\tfrac{1}{2\sigma^2}\Sigma_{i=1}^n (x_i - \mu)^2\}, \mu \in \Re \tag{11.2.1}$$

Observe that

$$\max_{\mu \in \Theta_0} L(\mu) = \{\sigma\sqrt{2\pi}\}^{-n} exp\{-\tfrac{1}{2\sigma^2}\Sigma_{i=1}^n (x_i - \mu_0)^2\} \qquad (11.2.2)$$

On the other hand, one has

$$\max_{\mu \in \Theta} L(\mu) = \{\sigma\sqrt{2\pi}\}^{-n} exp\{-\tfrac{1}{2\sigma^2}\Sigma_{i=1}^n (x_i - \overline{x})^2\}, \text{ since } \overline{x}, \qquad (11.2.3)$$
$$\text{the MLE of } \mu, \text{ maximizes } L(\mu) \text{ over } \Theta$$

Now, note that with any real number c, we can write

$$\Sigma_{i=1}^n (x_i - c)^2 = \Sigma_{i=1}^n (x_i - \overline{x})^2 + n(\overline{x} - c)^2 \qquad (11.2.4)$$

and hence the LR Λ from Equation (11.1.2) becomes:

$$exp\{-\tfrac{1}{2\sigma^2} [\Sigma_{i=1}^n (x_i - \mu_0)^2 - \Sigma_{i=1}^n (x_i - \overline{x})^2]\}$$
$$= exp\{-\tfrac{1}{2\sigma^2} [n(\overline{x} - \mu_0)^2]\}, \text{ in view of Equation (11.2.4)} \qquad (11.2.5)$$

The LRT from Equation (11.1.3) rejects H_0 if and only if Λ is "small" so that we decide as follows:

$$\text{Reject } H_0 \text{ if and only if } n(\overline{X} - \mu_0)^2/\sigma^2 > k \qquad (11.2.6)$$

where $k(> 0)$ is a *generic* constant. That is, we reject H_0 if and only if $\overline{X} - \mu_0$, when properly scaled, is too large ($> k$) or too small ($< -k$). The implementable form of the LRT will be:

$$\text{Reject } H_0 \text{ if and only if } |\sqrt{n}(\overline{X} - \mu_0)/\sigma| > z_{\alpha/2} \qquad (11.2.7)$$

The level of this two-sided Z test is:

$$P\{\text{Reject } H_0 \text{ when } \mu = \mu_0\}$$
$$= P\{|\sqrt{n}(\overline{X} - \mu_0)/\sigma| > z_{\alpha/2} \text{ when } \mu = \mu_0\}$$

which is α, since $\sqrt{n}(\overline{X} - \mu_0)/\sigma$ is distributed as $N(0,1)$ if $\mu = \mu_0$. Equation (11.2.7) gives the customary two-sided Z test with equal tails ($= \tfrac{1}{2}\alpha$).

From Equation (11.2.6), one could equivalently express the LRT as follows:

$$\text{Reject } H_0 \text{ if and only if } n(\overline{X} - \mu_0)^2/\sigma^2 > \chi_{1,\alpha}^2 \qquad (11.2.8)$$

where $\chi_{1,\alpha}^2$ is the upper $100\alpha\%$ point of a χ_1^2 distribution. Note that $z_{\alpha/2}^2 \equiv \chi_{1,\alpha}^2$!

> Section 8.5.1: No UMP test exists for this problem!

11.2.2 Mean: Unknown Mean and Variance

Assume that μ, σ^2 are both unknown. Given $\alpha \in (0, 1)$, we test $H_0 : \mu = \mu_0$ vs. $H_1 : \mu \neq \mu_0$ with level α. Here, μ_0 is a fixed real number. In this case, H_0, H_1 are not simple hypotheses. We have $\boldsymbol{\theta} = (\mu, \sigma^2), \Theta_0 = \{(\mu_0, \sigma^2) : \mu_0$ is fixed, $\sigma \in \Re^+\}$ and $\Theta = \{(\mu, \sigma^2) : \mu \in \Re, \sigma \in \Re^+\}$. Assume that the sample size $n \geq 2$. The likelihood function is:

$$L(\mu, \sigma^2) = \{\sigma\sqrt{2\pi}\}^{-n} exp\{-\Sigma_{i=1}^n (x_i - \mu)^2/(2\sigma^2)\}, (\mu, \sigma^2) \in \Re \times \Re^+ \tag{11.2.9}$$

Observe that:

$$\max_{(\mu, \sigma^2) \in \Theta_0} L(\mu, \sigma^2)$$

$$= \max_{\sigma^2 > 0} \{\sigma\sqrt{2\pi}\}^{-n} exp\{-\Sigma_{i=1}^n (x_i - \mu_0)^2/(2\sigma^2)\}$$

$$= \{\widehat{\sigma}\sqrt{2\pi}\}^{-n} exp\{-\Sigma_{i=1}^n (x_i - \mu_0)^2/(2\widehat{\sigma}^2)\}, \text{ since}$$

$$\widehat{\sigma}^2 = n^{-1}\Sigma_{i=1}^n (x_i - \mu_0)^2 \text{ is the MLE of } \sigma^2 \text{ if } \mu = \mu_0$$

$$= \{\widehat{\sigma}\sqrt{2\pi}\}^{-n} exp\{-n/2\} \tag{11.2.10}$$

On the other hand, one has:

$$\max_{(\mu, \sigma^2) \in \Theta} L(\mu, \sigma^2)$$

$$= \{\widehat{\sigma}^*\sqrt{2\pi}\}^{-n} exp\{-\Sigma_{i=1}^n (x_i - \overline{x})^2/(2\widehat{\sigma}^{*2})\},$$

$$\text{since } \overline{x}, \widehat{\sigma}^{*2} = n^{-1}\Sigma_{i=1}^n (x_i - \overline{x})^2 \text{ are the} \tag{11.2.11}$$

$$\text{MLEs of } \mu \text{ and } \sigma^2 \text{ respectively}$$

$$= \{\widehat{\sigma}^*\sqrt{2\pi}\}^{-n} exp\{-n/2\}$$

Now, combine Equations (11.1.3), (11.2.4), (11.2.10), and (11.2.11) to write:

$$\Lambda = \{\widehat{\sigma}^{*2}/\widehat{\sigma}^2\}^{n/2} = \left[1 + \frac{n(\overline{x} - \mu_0)^2}{\Sigma_{i=1}^n (x_i - \overline{x})^2}\right]^{-n/2} \tag{11.2.12}$$

Again, one would reject H_0 if and only if Λ is "small." So, we decide as follows:

Reject H_0 if and only if $n(\overline{X} - \mu_0)^2/\Sigma_{i=1}^n (X_i - \overline{X})^2 > k$ (11.2.13)

where $k(> 0)$ is a *generic* constant. That is, we reject H_0 if and only if $\overline{X} - \mu_0$, when properly scaled, becomes too large $(> k)$ or too small $(< -k)$. The implementable form of the LRT will be:

Reject H_0 if and only if $\left|\sqrt{n}(\overline{X} - \mu_0)/S\right| > t_{n-1, \alpha/2}$ (11.2.14)

The level of this two-sided t test is:

$$P\{\text{Reject } H_0 \text{ when } \mu = \mu_0\}$$

$$= P\{\left|\sqrt{n}(\overline{X} - \mu_0)/S\right| > t_{n-1, \alpha/2} \text{ when } \mu = \mu_0\} \tag{11.2.15}$$

which is α, since $\sqrt{n}(\overline{X} - \mu_0)/S$ has a Student's t_{n-1} distribution when $\mu = \mu_0$. This is the customary two-sided Student's t test with equal tails $(= \frac{1}{2}\alpha)$.

11.2.3   Variance: Unknown Mean and Variance

Assume that μ, σ^2 are both unknown. Given $\alpha \in (0,1)$, we test $H_0 : \sigma^2 = \sigma_0^2$ vs. $H_1 : \sigma^2 \neq \sigma_0^2$ with level α. Here, σ_0^2 is a fixed positive number. In this case, H_0, H_1 are not simple hypotheses. We have $\boldsymbol{\theta} = (\mu, \sigma^2), \Theta_0 = \{(\mu, \sigma_0^2) : \mu \in \mathfrak{R}, \sigma_0^2 \in \mathfrak{R}^+ \text{ is fixed}\}$ and $\Theta = \{(\mu, \sigma^2) : \mu \in \mathfrak{R}, \sigma^2 \in \mathfrak{R}^+\}$. We assume that sample size $n \geq 2$.

The likelihood function is given by Equation (11.2.9). Observe that:

$$
\begin{aligned}
\max_{(\mu,\sigma^2)\in\Theta_0} & L(\mu, \sigma^2) \\
&= \max_{\mu\in\mathfrak{R}}\{\sigma_0\sqrt{2\pi}\}^{-n} exp\{-\Sigma_{i=1}^n(x_i - \mu)^2/(2\sigma_0^2)\} \\
&= \{\sigma_0\sqrt{2\pi}\}^{-n} \ exp\{-\Sigma_{i=1}^n(x_i - \overline{x})^2/(2\sigma_0^2)\},
\end{aligned}
\tag{11.2.16}
$$

since $\overline{x}$ is the MLE of μ if $\sigma = \sigma_0$

As before, one has $\max_{(\mu,\sigma^2)\in\Theta} L(\mu, \sigma^2) = \{\hat{\sigma}^*\sqrt{2\pi}\}^{-n} exp\{-n/2\}$ from Equation (11.2.11) with $\hat{\sigma}^{*2} = n^{-1}\Sigma_{i=1}^n(x_i - \overline{x})^2$. Combine this with Equations (11.1.2) and (11.2.16) to obtain:

$$
\Lambda = \left[\left(\hat{\sigma}^{*2}/\sigma_0^2\right) exp\left\{-\left(\hat{\sigma}^{*2}/\sigma_0^2\right) +1\right\}\right]^{n/2}
\tag{11.2.17}
$$

One ought to reject H_0 if and only if Λ is "small." So, we decide as follows:

$$
\text{Reject } H_0 \text{ if and only if } \left(\hat{\sigma}^{*2}/\sigma_0^2\right) exp\left\{1 - (\hat{\sigma}^{*2}/\sigma_0^2)\right\} < k
\tag{11.2.18}
$$

where $k(> 0)$ is a *generic* constant.

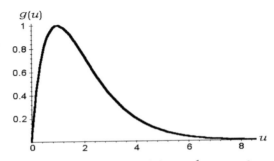

Figure 11.2.1. Plot of $g(u) = ue^{1-u}, u > 0$.

In order to obtain an implementable form, we proceed as follows: consider a function, $g(u) = ue^{1-u}$ for $u > 0$, and investigate when it is small

$(< k)$. Note that $g(1) = 1$ and $g'(u) = \{(1 - u)/u\}g(u)$ which is positive (negative) when $u < 1$ ($u > 1$). Hence, the function $g(u)$ is strictly increasing (decreasing) on the left- (right-) hand side of $u = 1$. Thus, $g(u)$ is going to be "small" for both small and large values of $u(> 0)$. This feature is clear from Figure 11.2.1. Thus, we rewrite the LRT from Equation (11.2.18) as follows:

$$\text{Reject } H_0 \text{ if and only if } \hat{\sigma}^{*2}/\sigma_0^2 < a \text{ or } \hat{\sigma}^{*2}/\sigma_0^2 > b$$

$$\text{for } 0 < a < b < \infty \tag{11.2.19}$$

as long as a, b are chosen so that the test has level α.

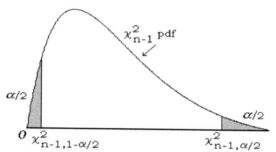

Figure 11.2.2. Two-sided χ_{n-1}^2 rejection region.

Recall that $(n - 1)S^2/\sigma_0^2 = \Sigma_{i=1}^n (X_i - \overline{X})^2/\sigma_0^2 = n\hat{\sigma}^{*2}/\sigma_0^2$ has a χ_{n-1}^2 distribution if $\sigma^2 = \sigma_0^2$ and hence a level α LRT is:

$$\text{Reject } H_0 \text{ if and only if } (n - 1)S^2/\sigma_0^2 > \chi_{n-1,\alpha/2}^2 \text{ or}$$

$$(n - 1)S^2/\sigma_0^2 < \chi_{n-1,1-\alpha/2}^2 \tag{11.2.20}$$

This is the customary two-sided Chi-square test with equal tails $(= \frac{1}{2}\alpha)$. See Figure 11.2.2.

11.3 Two-Sample LR Tests: Independent Normal

Now, we focus on some traditional two-sample problems for *independent* normal populations. The LR test procedures that we obtain do match with the two-sided tests that are customarily employed in practice. Suppose that $X_{i1}, ..., X_{in_i}$ are iid $N(\mu_i, \sigma_i^2), i = 1, 2$, and that X_1, X_2 are independent, $-\infty < \mu_1, \mu_2 < \infty, 0 < \sigma_1^2, \sigma_2^2 < \infty$. As usual, let

$$\overline{X}_i = n_i^{-1}\Sigma_{j=1}^{n_i} X_{ij}, S_i^2 = (n_i - 1)^{-1}\Sigma_{j=1}^{n_i}(X_{ij} - \overline{X}_i)^2$$

$$\text{and } S_P^2 = (n_1 + n_2 - 2)^{-1}\left\{(n_1 - 1)S_1^2 + (n_2 - 1)S_2^2\right\}, i = 1, 2$$

$$\tag{11.3.1}$$

denote the sample means, sample variances, and the pooled sample variance.

11.3.1 Compare Means: Unknown Means and Common Unknown Variance

Suppose that $\sigma_1^2 = \sigma_2^2 = \sigma^2$, but the common variance σ^2 is unknown and μ_1, μ_2 are unknown too. Given $\alpha \in (0,1)$, we test $H_0 : \mu_1 = \mu_2$ vs. $H_1 : \mu_1 \neq \mu_2$ with level α. Here, H_0, H_1 are not simple hypotheses. Assume that $n_i \geq 2, i = 1,2$.

Since H_0 specifies that the means are the same, we have $\Theta_0 = \{(\mu, \mu, \sigma^2) : \mu \in \Re, \sigma^2 \in \Re^+\}$ and $\Theta = \{(\mu_1, \mu_2, \sigma^2) : \mu_1 \in \Re, \mu_2 \in \Re, \sigma^2 \in \Re^+\}$. The likelihood function is:

$$L(\mu_1, \mu_2, \sigma^2) = \{\sigma\sqrt{2\pi}\}^{-(n_1+n_2)} exp\left\{-(2\sigma^2)^{-1}\Sigma_{i=1}^2\Sigma_{j=1}^{n_i}(x_{ij} - \mu_i)^2\right\},$$

$$(\mu_1, \mu_2, \sigma^2) \in \Re \times \Re \times \Re^+$$

$$(11.3.2)$$

so that

$$\max_{(\mu_1,\mu_2,\sigma^2)\in\Theta_0} L(\mu_1, \mu_2, \sigma^2)$$

$$= \max_{-\infty<\mu<\infty,0<\sigma^2<\infty} \{\sigma\sqrt{2\pi}\}^{-(n_1+n_2)} exp\{-\Sigma_{i=1}^2\Sigma_{j=1}^{n_i}(x_{ij} - \mu)^2/(2\sigma^2)\}$$

$$(11.3.3)$$

One may check that the MLEs of μ, σ^2 from this restricted likelihood function are:

$$\widehat{\mu} = (n_1\bar{x}_1 + n_2\bar{x}_2)/(n_1 + n_2), \widehat{\sigma^2} = \Sigma_{i=1}^2\Sigma_{j=1}^{n_i}(x_{ij} - \widehat{\mu})^2/(n_1 + n_2)$$

$$(11.3.4)$$

so that from Equations (11.3.3) and (11.3.4) we have:

$$\max_{(\mu_1,\mu_2,\sigma^2)\in\Theta_0} L(\mu_1, \mu_2, \sigma^2) = \{\widehat{\sigma}\sqrt{2\pi}\}^{-(n_1+n_2)} exp\{-(n_1 + n_2)/2\}$$

$$(11.3.5)$$

On the other hand, one has:

$$\max_{(\mu_1,\mu_2,\sigma^2)\in\Theta} L(\mu_1, \mu_2, \sigma^2)$$

$$= \{\widehat{\sigma}^*\sqrt{2\pi}\}^{-(n_1+n_2)} exp\{-\Sigma_{i=1}^2\Sigma_{j=1}^{n_i}(x_{ij} - \bar{x}_i)^2/(2\widehat{\sigma}^{*2})\},$$

since $\bar{x}_1, \bar{x}_2$ and $\widehat{\sigma}^{*2} = (n_1 + n_2)^{-1}\Sigma_{i=1}^2\Sigma_{j=1}^{n_i}(x_{ij} - \bar{x}_i)^2$

are the MLEs of μ_1, μ_2, σ^2

$$= \{\widehat{\sigma}^*\sqrt{2\pi}\}^{-(n_1+n_2)} exp\{-(n_1 + n_2)/2\}$$

$$(11.3.6)$$

Combine Equation (11.1.2) with Equations (11.3.5) and (11.3.6) to write:

$$\Lambda = \{\widehat{\sigma}^{*2}/\widehat{\sigma}^2\}^{(n_1+n_2)/2} = \left[\frac{\Sigma_{i=1}^2\Sigma_{j=1}^{n_i}(x_{ij} - \bar{x}_i)^2}{\Sigma_{i=1}^2\Sigma_{j=1}^{n_i}(x_{ij} - \widehat{\mu})^2}\right]^{(n_1+n_2)/2}$$

$$(11.3.7)$$

with $\widehat{\mu}$ from Equation (11.3.4). We reject H_0 if and only if Λ is "small." Thus, we decide as follows:

$$\text{Reject } H_0 \text{ if and only if } \frac{\Sigma_{i=1}^2 \Sigma_{j=1}^{n_i} (X_{ij} - \overline{X}_i)^2}{\Sigma_{i=1}^2 \Sigma_{j=1}^{n_i} (X_{ij} - \widehat{\mu})^2} < k \qquad (11.3.8)$$

where $k(> 0)$ is a *generic* constant. Again, we utilize Equation (11.2.4) to write

$$\Sigma_{i=1}^2 \Sigma_{j=1}^{n_i} (x_{ij} - \widehat{\mu})^2 = \Sigma_{i=1}^2 \Sigma_{j=1}^{n_i} (x_{ij} - \overline{x}_i)^2 + \Sigma_{i=1}^2 n_i (\overline{x}_i - \widehat{\mu})^2$$

$$= \Sigma_{i=1}^2 \Sigma_{j=1}^{n_i} (x_{ij} - \overline{x}_i)^2 + \left(\frac{n_1 n_2}{n_1 + n_2}\right) (\overline{x}_1 - \overline{x}_2)^2$$

so that one has:

$$\frac{\Sigma_{i=1}^2 \Sigma_{j=1}^{n_i} (X_{ij} - \widehat{\mu})^2}{\Sigma_{i=1}^2 \Sigma_{j=1}^{n_i} (X_{ij} - \overline{X}_i)^2} = 1 + \left(\frac{n_1 n_2}{n_1 + n_2}\right) \frac{(\overline{X}_1 - \overline{X}_2)^2}{\Sigma_{i=1}^2 \Sigma_{j=1}^{n_i} (X_{ij} - \overline{X}_i)^2} \qquad (11.3.9)$$

In other words, "small" values of $\Sigma_{i=1}^2 \Sigma_{j=1}^{n_i} (X_{ij} - \overline{X}_i)^2 / \Sigma_{i=1}^2 \Sigma_{j=1}^{n_i} (X_{ij} - \widehat{\mu})^2$ will correspond to "large" values of $(\overline{X}_1 - \overline{X}_2)^2 / \Sigma_{i=1}^2 \Sigma_{j=1}^{n_i} (X_{ij} - \overline{X}_i)^2$. Thus, we can rewrite the LRT from Equation (11.3.8) as:

$$\text{Reject } H_0 \text{ if and only if } \frac{|\overline{X}_1 - \overline{X}_2|}{\sqrt{\Sigma_{i=1}^2 \Sigma_{j=1}^{n_i} (X_{ij} - \overline{X}_i)^2}} > k \qquad (11.3.10)$$

Now, note that $(n_1 + n_2 - 2)S_P^2 = \Sigma_{i=1}^2 \Sigma_{j=1}^{n_i} (X_{ij} - \overline{X}_i)^2$ and from Example 4.5.2, recall that $\{n_1^{-1} + n_2^{-1}\}^{-1/2} (\overline{X}_1 - \overline{X}_2)S_P^{-1}$ has the $t_{n_1+n_2-2}$ distribution under H_0. In view of Equation (11.3.10), the implementable form of a level α LRT test is:

$$\text{Reject } H_0 \text{ if and only if}$$
$$\{n_1^{-1} + n_2^{-1}\}^{-1/2} |\overline{X}_1 - \overline{X}_2| S_P^{-1} > t_{n_1+n_2-2, \alpha/2} \qquad (11.3.11)$$

Observe that the LRT rejects H_0 when $\overline{X}_1 - \overline{X}_2$ is sizably different from zero with proper scaling. This is the customary two-sided Student's t test with the pooled sample variance and equal tails $(= \frac{1}{2}\alpha)$.

> Testing the equality of means of independent normal populations with unknown and unequal variances is complicated. This is referred to as: the *Behrens-Fisher problem*. A simple-minded confidence interval was suggested in Exercise 9.3.6.

11.3.2 Compare Variances: Unknown Means and Variances

We denote $\boldsymbol{\theta} = (\mu_1, \mu_2, \sigma_1^2, \sigma_2^2)$ and assume that all parameters are unknown. Given $\alpha \in (0,1)$, we test $H_0 : \sigma_1^2 = \sigma_2^2$ vs. $H_1 : \sigma_1^2 \neq \sigma_2^2$ with level α.

Again, H_0, H_1 are not simple hypotheses. Since H_0 specifies that two variances are the same, we write $\Theta_0 = \{(\mu_1, \mu_2, \sigma^2, \sigma^2) : \mu_1 \in \Re, \mu_2 \in \Re, \sigma^2 \in \Re^+\}$, and $\Theta = \{(\mu_1, \mu_2, \sigma_1^2, \sigma_2^2) : \mu_1 \in \Re, \mu_2 \in \Re, \sigma_1^2 \in \Re^+, \sigma_2^2 \in \Re^+\}$. The likelihood function is given by Equation (11.3.2) so that

$$\max_{(\mu_1,\mu_2,\sigma^2,\sigma^2)\in\Theta_0} L(\mu_1,\mu_2,\sigma_1^2,\sigma_2^2)$$

$$= \max_{-\infty<\mu_1,\mu_2<\infty,0<\sigma^2<\infty} \{\sigma\sqrt{2\pi}\}^{-(n_1+n_2)} exp\{-\tfrac{1}{2\sigma^2}\Sigma_{i=1}^2\Sigma_{j=1}^{n_i}(x_{ij}-\mu_i)^2\}$$

$$(11.3.12)$$

One may check that the MLEs of μ_1, μ_2, σ^2 obtained from this restricted likelihood function are

$$\widehat{\mu}_1 = \overline{x}_1, \widehat{\mu}_2 = \overline{x}_2, \widehat{\sigma^2} = \Sigma_{i=1}^2\Sigma_{j=1}^{n_i}(x_{ij}-\overline{x}_i)^2/(n_1+n_2) \qquad (11.3.13)$$

so that from Equations (11.3.12) and (11.3.13), we have:

$$\max_{(\mu_1,\mu_2,\sigma^2,\sigma^2)\in\Theta_0} L(\mu_1,\mu_2,\sigma_1^2,\sigma_2^2) = \{\widehat{\sigma}\sqrt{2\pi}\}^{-(n_1+n_2)} exp\{-(n_1+n_2)/2\}$$

$$(11.3.14)$$

On the other hand, one has

$$\max_{(\mu_1,\mu_2,\sigma_1^2,\sigma_2^2)\in\Theta} L(\mu_1,\mu_2,\sigma_1^2,\sigma_2^2)$$

$$= \{\sqrt{2\pi}\}^{-(n_1+n_2)}(\widehat{\sigma}_1^{n_1}\widehat{\sigma}_2^{n_2})^{-1} exp\{-\tfrac{1}{2}\Sigma_{i=1}^2\tfrac{1}{\sigma_i^2}\Sigma_{j=1}^{n_i}(x_{ij}-\overline{x}_i)^2\},$$

since $\overline{x}_i, \widehat{\sigma}_i^2 = n_i^{-1}\Sigma_{j=1}^{n_i}(x_{ij}-\overline{x}_i)^2$ are the MLEs of

$$\mu_i, \sigma_i^2, i = 1, 2$$

$$= \{\sqrt{2\pi}\}^{-(n_1+n_2)}(\widehat{\sigma}_1^{n_1}\widehat{\sigma}_2^{n_2})^{-1} exp\{-(n_1+n_2)/2\}$$

$$(11.3.15)$$

Combine Equation (11.1.2) with Equations (11.3.14) and (11.3.15) to write

$$\Lambda = \frac{\widehat{\sigma}_1^{n_1}\widehat{\sigma}_2^{n_2}}{\widehat{\sigma}^{(n_1+n_2)}} = \frac{aS_1^{n_1}S_2^{n_2}}{[S_1^2+bS_2^2]^{(n_1+n_2)/2}} \qquad (11.3.16)$$

where $a \equiv a(n_1, n_2), b \equiv b(n_1, n_2)$ are positive numbers involving n_1, n_2 only. Now, one will reject H_0 if and only if Λ is "small." So, we decide as follows:

$$\text{Reject } H_0 \text{ if and only if } \frac{S_1^{n_1}S_2^{n_2}}{[S_1^2+bS_2^2]^{(n_1+n_2)/2}} < k$$

or equivalently,

$$\text{Reject } H_0 \text{ if and only if } \frac{(S_1^2/S_2^2)^{n_1/2}}{[(S_1^2/S_2^2)+b]^{(n_1+n_2)/2}} < k \qquad (11.3.17)$$

where $k(> 0)$ is a *generic* constant.

In order to express the LRT in an implementable form, we proceed as follows: consider the function

$$g(u) = u^{n_1/2}(u + b)^{-(n_1+n_2)/2} \text{ for } u > 0$$

and investigate its behavior in order to check when it is small $(< k)$. Note that

$$g'(u) = \tfrac{1}{2}u^{(n_1-2)/2}(u + b)^{-(n_1+n_2+2)/2}\{n_1 b - n_2 u\}$$

which is positive (negative) when $u < (>)n_1 b/n_2$. Hence, $g(u)$ is strictly increasing (decreasing) on the left (right) hand side of $u = n_1 b/n_2$. Thus, $g(u)$ ought to be "small" for both small or large values of $u(> 0)$.

So, we rewrite the LRT from Equation (11.3.17) as follows:

$$\text{Reject } H_0 \text{ if and only if } S_1^2/S_2^2 < c \text{ or } S_1^2/S_2^2 > d$$
$$\text{for } 0 < c < d < \infty \tag{11.3.18}$$

where the numbers c, d are to be chosen in such a way that the test has level α.

Recall that S_1^2/S_2^2 has a F_{n_1-1,n_2-1} distribution when $\sigma_1^2 = \sigma_2^2$. Hence, a level α LRT would be:

$$\text{Reject } H_0 \text{ if and only if } S_1^2/S_2^2 > F_{n_1-1,n_2-1,\alpha/2} \text{ or}$$
$$S_1^2/S_2^2 < F_{n_1-1,n_2-1,1-\alpha/2} \tag{11.3.19}$$

Thus, this test rejects H_0 when S_1^2/S_2^2 is sizably different from one. The LRT is the customary two-sided F test with equal tails $(= \tfrac{1}{2}\alpha)$.

11.4   Bivariate Normal

Let pairs of random variables (X_{1j}, X_{2j}) be iid $N_2(\mu_1, \mu_2, \sigma_1^2, \sigma_2^2, \rho), j = 1, ..., n$ with all parameters unknown, $(\mu_l, \sigma_l^2) \in \Re \times \Re^+, l = 1, 2$ and $-1 < \rho < 1$. Test procedures are summarized for comparing means μ_1, μ_2, for the correlation coefficient ρ, and also for comparing the variances σ_1^2, σ_2^2. Among these, the test for ρ qualifies as a bonafide LRT.

11.4.1   Compare Means: Paired t Method

Given $\alpha \in (0, 1)$, we test $H_0 : \mu_1 = \mu_2$ vs. an upper-, lower-, or two-sided alternative hypothesis H_1 with level α when $n \geq 2$. Denote:

$$Y_j = X_{1j} - X_{2j}, j = 1, ..., n, \text{ and } \overline{Y} = n^{-1}\Sigma_{j=1}^n Y_j,$$
$$S^2 = (n - 1)^{-1}\Sigma_{j=1}^n (Y_j - \overline{Y})^2 \tag{11.4.1}$$

Observe that the paired "differences" $Y_1, ..., Y_n$ are iid $N(\mu_1 - \mu_2, \sigma^2)$ where $\sigma^2 = \sigma_1^2 + \sigma_2^2 - 2\rho\sigma_1\sigma_2$. Since $\mu_1 - \mu_2$ and σ^2 are unknown, the two-sample problem on hand has reduced to a one-sample problem in terms of the pivot:

$$U \equiv \frac{n^{1/2}[\overline{Y} - (\mu_1 - \mu_2)]}{S} \overset{\text{Under}}{\underset{H_0}{\Rightarrow}} U_{calc} = \frac{n^{1/2}\overline{Y}}{S} \qquad (11.4.2)$$

Under $H_0 : \mu_1 = \mu_2$, the test statistic U has the Student's t_{n-1} distribution.

Upper-Sided Alternative:

Test $H_0 : \mu_1 = \mu_2$ vs. $H_1 : \mu_1 > \mu_2$. In the spirit of Section 8.4, we propose the following level α test:

Reject H_0 in favor of $H_1 : \mu_1 > \mu_2$ if and only if $U_{calc} > t_{n-1,\alpha}$
$$(11.4.3)$$

Lower-Sided Alternative:

Test $H_0 : \mu_1 = \mu_2$ vs. $H_1 : \mu_1 < \mu_2$. In the spirit of Section 8.4, we propose the following level α test:

Reject H_0 in favor of $H_1 : \mu_1 < \mu_2$ if and only if $U_{calc} < -t_{n-1,\alpha}$
$$(11.4.4)$$

Two-Sided Alternative:

Test $H_0 : \mu_1 = \mu_2$ vs. $H_1 : \mu_1 \neq \mu_2$. In the spirit of Equation (11.2.14), we propose the following level α test:

Reject H_0 in favor of $H_1 : \mu_1 \neq \mu_2$ if and only if
$$(11.4.5)$$
$$U_{calc} < -t_{n-1,\alpha/2} \text{ or } U_{calc} > t_{n-1,\alpha/2}$$

11.4.2 LR Test for Correlation Coefficient

Given $\alpha \in (0,1)$, we test $H_0 : \rho = 0$ vs. $H_1 : \rho \neq 0$ with level α when $n \geq 3$. Denote $\theta = (\mu_1, \mu_2, \sigma_1^2, \sigma_2^2, \rho)$, write $\Theta_0 = \{(\mu_1, \mu_2, \sigma_1^2, \sigma_2^2, 0) : \mu_1 \in \Re, \mu_2 \in \Re, \sigma_1^2 \in \Re^+, \sigma_2^2 \in \Re^+\}$, and $\Theta = \{(\mu_1, \mu_2, \sigma_1^2, \sigma_2^2, \rho) : \mu_1 \in \Re, \mu_2 \in \Re, \sigma_1^2 \in \Re^+, \sigma_2^2 \in \Re^+, \rho \in (-1,1)\}$. The likelihood function is:

$$L(\mu_1, \mu_2, \sigma_1^2, \sigma_2^2, \rho)$$

$$= \{2\pi\sigma_1\sigma_2\sqrt{(1-\rho^2)}\}^{-n} exp \left\{ -\frac{1}{2(1-\rho^2)} \left[\frac{\sum_{j=1}^n (x_{1j} - \mu_1)^2}{\sigma_1^2} \right. \right.$$

$$\left. \left. + \frac{\sum_{j=1}^n (x_{2j} - \mu_2)^2}{\sigma_2^2} - \frac{2\rho\sum_{j=1}^n (x_{1j} - \mu_1)(x_{2j} - \mu_2)}{\sigma_1\sigma_2} \right] \right\} \qquad (11.4.6)$$

for all $\theta \in \Theta$. We leave it as an exercise to show that the MLEs for $\mu_1, \mu_2, \sigma_1^2, \sigma_2^2$ and ρ are respectively given by:

$$\overline{x}_1, \overline{x}_2, u_1^2 = n^{-1}\sum_{i=1}^n (x_{1i} - \overline{x}_1)^2, u_2^2 = n^{-1}\sum_{i=1}^n (x_{2i} - \overline{x}_2)^2$$

$$\text{and } r = n^{-1}\sum_{i=1}^n (x_{1i} - \overline{x}_1)(x_{2i} - \overline{x}_2)/(u_1 u_2)$$

These stand for the customary sample means, sample variances (not unbiased), and the sample correlation coefficient. From Equation (11.4.6), one has

$$\max_{\theta \in \Theta} L(\mu_1, \mu_2, \sigma_1^2, \sigma_2^2, \rho)$$
$$= \{2\pi u_1 u_2 \sqrt{(1 - r^2)}\}^{-n} exp\{-n\} \tag{11.4.7}$$

Under H_0, that is, when $\rho = 0$, the likelihood function reduces to

$$L(\mu_1, \mu_2, \sigma_1^2, \sigma_2^2)$$
$$= \{2\pi\sigma_1\sigma_2\}^{-n} exp\left\{-\frac{1}{2}\left[\frac{\sum_{j=1}^{n}(x_{1j} - \mu_1)^2}{\sigma_1^2} + \frac{\sum_{j=1}^{n}(x_{2j} - \mu_2)^2}{\sigma_2^2}\right]\right\} \tag{11.4.8}$$

Again, we leave it as an exercise to show that the MLEs for μ_1, μ_2, σ_1^2 and σ_2^2 are respectively given by $\bar{x}_1, \bar{x}_2, u_1^2$ and u_2^2. So, one has:

$$\max_{\theta \in \Theta_0} L(\mu_1, \mu_2, \sigma_1^2, \sigma_2^2) = \{2\pi u_1 u_2\}^{-n} exp\{-n\} \tag{11.4.9}$$

Now, one will reject H_0 if and only if $\Lambda \equiv (1 - r^2)^{n/2}$ is "small." So, we decide as follows:

$$\text{Reject } H_0 \text{ if and only if } (1 - r^2)^{n/2} < k$$

or equivalently

$$\text{Reject } H_0 \text{ if and only if } r^2/(1 - r^2) > k \tag{11.4.10}$$

where $k(> 0)$ is a *generic* constant. Note that $r^2/(1 - r^2)$ is a one-to-one function of r^2. It is easy to see that this test rejects H_0 when r is sizably different from zero.

Now, recall from Equation (4.6.9) that $r\sqrt{n-2}/\sqrt{1-r^2}$ has the Student's t_{n-2} distribution when $\rho = 0$. This is why we assumed that n was at least three. The derivation of this sampling distribution was one of the earliest fundamental contributions of Fisher (1915). From Equation (11.4.10), a level α LR test can be expressed as follows:

$$\text{Reject } H_0 \text{ if and only if } \left|\frac{r\sqrt{n-2}}{\sqrt{1-r^2}}\right| > t_{n-2,\alpha/2} \tag{11.4.11}$$

Upper-Sided Alternative:

Test $H_0 : \rho = 0$ vs. $H_1 : \rho > 0$. We propose the following upper-sided level α test:

$$\text{Reject } H_0 \text{ in favor of } H_1 : \rho > 0 \text{ if and only if } \frac{r\sqrt{n-2}}{\sqrt{1-r^2}} > t_{n-2,\alpha} \tag{11.4.12}$$

Lower-Sided Alternative:

Test $H_0 : \rho = 0$ vs. $H_1 : \rho < 0$. We propose the following lower-sided level α test:

$$\text{Reject } H_0 \text{ in favor of } H_1 : \rho < 0 \text{ if and only if } \frac{r\sqrt{n-2}}{\sqrt{1-r^2}} < -t_{n-2,\alpha}$$

$$(11.4.13)$$

11.4.3 Compare Variances

Given $\alpha \in (0,1)$, we test $H_0 : \sigma_1^2 = \sigma_2^2$ against an upper-, lower-, or two-sided alternative hypothesis H_1 with level α when $n \geq 3$. Denote:

$$Y_{1j} = X_{1j} + X_{2j}, Y_{2j} = X_{1j} - X_{2j}, j = 1, ..., n$$

$$\overline{Y}_1 = n^{-1}\Sigma_{j=1}^n Y_{1j}, \overline{Y}_2 = n^{-1}\Sigma_{j=1}^n Y_{2j}$$

$$S_1^2 = (n-1)^{-1}\Sigma_{j=1}^n (Y_{1j} - \overline{Y}_1)^2, S_2^2 = (n-1)^{-1}\Sigma_{j=1}^n (Y_{2j} - \overline{Y}_2)^2$$

$$r^* = (n-1)^{-1}\Sigma_{j=1}^n (Y_{1j} - \overline{Y}_1)(Y_{2j} - \overline{Y}_2)/(S_1 S_2)$$

$$(11.4.14)$$

Note that the observed pairs (Y_{1i}, Y_{2i}) are iid $N_2(\nu_1, \nu_2, \tau_1^2, \tau_2^2, \rho^*), i = 1, ..., n(\geq 3)$ where $\nu_1 = \mu_1 + \mu_2, \nu_2 = \mu_1 - \mu_2, \tau_1^2 = \sigma_1^2 + \sigma_2^2 + 2\rho\sigma_1\sigma_2, \tau_2^2 = \sigma_1^2 + \sigma_2^2 - 2\rho\sigma_1\sigma_2$ and $Cov(Y_{1i}, Y_{2i}) = \sigma_1^2 - \sigma_2^2$ so that $\rho^* = (\sigma_1^2 - \sigma_2^2)/(\tau_1\tau_2)$. Of course, these newly defined parameters $\nu_1, \nu_2, \tau_1^2, \tau_2^2, \rho^*$ are all unknown, $(\nu_l, \tau_l^2) \in \Re \times \Re^+, l = 1, 2$ and $-1 < \rho^* < 1$.

It is clear that the testing problem for $H_0 : \sigma_1^2 = \sigma_2^2$ reduces to that for $H_0 : \rho^* = 0$, whereas the upper-, lower-, or two-sided alternative hypothesis regarding σ_1^2, σ_2^2 will surely translate into an upper-, lower-, or two-sided alternative hypothesis regarding ρ^*, namely, $\rho^* > 0, \rho^* < 0$, and $\rho^* \neq 0$. So, a level α test procedure can be derived by mimicking the proposed methodologies from Section 11.4.2, by replacing r with r^* obtained from the transformed data $(Y_{1j}, Y_{2j}), j = 1, ..., n(\geq 3)$.

The final test procedures will be identical with those given by Equations (11.4.11) and (11.4.13) respectively for a two-sided, upper-sided, and lower-sided H_1, as long as r^* is substituted in place of r.

11.5 Exercises and Complements

11.2.1 Verify Equation (11.2.4).

11.2.2 Let $X_1, ..., X_n$ be iid $N(0, \sigma^2)$ where $\sigma^2(> 0)$ is an unknown parameter. Given $\alpha \in (0,1)$, derive a level α LRT for $H_0 : \sigma^2 = \sigma_0^2(> 0)$ vs. $H_1 : \sigma^2 \neq \sigma_0^2$ in an implementable form.

11.2.3 Let $X_1, ..., X_n$ be iid $N(\mu, \sigma^2)$ where $\sigma^2(\in \Re^+)$ is unknown but $\mu(\in \Re)$ is assumed known. Given $\alpha \in (0,1)$, derive a level α LRT for $H_0 : \sigma^2 = \sigma_0^2(> 0)$ vs. $H_1 : \sigma^2 \neq \sigma_0^2$ in an implementable form.

11.2.4 Suppose that X_1, X_2 are iid $N(\mu, \sigma^2)$ where $\mu(\in \Re), \sigma^2(\in \Re^+)$ are both assumed unknown. Given $\alpha \in (0, 1)$, reconsider the level α LRT from Equation (11.2.20) for $H_0 : \sigma^2 = \sigma_0^2(> 0)$ vs. $H_1 : \sigma^2 \neq \sigma_0^2$. Show that the same test can be expressed as follows:

$$\text{Reject } H_0 \text{ if and only if } |X_1 - X_2| > \sqrt{2}\sigma_0 z_{\alpha/2}$$

11.2.5 Suppose $X_1, ..., X_n$ are iid with a common pdf:

$$\theta^{-1} exp\{-x/\theta\}I(x > 0)$$

where $\theta(> 0)$ is unknown. Given $\alpha \in (0, 1)$, derive a level α LRT for $H_0 : \theta = \theta_0(> 0)$ vs. $H_1 : \theta \neq \theta_0$ in an implementable form.

11.2.6 Suppose $X_1, ..., X_n$ are iid with a common pdf

$$2\theta^{-1} x\, exp(-x^2/\theta)I(x > 0)$$

where $\theta(> 0)$ is unknown. Given $\alpha \in (0, 1)$, derive a level α LRT for $H_0 : \theta = \theta_0(> 0)$ vs. $H_1 : \theta \neq \theta_0$ in an implementable form.

11.2.7 Suppose $X_1, ..., X_n$ are iid having a common pdf:

$$\{x\sigma\sqrt{2\pi}\}^{-1} exp\left\{-[log(x) - \mu]^2/(2\sigma^2)\right\}$$

with $x > 0, -\infty < \mu < \infty, 0 < \sigma < \infty$. Here, μ is the only unknown parameter. Given $\alpha \in (0, 1)$, derive a level α LRT for $H_0 : \mu = \mu_0$ vs. $H_1 : \mu \neq \mu_0$ in an implementable form.

11.2.8 (Pareto Distribution) Suppose $X_1, ..., X_n$ are iid from a *Pareto population* with its pdf:

$$\theta\gamma^\theta x^{-\theta-1}I(\gamma < x < \infty)$$

where $\theta \in \Re^+, \gamma \in \Re^+$ are unknown parameters.

(i)   Show that the MLEs of γ and θ are given by $\hat{\gamma} = X_{n:1}$, $\hat{\theta} = n/T$ where $T = log(\Pi_{i=1}^n X_i/X_{n:1}^n)$;

(ii)   Show that the likelihood ratio test for $H_0 : \theta = 1$ vs. $H_1 : \theta \neq 1$ would reject H_0 if and only if $T \leq a$ or $T \geq b$, for $0 < a < b < \infty$;

(iii)   Give the equal-tailed level α likelihood ratio test from part (ii) in an implementable form.

11.3.1 Verify the expressions of the MLEs in Equation (11.3.4).

11.3.2 Suppose that $X_{i1}, ..., X_{in_i}$ are iid $N(\mu_i, \sigma_i^2), i = 1, 2$, and X_1, X_2 are independent. Assume that $(\mu_1, \mu_2) \in \Re \times \Re$ are unknown and $(\sigma_1^2, \sigma_2^2) \in \Re^+ \times \Re^+$ are known. Given $\alpha \in (0, 1)$, derive a level α LRT for $H_0 : \mu_1 = \mu_2$ vs. $H_1 : \mu_1 \neq \mu_2$ in an implementable form. {*Hint:* The LRT rejects H_0 if and only if $|\overline{X}_1 - \overline{X}_2| > z_{\alpha/2}\sqrt{n_1^{-1}\sigma_1^2 + n_2^{-1}\sigma_2^2}$.}

11.3.3 Suppose $X_{i1}, ..., X_{in_i}$ are iid $N(\mu_i, \sigma^2), i = 1, 2$, and X_1, X_2 are independent. Assume that $(\mu_1, \mu_2) \in \Re \times \Re$ are unknown and $\sigma^2 \in \Re^+$ is known. Given $\alpha \in (0, 1)$, derive a level α LRT for $H_0 : \mu_1 = \mu_2$ vs. $H_1 : \mu_1 \neq \mu_2$ in an implementable form. {*Hint:* The LRT rejects H_0 if and only if $|\overline{X}_1 - \overline{X}_2| > \sigma z_{\alpha/2}\sqrt{n_1^{-1} + n_2^{-1}}$.}

11.3.4 Verify the expressions of the MLEs in Equation (11.3.13).

11.3.5 Suppose $X_{i1}, ..., X_{in_i}$ are iid $N(0, \sigma_i^2), i = 1, 2$, and X_1, X_2 are independent. Assume that $(\sigma_1^2, \sigma_2^2) \in \Re^+ \times \Re^+$ are unknown. Given $\alpha \in (0, 1)$, derive a level α LRT $H_0 : \sigma_1^2 = \sigma_2^2$ vs. $H_1 : \sigma_1^2 \neq \sigma_2^2$ in an implementable form.

11.3.6 Let $X_{i1}, ..., X_{in_i}$ be iid Exponential(θ_i), $i = 1, 2$, and X_1, X_2 are independent, with $(\theta_1, \theta_2) \in \Re^+ \times \Re^+$ unknown. Given $\alpha \in (0, 1)$, derive a level α LRT for $H_0 : \theta_1 = \theta_2$ vs. $H_1 : \theta_1 \neq \theta_2$ in an implementable form. {*Hint:* Proceed along Section 11.3.2 in principle. With $0 < c < d < \infty$, a LRT rejects H_0 if and only if $\overline{X}_1/\overline{X}_2 < c$ or $> d$.}

11.4.1 A nutritional science project had involved eight overweight men of comparable background which *included* eating habits, family traits, health condition, and job-related stress. An experiment was conducted to study the average reduction in weight for overweight men following a regimen of nutritional diet and exercise. The technician weighed in each individual before they were to enter this program. At the conclusion of the study which took two months, each individual was weighed in again. It was believed that the assumption of a bivariate normal distribution would be reasonable to use for (X_1, X_2). The adjoining table gives the observed data.

ID# of Individual	Weight (x_1, pounds) before study	Weight (x_2, pounds) after study
1	235	220
2	189	175
3	156	150
4	172	160
5	165	169
6	180	170
7	170	173
8	195	180

Test at 5% level whether the true average weights taken before and after going through the regimen are significantly different.

11.4.2 Let (X_{1i}, X_{2i}) be iid $N_2(\mu_1, \mu_2, \sigma_1^2, \sigma_2^2, \rho), i = 1, ..., n(\geq 2)$. Assume that μ_1, μ_2 are unknown but $\sigma_1^2, \sigma_2^2, \rho$ are known where $(\mu_l, \sigma_l) \in \Re \times \Re^+, l = 1, 2$ and $-1 < \rho < 1$. Given $\alpha \in (0, 1)$, construct a level α test for $H_0 : \mu_1 = \mu_2$ vs. $H_1 : \mu_1 \neq \mu_2$ in an implementable form. {*Hint:* Improvise with the methodology from Section 11.4.1.}

11.4.3 Suppose that pairs of random variables (X_{1i}, X_{2i}) are iid bivariate normal, $N_2(\mu_1, \mu_2, \sigma^2, \sigma^2, \rho), i = 1, ..., n(\geq 2)$. Assume that all the parameters μ_1, μ_2, σ^2 and ρ are unknown where $(\mu_1, \mu_2) \in \Re \times \Re, \sigma^2 \in \Re^+$ and $-1 < \rho < 1$. With fixed $\alpha \in (0, 1)$, construct a level α test for $H_0 : \mu_1 = \mu_2$ against $H_1 : \mu_1 \neq \mu_2$ in an implementable form. {*Hint*: Improvise with the methodology from Section 11.4.1.}

11.4.4 In what follows, the data on systolic blood pressure (X_1) and age (X_2) for a sample of 10 women of similar health conditions are given.

ID #	X_1	X_2	ID #	X_1	X_2
1	122	41	6	144	44
2	148	52	7	138	51
3	146	54	8	138	56
4	162	60	9	145	49
5	135	45	10	144	58

At 5% level, test whether the population correlation coefficient ρ_{X_1, X_2} is significantly different from zero. We are told to assume bivariate normality of (X_1, X_2).

12

Large-Sample Methods

12.1 Introduction

Thus far, we have presented many standard topics and approaches, both in probability theory and statistical inference. One would have noticed that, except in Chapter 5, the proposed methodologies were predominantly exact whatever the sample size n. In this chapter, we summarize some *approximate* confidence interval and test procedures which work well when n is "large." These methods help one to achieve an *approximate* confidence coefficient or *approximate* level for some kinds of problems that cannot be handled exactly.

Section 12.2 briefly touches upon useful large-sample properties of a *maximum likelihood estimator* (MLE). We discuss Fisher's notion of *asymptotic relative efficiency* (ARE) along with an example in Section 12.3. In Section 12.4.1, we introduce large-sample confidence interval and test procedures for the mean of a population having an unknown distribution. Analogous methods for the success probability p in a Bernoulli distribution are briefly included in Section 12.4.2. The notion of a *variance stabilizing transformation* is mentioned in Section 12.5.

12.2 Maximum Likelihood Estimation

In this section, we provide some useful large-sample properties of an MLE. Consider $X_1, ..., X_n$ which are *independent and identically distributed* (iid) with a common *probability mass function* (pmf) or *probability density function* (pdf) $f(x; \theta)$, $x \in \mathcal{X} \subseteq \Re$ and $\theta \in \Theta \subseteq \Re$. Clearly, θ is assumed real valued. We also assume that the parameter space Θ is a subinterval of $\Re$. Now, having observed $\mathbf{X} = \mathbf{x}$, recall that a likelihood function is:

$$L(\theta) = \Pi_{i=1}^{n} f(x_i; \theta), \theta \in \Theta \qquad (12.2.1)$$

We mention some standing assumptions. These assumptions are similar in spirit to those used in the derivation of the Cramér-Rao Inequality (Section 7.5.1).

A1: $\frac{\partial}{\partial \theta}[log f(x; \theta)]$, $\frac{\partial^2}{\partial \theta^2}[log f(x; \theta)]$, and $\frac{\partial^3}{\partial \theta^3}[log f(x; \theta)]$ are assumed finite for all $x \in \mathcal{X}$ and $\theta \in \Theta$, an interval around true unknown θ.

A2: Consider $\int \frac{\partial}{\partial \theta}[f(x; \theta)]dx$, $\int \frac{\partial^2}{\partial \theta^2}[f(x; \theta)]dx$ and $\int \left\{ \frac{\partial}{\partial \theta}[f(x; \theta)] \right\}^2 dx$.

The first two integrals equal zero whereas the third integral is positive for true unknown $\theta \in \Theta$.

A3: For every θ in an interval around true unknown θ, let

$$\left| \frac{\partial^3}{\partial \theta^3} [\log f(x; \theta)] \right| < a(x)$$

such that $E_\theta[a(X_1)] < b$, a constant not involving θ.

Assumptions A1 through A3 are routinely satisfied by many standard distributions, for example, binomial, Poisson, normal, gamma, and exponential. In order to find the MLE for θ in regular cases, one frequently starts with a derivative of the log-likelihood function and solves the *likelihood equation*:

$$\frac{\partial}{\partial \theta} [\log L(\theta)] = 0 \qquad (12.2.2)$$

Will this equation necessarily have a solution? If it does, will the solution be unique? Assumptions A1 through A3 will guarantee that we can answer both questions in the affirmative.

> Likelihood Equation (12.2.2) has a solution, denoted by $\widehat{\theta}_n(\mathbf{x})$, such that $\widehat{\theta}_n(\mathbf{X})$ is a consistent estimator for θ. $\qquad (12.2.3)$

That is, an MLE of θ will stay close to the true value of θ with high probability if the sample size n is sufficiently large. One should also note the following crucial property of an MLE:

> Let $\widehat{\theta}_n(\mathbf{x})$ be a solution of Equation (12.2.2) and $\widehat{\theta}_n(\mathbf{X})$ be consistent for θ. Then, $\sqrt{n} \left[\widehat{\theta}_n(\mathbf{X}) - \theta \right] \overset{\mathcal{L}}{\to} N(0, 1/\mathcal{I}(\theta))$ as $n \to \infty$ if θ is the true value. Here, $\mathcal{I}(\theta)$ is the Fisher-information in a single observation.

$$(12.2.4)$$

In a variety of situations, the asymptotic variance $\sqrt{n}\widehat{\theta}_n(\mathbf{X})$ coincides with $1/\mathcal{I}(\theta)$. From Section 7.5.1, one may recall that $1/\mathcal{I}(\theta)$ was the Cramér-Rao lower bound (CRLB) for the variance of unbiased estimators of θ. What we are claiming now is then this: in many routine problems, the variance of $\widehat{\theta}_n(\mathbf{X})$, an MLE of θ, has asymptotically the smallest possible value.

This phenomenon was referred to as the *asymptotic efficiency* property of an MLE by Fisher (1922, 1925a,b) where he laid foundations of likelihood theory. Cramér (1946a, Chapter 33) pioneered mathematical treatments under full generality with assumptions along the lines of A1 through A3. One may note that the assumptions A1 through A3 are merely *sufficient* conditions for the stated large-sample properties of an MLE to hold. In the decades that followed Cramér's (1946a) numerous path-breaking contributions, many researchers including Cramér derived properties of the MLEs comparable to those stated in Equations (12.2.3) and (12.2.4), but under less restrictive assumptions.

12.3   Asymptotic Relative Efficiency

Suppose that there are two competing *consistent* estimators T_{1n} and T_{2n} for the same parametric function $\tau(\theta)$. Let us also suppose that

$$\sqrt{n}\left(T_{in} - \tau(\theta)\right) \overset{\mathcal{L}}{\to} N\left(0, \sigma_i^2\right) \text{ as } n \to \infty \qquad (12.3.1)$$

where $\sigma_i^2 \equiv \sigma_i^2(\theta), i = 1, 2$. Which estimator should one favor more? Fisher introduced a notion called *asymptotic relative efficiency* (ARE) of one estimator relative to another. One defines the asymptotic efficiency of T_{1n} relative to T_{2n} as follows:

$$ARE_{T_{1n}, T_{2n}}(\theta) \equiv \sigma_2^2 / \sigma_1^2 \qquad (12.3.2)$$

That is, if $\sigma_2^2 / \sigma_1^2 > (<)1$, $ARE_{T_{1n}, T_{2n}}$ will (not) exceed one so that T_{1n} would be considered more (less) efficient than T_{2n}. If $ARE_{T_{1n}, T_{2n}} = 1$, then according to this efficiency measure, the estimators T_{1n}, T_{2n} would have similar large-sample performance in the sense of ARE. Note that, in general, $ARE_{T_{1n}, T_{2n}}(\theta)$ may involve θ.

Example 12.3.1 Let $X_1, ..., X_n$ be iid $N(\theta, 1)$ with an unknown parameter $\theta \in \Re$. We wish to estimate:

$$\tau(\theta) \equiv P_\theta(X_1 > 0) = 1 - \Phi(-\theta) = \Phi(\theta)$$

Let $T_{1n} \equiv \Phi(\overline{X}_n)$, the MLE for $\tau(\theta)$, in view of the MLE's invariance property (Theorem 7.2.1). Obviously, $\sqrt{n}\left(\overline{X}_n - \theta\right)$ is distributed as $N(0, 1)$ when θ is true. Now, the Mann-Wald Theorem 5.3.4 implies:

$$\sqrt{n}\left(T_{1n} - \tau(\theta)\right) \overset{\mathcal{L}}{\to} N\left(0, \phi^2(\theta)\right) \text{ as } n \to \infty \qquad (12.3.3)$$

Consider another estimator of $\tau(\theta)$, defined as $T_{2n} \equiv n^{-1}\Sigma_{i=1}^n I(X_i > 0)$, which is the sample fraction of the observations that are positive. Both T_{1n}, T_{2n} are consistent estimators of $\tau(\theta)$. Also, T_{2n} is an unbiased estimator of $\tau(\theta)$. Note that $\Sigma_{i=1}^n I(X_i > 0)$ is distributed as Binomial$(n, \Phi(\theta))$ when θ obtains so that the *central limit theorem* (CLT) will give:

$$\sqrt{n}\left(T_{2n} - \tau(\theta)\right) \overset{\mathcal{L}}{\to} N\left(0, \Phi(\theta)[1 - \Phi(\theta)]\right) \text{ as } n \to \infty \qquad (12.3.4)$$

Now, combining Equations (12.3.3) and (12.3.4) with Equation (12.3.2), we have:

$$ARE_{T_{1n}, T_{2n}}(\theta) \equiv \frac{\Phi(\theta)[1 - \Phi(\theta)]}{\phi^2(\theta)} \qquad (12.3.5)$$

Here, ARE clearly depends upon θ, and the expression in Equation (12.3.5) is obviously symmetric in θ! *Without any loss of generality*, we assume that θ is nonnegative. Surely, $\lim_{\theta \to 0} ARE_{T_{1n}, T_{2n}}(\theta) = \frac{1}{2}\pi > 1$, which shows that the MLE of $\tau(\theta)$, T_{1n} is more efficient than T_{2n} when θ is near zero. Also, observe that $ARE_{T_{1n},T_{2n}}(\theta)$ is strictly increasing in $\theta(> 0)$ since

$$\frac{d}{d\theta}\left(\frac{\Phi(\theta)[1-\Phi(\theta)]}{\phi^2(\theta)}\right) = \frac{\phi(\theta)+2\theta\Phi^2(\theta)}{\phi^2(\theta)} > 0 \text{ for all } \theta(>0) \qquad (12.3.6)$$

That is, asymptotically, T_{1n} is a "more efficient" estimator for $\tau(\theta)$ than T_{2n} whatever be $\theta \geq 0$. Even if θ is near zero, $ARE_{T_{1n},T_{2n}}(\theta)$ is $1.5708(\approx \frac{1}{2}\pi)$ so that T_{1n} is already "more efficient" than T_{2n}! As $\theta(> 0)$ is chosen larger and larger, T_{1n} steadily becomes more and more efficient than T_{2n}. Similar arguments work if $\theta < 0$. According to the ARE criterion, asymptotically, T_{1n} wins over T_{2n} hands down, whatever be $\theta \in \Re$. ▲

12.4   Confidence Intervals and Tests of Hypotheses

We begin with *one-sample* problems and give confidence intervals and tests for the unknown mean μ of an arbitrary population having a finite variance. Next, such methodologies are mentioned very briefly for the success probability p in Bernoulli trials. Analogously, we also include *two-sample* problems. These methodologies rely heavily upon the CLT introduced in Chapter 5.

> In practice, a sample of size **thirty or more** is often considered **large**. The readers should note, however, that this is only a practical guideline. This should not be misunderstood as a mathematical result.

12.4.1   Distribution-Free Mean

Suppose that $X_1, ..., X_n$ are iid real valued random variables from a population with a common pmf or pdf $f(x; \boldsymbol{\theta})$ where the parameter vector $\boldsymbol{\theta}$ or f or both are unknown. Assume that the population variance $\sigma^2 \equiv \sigma^2(\boldsymbol{\theta})$ is positive and finite, which would imply that the population mean $\mu \equiv \mu(\boldsymbol{\theta})$ is also finite. Customarily, μ and σ^2 are both assumed unknown. Let

$$\overline{X}_n = n^{-1}\Sigma_{i=1}^n X_i, S_n^2 = (n-1)^{-1}\Sigma_{i=1}^n(X_i - \overline{X}_n)^2$$

respectively be the sample mean and variance, $n \geq 2$. Apply the CLT and write:

$$\sqrt{n}(\overline{X}_n - \mu)/S_n \xrightarrow{\mathcal{L}} N(0,1) \text{ as } n \to \infty \qquad (12.4.1)$$

For large n, we use

$$\sqrt{n}(\overline{X}_n - \mu)/S_n$$

as an *approximate pivot* since its asymptotic distribution is free from μ, σ, and f.

Confidence Interval:

Let us first handle a **one-sample** problem. Given $\alpha \in (0,1)$, we claim:

$$P_{\boldsymbol{\theta}}\{|\sqrt{n}(\overline{X}_n - \mu)/S_n| < z_{\alpha/2}\} \approx 1 - \alpha \qquad (12.4.2)$$

so that the confidence interval

$$[\,\overline{X}_n \pm z_{\alpha/2}n^{-1/2}S_n]\ \text{for the population mean}\ \mu \qquad (12.4.3)$$

will have an *approximate* confidence coefficient $1 - \alpha$ whatever be $\boldsymbol{\theta}$ or f.

Next, let us discuss a **two-sample** problem. Suppose that $X_{i1}, ..., X_{in_i}$ are iid from the ith population with a common pmf or pdf $f_i(x_i; \boldsymbol{\theta}_i)$ where the parameter vector $\boldsymbol{\theta}_i$ or f_i or both are unknown, $i = 1, 2$. Assume that X_1, X_2 are independent. Let $\sigma_i^2 \equiv \sigma_i^2(\boldsymbol{\theta}_i)$ from the ith population be positive and finite so that the mean $\mu_i \equiv \mu_i(\boldsymbol{\theta}_i)$ is also finite, $i = 1, 2$. Let μ_i, σ_i^2 be unknown, $i = 1, 2$. Denote:

$$\overline{X}_{in_i} = n_i^{-1}\Sigma_{j=1}^{n_i}X_{ij}, S_{in_i}^2 = (n_i - 1)^{-1}\Sigma_{j=1}^{n_i}(X_{ij} - \overline{X}_i)^2$$
$$\text{for}\ n_i \geq 2, i = 1, 2$$

Now, we invoke the following form of the CLT: if $n_1 \to \infty, n_2 \to \infty$ such that $n_1/n_2 \to \delta$ for some $0 < \delta < \infty$, then

$$\frac{(\overline{X}_{1n_1} - \overline{X}_{2n_2}) - (\mu_1 - \mu_2)}{\sqrt{\sigma_1^2 n_1^{-1} + \sigma_2^2 n_2^{-1}}} \xrightarrow{\pounds} N(0, 1) \qquad (12.4.4)$$

Using Slutsky's Theorem 5.3.2, one immediately concludes: if $n_1 \to \infty, n_2 \to \infty$ such that $n_1/n_2 \to \delta$ for some $0 < \delta < \infty$, then

$$\frac{(\overline{X}_{1n_1} - \overline{X}_{2n_2}) - (\mu_1 - \mu_2)}{\sqrt{S_{1n_1}^2 n_1^{-1} + S_{2n_2}^2 n_2^{-1}}} \xrightarrow{\pounds} N(0, 1) \qquad (12.4.5)$$

So, for large sample sizes n_1 and n_2, we use

$$\{(\overline{X}_{1n_1} - \overline{X}_{2n_2}) - (\mu_1 - \mu_2)\}/\sqrt{S_{1n_1}^2 n_1^{-1} + S_{2n_2}^2 n_2^{-1}} \qquad (12.4.6)$$

as an *approximate pivot* since its asymptotic distribution does not involve μ_i, σ_i and $f_i, i = 1, 2$. Then, given $\alpha \in (0, 1)$, we claim:

$$P_{\boldsymbol{\theta}_1, \boldsymbol{\theta}_2}\left\{\left|\frac{(\overline{X}_{1n_1} - \overline{X}_{2n_2}) - (\mu_1 - \mu_2)}{\sqrt{S_{1n_1}^2 n_1^{-1} + S_{2n_2}^2 n_2^{-1}}}\right| < z_{\alpha/2}\right\} \approx 1 - \alpha \qquad (12.4.7)$$

and hence the confidence interval

$$[\,(\overline{X}_{1n_1} - \overline{X}_{2n_2}) \pm z_{\alpha/2}\sqrt{S_{1n_1}^2 n_1^{-1} + S_{2n_2}^2 n_2^{-1}}\,]\ \text{for}\ (\mu_1 - \mu_2) \qquad (12.4.8)$$

will have an *approximate* confidence coefficient $1 - \alpha$ whatever be $\boldsymbol{\theta}_i$ or $f_i, i = 1, 2$.

Tests of Hypotheses:

Again, we first handle a **one-sample** problem. Given $\alpha \in (0, 1)$, we test $H_0 : \mu = \mu_0$ with an *approximate* level α where μ_0 is a fixed number. The alternative hypotheses will be stated shortly. We consider the test statistic:

$$U \equiv \sqrt{n}(\overline{X}_n - \mu_0)/S_n, \text{ approximately distributed as } N(0, 1) \quad (12.4.9)$$

for large n when $\mu = \mu_0$.

Thus, we propose the following *approximate* level α tests:

> Reject H_0 in favor of $H_1 : \mu > \mu_0$ if and only if $U > z_\alpha$
> Reject H_0 in favor of $H_1 : \mu < \mu_0$ if and only if $U < -z_\alpha$ $\quad (12.4.10)$
> Reject H_0 in favor of $H_1 : \mu \neq \mu_0$ if and only if $|U| > z_{\alpha/2}$

Next, we address a **two-sample** problem. Given $\alpha \in (0, 1)$, we test $H_0 : \mu_1 = \mu_2$ with an *approximate* level α. The alternative hypotheses will be stated shortly. Using Equation (12.4.6), we consider the test statistic:

$$W \equiv \frac{(\overline{X}_{1n_1} - \overline{X}_{2n_2})}{\sqrt{S_{1n_1}^2 n_1^{-1} + S_{2n_2}^2 n_2^{-1}}}, \text{ approximately distributed as } N(0, 1)$$

$$(12.4.11)$$

for large n_1, n_2 when $\mu_1 = \mu_2$.

Again, we propose the following *approximate* level α tests:

> Reject H_0 in favor of $H_1 : \mu_1 > \mu_2$ if and only if $W > z_\alpha$
> Reject H_0 in favor of $H_1 : \mu_1 < \mu_2$ if and only if $W < -z_\alpha$ $\quad (12.4.12)$
> Reject H_0 in favor of $H_1 : \mu_1 \neq \mu_2$ if and only if $|W| > z_{\alpha/2}$

12.4.2 Binomial Proportion

Inferences for a binomial proportion p may be carried out essentially in the same way when n is large. Suppose that $X_1, ..., X_n$ are iid Bernoulli(p) where $0 < p < 1$ is unknown. A minimal sufficient estimator of p is $\overline{X}_n$, which is the same as the sample proportion $\widehat{p}_n$ of successes. For a confidence interval of p, one will use

$$\sqrt{n}(\widehat{p}_n - p)/\sqrt{\widehat{p}_n(1 - \widehat{p}_n)}$$

as an *approximate pivot*. In a *two-sample* case, analogous to Equation (12.4.6), one will use

$$\frac{(\widehat{p}_{1n_1} - \widehat{p}_{2n_2}) - (p_1 - p_2)}{\sqrt{\widehat{p}_{1n_1}(1 - \widehat{p}_{1n_1})n_1^{-1} + \widehat{p}_{2n_2}(1 - \widehat{p}_{2n_2})n_2^{-1}}} \quad (12.4.13)$$

as an *approximate pivot* to obtain a confidence interval for $p_1 - p_2$.
   In order to test $H_0 : p = p_0$, one will use the test statistic

$$U \equiv \sqrt{n}(\widehat{p}_n - p_0)/\sqrt{p_0(1 - p_0)} \qquad (12.4.14)$$

which is *approximately* distributed as $N(0,1)$ for large n when $p = p_0$. A
rejection region can be set up as in Equation (12.4.10) depending upon the
nature of H_1. In a *two-sample* case, analogous to Equation (12.4.11), one
will use

$$W \equiv \frac{(\widehat{p}_{1n_1} - \widehat{p}_{2n_2})}{\sqrt{\widehat{p}_n(1 - \widehat{p}_n)\{n_1^{-1} + n_2^{-1}\}}} \qquad (12.4.15)$$

as an *approximate pivot* to test $H_0 : p_1 = p_2$ where $\mathbf{n} = (n_1, n_2)$ and the
pooled estimator of p is $\widehat{p}_{\mathbf{n}} = (n_1\widehat{p}_{1n_1} + n_2\widehat{p}_{2n_2})/(n_1 + n_2)$ under H_0. A
rejection region can be easily set up in the spirit of Equation (12.4.12)
depending upon the nature of H_1.

12.5   Variance Stabilizing Transformation

Consider a sequence of real valued statistics $\{T_n; n \geq 1\}$ such that

$$n^{1/2}(T_n - \theta) \overset{\pounds}{\to} N(0, \sigma^2) \text{ as } n \to \infty$$

Then, the Mann-Wald Theorem 5.3.4 implies:

$$n^{1/2}\left[g(T_n) - g(\theta)\right] \overset{\pounds}{\to} N\left(0, [\sigma g'(\theta)]^2\right) \text{ as } n \to \infty \qquad (12.5.1)$$

if $g(.)$ is a continuous real valued function and $g'(\theta)$ is finite and nonzero.
   When σ^2 itself involves θ, one may like to determine an appropriate function $g(.)$ such that for large n, the *approximate variance* of a transformed
statistic $g(T_n)$ becomes free from θ. Such a function $g(.)$ is called a *variance
stabilizing transformation*. We explain this with examples.
   Let us go back to Section 12.4.1. Suppose $X_1, ..., X_n$ are iid Bernoulli(p)
random variables where $0 < p < 1$ is unknown. Let $\widehat{p}_n$ be the sample
proportion of successes in n independent runs. One has $E_p[\widehat{p}_n] = p$ and
$V_p[\widehat{p}_n] = p(1 - p)/n$. Even though

$$\sqrt{n}\,(\widehat{p}_n - p) \overset{\pounds}{\to} N(0, p(1 - p)) \text{ as } n \to \infty$$

the expression

$$\sqrt{n}(\widehat{p}_n - p)/\sqrt{\widehat{p}_n(1 - \widehat{p}_n)}$$

is rather unstable as a pivot!
   We invoke the Mann-Wald Theorem from Equation (12.5.1) and seek out
a suitable function $g(.)$ such that the asymptotic variance of

$$\sqrt{n}\,[g(\widehat{p}_n) - g(p)]$$

does not involve p. In other words, we must have:

$$g'(p)\sqrt{p(1-p)} = k, \text{ a constant} \qquad (12.5.2)$$

That is, we can write:

$$g(p) \equiv \int g'(p)dp = k \int p^{-1/2}(1-p)^{-1/2}dp$$

Nest, we substitute $p = \sin^2(\theta)$ to rewrite:

$$g(p) = 2k \int d\theta = 2k \sin^{-1}(\sqrt{p}) + \text{constant} \qquad (12.5.3)$$

From Equation (12.5.3), it is clear that $\sin^{-1}\left(\sqrt{\widehat{p}_n}\right)$ should be the function! Let us consider the asymptotic distribution of

$$\sqrt{n}\left[\sin^{-1}\left(\sqrt{\widehat{p}_n}\right) - \sin^{-1}\left(\sqrt{p}\right)\right]$$

Again, we apply Equation (12.5.1) to write:

$$\sqrt{n}\left[\sin^{-1}\left(\sqrt{\widehat{p}_n}\right) - \sin^{-1}\left(\sqrt{p}\right)\right] \xrightarrow{\mathcal{L}} N(0, \tfrac{1}{4}) \text{ as } n \to \infty$$

since we have $g'(p) = \dfrac{1}{2\sqrt{p(1-p)}}$. That is, for large n, we may consider a pivot:

$$2\sqrt{n}\left[\sin^{-1}\left(\sqrt{\widehat{p}_n}\right) - \sin^{-1}\left(\sqrt{p}\right)\right], \text{ which is } \textit{approximately } N(0,1)$$

$$\qquad (12.5.4)$$

In the literature, for moderate values of n, the following fine-tuned approximation is frequently used:

$$2\sqrt{n}\left[\sin^{-1}\left(\sqrt{\{\widehat{p}_n + \tfrac{3}{8n}\}\{1 + \tfrac{3}{4n}\}^{-1}}\right) - \sin^{-1}\left(\sqrt{p}\right)\right] \qquad (12.5.5)$$

See Johnson and Kotz (1969, Chapter 3, Section 12.1) for a variety of other related approximations.

For large n, one may use either Equation (12.5.4) or Equation (12.5.5) to derive an *approximate* $100(1 - \alpha)\%$ confidence interval for p. Also, in order to test $H_0 : p = p_0$, for large n, one may use the test statistic:

$$Z_{calc} \equiv 2\sqrt{n}\left[\sin^{-1}\left(\sqrt{\widehat{p}_n}\right) - \sin^{-1}\left(\sqrt{p_0}\right)\right] \qquad (12.5.6)$$

to come up with an *approximate* level α test against an appropriate alternative hypothesis. The details are left out for brevity.

Remark 12.5.1 In a Poisson(λ) case, one may check that the associated variance stabilizing transformation turns out to be $g(u) = \sqrt{u}, u > 0$. See Johnson and Kotz (1969, Chapter 4, Section 7). In the case of r, the sample correlation coefficient, obtained from a random sample having a bivariate normal distribution, the associated variance stabilizing transformation turns out to be

$$g(u) \equiv \tanh^{-1}(r_n) = \frac{1}{2}\log\left\{\frac{1+r_n}{1-r_n}\right\}$$

This is especially well known as Fisher's Z transformation that was introduced in Fisher (1925a). Gayen (1951) updated Fisher's approximations and showed that for moderate sample size n,

$$\sqrt{n-3}\left[\tanh^{-1}(r_n) - \tanh^{-1}(\rho)\right] \tag{12.5.7}$$

approximated a standard normal variable a little better than

$$\sqrt{n}\left[\tanh^{-1}(r_n) - \tanh^{-1}(\rho)\right]$$

12.6   Exercises and Complements

12.2.1 Suppose that X has a Cauchy pdf:

$$f(x;\theta) = \pi^{-1}\{1 + (x-\theta)^2\}^{-1}I(x \in \Re)$$

where $\theta(\in \Re)$ is unknown. Show that the Fisher-information $\mathcal{I}(\theta) = \frac{1}{2}$. {*Hint*: With $y = x - \theta$, first write $\mathcal{I}(\theta) = \pi^{-1}\int_{-\infty}^{\infty} 4y^2(1+y^2)^{-3}dy$. Substitute $y = \tan(u)$.}

12.2.2 (Exercise 12.2.1 Continued) Show that $\sqrt{n}[\widehat{\theta}_n(\mathbf{X}) - \theta] \xrightarrow{\mathcal{L}} N(0,2)$ as $n \to \infty$ where $\widehat{\theta}_n(\mathbf{X})$ is the MLE of θ based on data $\mathbf{X} \equiv (X_1, ..., X_n)$ drawn from a Cauchy population.

12.2.3 Let $\widehat{\theta}_n \equiv \widehat{\theta}_n(\mathbf{X})$ be the MLE of $\theta(\in \Re - \{0\})$ satisfying Equations (12.2.3) and (12.2.4). Find the asymptotic distributions of

$$\sqrt{n}\left[\theta^{-1}\widehat{\theta}_n(\mathbf{X}) - 1\right] \text{ and } \sqrt{n}\left[1 - \theta\widehat{\theta}_n^{-1}(\mathbf{X})\right]$$

12.2.4 (**Logistic Distribution**) Suppose that X has a *logistic distribution* with its pdf:

$$f(x;\theta) = e^{-(x-\theta)}\{1 + e^{-(x-\theta)}\}^{-2}I(x \in \Re)$$

where $\theta(\in \Re)$ is unknown. Show that Fisher-information $\mathcal{I}(\theta) = 1/3$. {*Hint*: With $y = x - \theta$, express $\mathcal{I}(\theta)$ as

$$\int_{-\infty}^{\infty}[e^{-y}\{1+e^{-y}\}^{-2} + 4e^{-3y}\{1+e^{-y}\}^{-4} - 4e^{-2y}\{1+e^{-y}\}^{-3}]dy$$

Substitute $u = \{1 + e^{-y}\}^{-1}$.}

12.2.5 (Exercise 12.2.4 Continued) Show that $\sqrt{n}\left[\widehat{\theta}_n(\mathbf{X}) - \theta\right]$ is asymptotically distributed as $N(0,3)$.

12.3.1 Let $X_1, ..., X_n$ be iid Poisson(λ) with $\lambda(> 0)$ unknown. We wish to estimate $\tau(\lambda) \equiv P_\lambda(X_1 = 0)$. Consider T_{1n}, the MLE for $\tau(\lambda)$, and another estimator, $T_{2n} \equiv n^{-1}\Sigma_{i=1}^n I(X_i = 0)$ for $\tau(\lambda)$. Obtain $ARE_{T_{1n},T_{2n}}(\lambda)$ and investigate its behavior as a function of λ. Between the estimators T_{1n} and T_{2n}, which one stands out, if any?

12.3.2 Let $X_1, ..., X_n$ be iid $N(\theta, 1)$ with an unknown parameter $\theta \in \Re$. We wish to estimate $\tau(\theta) \equiv e^\theta$. Consider T_{1n}, the MLE for $\tau(\theta)$, and another estimator, T_{2n}, the *uniformly minimum variance unbiased estimator* (UMVUE) for $\tau(\theta)$. Obtain $ARE_{T_{1n},T_{2n}}(\theta)$ and investigate its behavior as a function of θ. Between the estimators T_{1n} and T_{2n}, which one stands out, if any?

12.3.3 Suppose $X_1, ..., X_n$ are iid $N(\theta, 1)$ with an unknown parameter $\theta \in \Re$. We wish to estimate $\tau(\theta) \equiv \theta^2$. Consider T_{1n}, the MLE for $\tau(\theta)$, and another estimator, T_{2n}, the UMVUE for $\tau(\theta)$. Obtain $ARE_{T_{1n},T_{2n}}(\theta)$ and investigate its behavior as a function of θ. Between the estimators T_{1n} and T_{2n}, which one stands out, if any?

12.3.4 Suppose that $X_1, ..., X_n$ are iid $N(\theta, \theta^2)$ with an unknown parameter $\theta \in \Re^+$. We wish to estimate $\tau(\theta) \equiv \theta^2$. Consider $T_{1n} = \overline{X}_n^2$ and another estimator, $T_{2n} = S_n^2$, for $\tau(\theta)$. Obtain the expression for $ARE_{T_{1n},T_{2n}}(\theta)$ and investigate its behavior as a function of θ. Between the estimators T_{1n} and T_{2n}, which one stands out, if any?

12.4.1 A coffee machine automatically fills cups placed underneath the dispenser. The average (μ) amount of fill must not vary too much from the target (4 fl. ounces) because the overfill will add extra cost to the manufacturer while the underfill will generate complaints from the customers. A random sample of fills for 35 cups gave $\overline{x} = 3.98$ ounces and $s = 0.008$ ounces. Test $H_0 : \mu = 4$ vs. $H_1 : \mu \neq 4$ at an *approximate* 1% level.

12.4.2 (Exercise 12.4.1 Continued) Obtain an *approximate* 95% confidence interval for the average (μ) amount of fill per cup.

12.4.3 Sixty automobiles of the same make and model were tested by drivers with similar road habits, and the gas mileage for each was recorded over a week. The summary results were $\overline{x} = 19.28$ MPG and $s = 2.53$ MPG. Construct an *approximate* 90% confidence interval for the average (μ) MPG.

12.4.4 A company has been experimenting with a new liquid diet. The investigator wants to test whether the average (μ) weight loss for individuals on this diet is more than five pounds over the initial two-week period. Fifty individuals with similar age, height, weight, and metabolic structure were started on this diet. After two weeks, the weight reduction for each individual was recorded which gave the sample average 5.78 pounds and standard deviation 1.12 pounds. Test $H_0 : \mu = 5$ vs. $H_1 : \mu > 5$ at an *approximate* 5% level.

12.4.5 A large tire manufacturing company has two factories, A and B. It is believed that employees in Factory A are paid less monthly salaries on an average than the employees in Factory B for comparable jobs and job-related performances. The local union for Factory A randomly sampled 35 employees from each factory and recorded each individual's monthly gross salary. The summary data follows:

Factory A	Factory B
$n_1 = 35$	$n_2 = 35$
$\overline{x}_{1n_1} = \$2854.72$	$\overline{x}_{2n_2} = \$3168.27$
$s_{1n_1} = \$105.29$	$s_{2n_2} = \$53.55$

At an *approximate* 5% level, test whether the data validate a belief that employees in Factory A are paid less on an average than employees in Factory B.

12.4.6 Let $X_1, ..., X_n$ be iid Bernoulli(p) random variables with $p \in (0,1)$ unknown. Consider the minimal sufficient estimator $\hat{p}_n$, the sample proportion of successes in n independent replications, for p. For large n, one has:

$$P_p \left\{ \left| \sqrt{n}(\hat{p}_n - p)/\sqrt{p(1-p)} \right| < z_{\alpha/2} \right\} \approx 1 - \alpha$$
$$\Rightarrow P_p\{p : n(\hat{p}_n - p)^2 < [p(1-p)]z_{\alpha/2}^2\} \approx 1 - \alpha$$

Now, solve the quadratic equation in p to derive an *approximate* $100(1-\alpha)\%$ confidence interval for p.

12.4.7 A critic of the insurance industry claimed that less than 30% of working women in a city carried employer-provided health benefits. In a random sample of 2000 working women in the city, we observed 20% with employer-provided health benefits. At an *approximate* 5% level, test the validity of the critic's claim.

12.4.8 (Sign Test) Let $X_1, ..., X_n$ be iid continuous random variables with a common pdf $f(x; \theta)$, symmetric around $x = \theta$, with $x \in \Re, \theta \in \Re$. The problem is to test $H_0 : \theta = 0$ vs. $H_1 : \theta > 0$ with given level $\alpha \in (0,1)$. Define $Y_i = 1$ or 0 according as $X_i > 0$ or $X_i \leq 0, i = 1, ..., n$.

(i)   Show that $Y_1, ..., Y_n$ are iid Bernoulli(p) where $p = \frac{1}{2}$ or $p > \frac{1}{2}$ according as $\theta = 0$ or $\theta > 0$ respectively whatever be f;

(ii) Argue that one can equivalently test $H_0 : p = \frac{1}{2}$ vs. $H_1 : p > \frac{1}{2}$ based on the observations $Y_1, ..., Y_n$;

(iii) Propose an *appropriate* level α test for the problem introduced in part (ii). It is called a *sign test*.

12.4.9 (Comparing Sign Test with Z Test: Normal Distribution)
Let $X_1, ..., X_n$ be iid $N(\theta, \sigma^2)$ random variables where $\theta \in \Re$ is unknown, but $\sigma^2 \in \Re^+$ is known. The problem is to test $H_0 : \theta = 0$ vs. $H_1 : \theta > 0$ with given level $\alpha \in (0, 1)$. Define $Y_i = 1$ or 0 according as $X_i > 0$ or $X_i \leq 0, i = 1, ..., n$. The *uniformly most powerful* (UMP) level α test is expressed as follows:

Test #1: Reject H_0 if and only if $\sqrt{n}\overline{X}_n/\sigma > z_\alpha$

An *approximate* level α sign test from Exercise 12.4.8 looks like:

Test #2: Reject H_0 if and only if $2\sqrt{n}\left[\overline{Y}_n - \frac{1}{2}\right] > z_\alpha$

for large n. Now, address the following problems. Assume $\sigma^2 = 1$.

(i)   Compare powers of Test #1 and Test #2 when $n = 30, 40$, $50, \alpha = 0.05$ and $\theta = 0.03, 0.02, 0.01$ (which are close to $\theta = 0$);

(ii)   Fix $n = 50$. Evaluate the sample sizes n_1, n_2, n_3 such that Test #2 with sample size $n = n_1, n = n_2, n = n_3$ has respectively the same power as that of Test #1, used with $n = 50$, at $\theta = 0.03, 0.02, 0.01$;

(iii) How does n_1, n_2, n_3 compare with $n = 50$? Any comments?
{*Note*: Such investigations lead to *Bahadur efficiency*. Refer to Bahadur (1971).}

12.5.1 Suppose that $X_1, ..., X_n$ are iid with a common $N(\mu, \sigma^2)$ distribution where the parameters are $\mu \in \Re, \sigma^2 \in \Re^+$. Asymptotically,

$$\sqrt{n}\left[S_n^2 - \sigma^2\right] \text{ is distributed as } N(0, 2\sigma^4)$$

where S_n^2 is the sample variance. Find the variance stabilizing transformation $g(S_n^2)$. Obtain explicitly the asymptotic distribution of

$$\sqrt{n}\left[g(S_n^2) - g(\sigma^2)\right]$$

12.5.2 Suppose that $X_1, ..., X_n$ are iid with a common exponential distribution having mean $\beta(> 0)$. Asymptotically,

$$\sqrt{n}\left[\overline{X}_n - \beta\right] \text{ is distributed as } N(0, \beta^2)$$

where $\overline{X}_n$ is the sample mean. Find the variance stabilizing transformation $g(\overline{X}_n)$. Obtain explicitly the asymptotic distribution of

$$\sqrt{n}\left[g(\overline{X}_n) - g(\beta)\right]$$

12.5.3 Suppose $X_1, ..., X_n$ are iid with a common $N(0, \theta^2)$ distribution where $\theta \in \Re^+$. Denote $T_n = \frac{1}{2}n^{-1}\Sigma_{i=1}^n X_i^2$. Derive the asymptotic distribution of $\sqrt{n}\left[T_n - \theta^2\right]$. Find the variance stabilizing transformation $g(T_n)$. Obtain explicitly the asymptotic distribution of

$$\sqrt{n}\left[g(T_n) - g(\theta^2)\right]$$

13

Abbreviations, Historical Notes, and Tables

Section 13.1 begins with a summary of frequently used abbreviations and notation. Section 13.2 provides brief historical notes. In the end, some standard statistical tables are included. These were prepared with the help of MAPLE.

13.1   Abbreviations and Notations

Some abbreviations and notations are summarized here.

$\mathcal{X}$	space of the values x for the random variable X
$\Re$	real line; $(-\infty, \infty)$
$\Re^+$	positive half of $\Re$; $(0, \infty)$
$\in$; $x \in A$	belongs to; x belongs to the set A
$A(\subset) \subseteq B$	A is a (proper) subset of B
▲	denotes end of an example
■	denotes end of a proof in a theorem
♦	denotes end of a lengthy proof in a subcase
$\varphi, \emptyset$	empty set
lhs	left-hand side
rhs	right-hand side
log, log	natural (base e) logarithm
pmf; PMF	probability mass function
pdf; PDF	probability density function
df; DF	distribution function
cdf; CDF	cumulative distribution function
mgf; MGF	moment generating function
iid	independent and identically distributed
$\Pi_{i=1}^n a_i$	product of $a_1, a_2, ..., a_n$
w.p. 1	with probability one
$\mathbf{x}, \mathbf{X}, \boldsymbol{\theta}$	vectors of dimension more than one
$\Gamma(\alpha), \alpha > 0$	gamma function; $\int_0^\infty e^{-x} x^{\alpha-1} dx$

$\text{beta}(\alpha, \beta), \alpha > 0, \beta > 0$	beta function; $\Gamma(\alpha)\Gamma(\beta)/\Gamma(\alpha + \beta)$
$[f(x)]_{x=a}^{x=b}$	stands for $f(b) - f(a)$
$I(A)$ or I_A	indicator function of a set A
$\mathcal{I}(\theta)$	Fisher-information about θ
$X_n \xrightarrow{P} a$	X_n converges to a in probability as $n \to \infty$
$X_n \xrightarrow{\mathcal{L}} X$	X_n converges to X in law (or distribution) as $n \to \infty$
$a(x) \backsim b(x)$	$a(x)/b(x) \to 1$ as $x \to \infty$
χ_ν^2	Chi-square distribution with ν degrees of freedom
$\phi(.), \Phi(.)$	standard normal pdf and cdf, respectively
$\overline{X}$ or $\overline{X}_n$	sample mean from $X_1, ..., X_n$
S^2 or S_n^2	sample variance from $X_1, ..., X_n$; divisor $n - 1$
$X_{n:i}$	i^{th}-order statistic in $X_{n:1} \le X_{n:2} \le ... \le X_{n:n}$
$det(A)$	determinant of a square matrix A
A'	transposed matrix obtained from $A_{m \times n}$
$I_{k \times k}$	identity matrix of order $k \times k$
WLLN	weak law of large numbers
CLT	central limit theorem
$a \approx b$	a and b are approximately the same
p.d.	positive definite
n.d.	negative definite
p.s.d.	positive semidefinite
$B(\theta)$	bias of an estimator
MSE	mean squared error
MLE	maximum likelihood estimator
CRLB	Cramér-Rao lower bound
UMVUE	uniformly minimum variance unbiased estimator
MP test	most powerful test
UMP test	uniformly most powerful test
UMPU test	uniformly most powerful unbiased test
LR	likelihood ratio
LRT	likelihood ratio test
$a \propto b$	a, b are proportional
z_α	upper $100\alpha\%$ point of $N(0, 1)$
$\chi_{\nu,\alpha}^2$	upper $100\alpha\%$ point of χ_ν^2
$t_{\nu,\alpha}$	upper $100\alpha\%$ point of Student's t_ν
$F_{\nu_1,\nu_2,\alpha}$	upper $100\alpha\%$ point of F_{ν_1,ν_2}

13.2   Historical Notes

In this section, we add some historical notes briefly and point toward selected sources where one could get more information. The aim is to let our
readers appreciate statistics knowing its heritage and how this field has
continued to flourish.

An edited volume of Pearson and Kendall (1970) included valuable selected articles. This is an excellent resource. *Statistical Science*, a quarterly
journal published by the Institute of Mathematical Statistics, includes con-
versation pieces of eminent statisticians. These articles are filled with illuminating historical material. Bellhouse and Genest (1999) as well as their
discussants gave an account of the rich history of statistics in Canada.

The archive of the American Statistical Association houses films of lec-
tures and interviews of many eminent statisticians. The list includes (alpha-
betically): T. W. Anderson, D. Blackwell, R. C. Bose, G. Box, D. Brillinger,
H. Chernoff, W. G. Cochran, D. R. Cox, H. Cramér, E. W. Deming, B.
F. Efron, I. J. Good, F. Graybill, M. Hansen, R. V. Hogg, S. Hunter,
O. Kempthorne, E. L. Lehmann, F. Mosteller, J. Neyman, G. Noether, I.
Olkin, E. J. G. Pitman, C. R. Rao, H. Robbins, E. L. Scott, J. W. Tukey,
and M. Zelen. These wonderful films are filled with fascinating stories.

For a reader who is beginning to appreciate the *history of statistics*, the
following short list (not arranged in any particular order) may be helpful.
Consider reading:

R. A. Fisher, The Life of a Scientist, by Joan Fisher Box (1978)

Neyman from Life, by Constance Reid (1982)

The Making of Statisticians, edited by Joe Gani (1982)

*American Contributions to Mathematical Statistics in the Nineteenth
Century,* edited by Steve Stigler (1980)

The History of Statistics: The Measurement of Uncertainty Before 1900,
by Steve Stigler (1986)

Glimpses of India's Statistical Heritage, edited by J. K. Ghosh, S. K.
Mitra, and K. R. Parthasarathy (1992)

> **Enjoy and Celebrate Statistics by Learning Its History**

13.3   Selected Statistical Tables

This section provides some of the standard statistical tables. These were
prepared with the help of MAPLE.

Tables 13.3.1a and 13.3.1b correspond to the *distribution function* (df)
of a standard normal distribution. Tables 13.3.2, 13.3.3, and 13.3.4 respectively correspond to the percentage points of the Chi-square, Student's t,
and F distribution. One may look at Lindley and Scott (1995) or other
sources for more extensive sets of tables like these.

13.3.1 *Standard Normal Distribution Function*

In Tables 13.3.1a and 13.3.1b, the first column and row respectively indicate the "first" and "second" decimal points of z. Look at Figures 13.3.1 and 13.3.2.

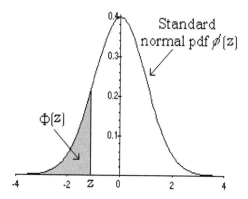

Figure 13.3.1. $N(0,1)$ df $\Phi(z)$ is the shaded area when $z < 0$.

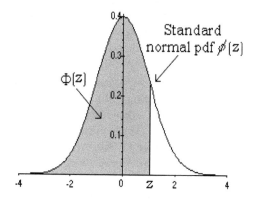

Figure 13.3.2. $N(0,1)$ dfF $\Phi(z)$ is the shaded area when $z > 0$.

Example 13.3.1 Consider the following probability calculations:

$P\{Z < 2\} = \Phi(2) = 0.97725;$

$P\{Z < 2.1\} = \Phi(2.1) = 0.98214;$

$P\{Z < 1.21\} = \Phi(1.21) = 0.88686;$

$P\{Z < -2\} = 1 - \Phi(2) = 0.02275;$

$P\{Z < -1.21\} = 1 - 0.88686 = 0.11314;$

$P\{-1.28 < Z < 2.36\} = 0.98214 - 0.11314 = 0.869.$

These are some examples of how one can use the standard normal table (Table 13.3.1a and Table 13.3.1b) to evaluate various kinds of normal probabilities. ▲

Table 13.3.1a. Values of the $N(0,1)$ Distribution
Function: $\Phi(z) = \int_{-\infty}^{z} (\sqrt{2\pi})^{-1} e^{-x^2/2} dx$

z	0.00	0.01	0.02	0.03	0.04
0.0	0.50000	0.50399	0.50798	0.51197	0.51595
0.1	0.53983	0.54380	0.54776	0.55172	0.55567
0.2	0.57926	0.58317	0.58706	0.59095	0.59483
0.3	0.61791	0.62172	0.62552	0.62930	0.63307
0.4	0.65542	0.65910	0.66276	0.66640	0.67003
0.5	0.69146	0.69497	0.69847	0.70194	0.70540
0.6	0.72575	0.72907	0.73237	0.73565	0.73891
0.7	0.75804	0.76115	0.76424	0.76730	0.77035
0.8	0.78814	0.79103	0.79389	0.79673	0.79955
0.9	0.81594	0.81859	0.82121	0.82381	0.82639
1.0	0.84134	0.84375	0.84614	0.84849	0.85083
1.1	0.86433	0.86650	0.86864	0.87076	0.87286
1.2	0.88493	0.88686	0.88877	0.89065	0.89251
1.3	0.90320	0.90490	0.90658	0.90824	0.90988
1.4	0.91924	0.92073	0.92220	0.92364	0.92507
1.5	0.93319	0.93448	0.93574	0.93699	0.93822
1.6	0.94520	0.94630	0.94738	0.94845	0.94950
1.7	0.95543	0.95637	0.95728	0.95818	0.95907
1.8	0.96407	0.96485	0.96562	0.96638	0.96712
1.9	0.97128	0.97193	0.97257	0.97320	0.97381
2.0	0.97725	0.97778	0.97831	0.97882	0.97932
2.1	0.98214	0.98257	0.98300	0.98341	0.98382
2.2	0.98610	0.98645	0.98679	0.98713	0.98745
2.3	0.98928	0.98956	0.98983	0.99010	0.99036
2.4	0.99180	0.99202	0.99224	0.99245	0.99266
2.5	0.99379	0.99396	0.99413	0.99430	0.99446
2.6	0.99534	0.99547	0.99560	0.99573	0.99585
2.7	0.99653	0.99664	0.99674	0.99683	0.99693
2.8	0.99744	0.99752	0.99760	0.99767	0.99774
2.9	0.99813	0.99819	0.99825	0.99831	0.99836
3.0	0.99865	0.99869	0.99874	0.99878	0.99882
3.1	0.99903	0.99906	0.99910	0.99913	0.99916
3.2	0.99931	0.99934	0.99936	0.99938	0.99940
3.3	0.99952	0.99953	0.99955	0.99957	0.99958
3.4	0.99966	0.99968	0.99969	0.99970	0.99971
3.5	0.99977	0.99978	0.99978	0.99979	0.99980

Table 13.3.1b. Values of the N(0,1) Distribution
Function: $\Phi(z) = \int_{-\infty}^{z} (\sqrt{2\pi})^{-1} e^{-x^2/2} dx$

z	0.05	0.06	0.07	0.08	0.09
0.0	0.51994	0.52392	0.52790	0.53188	0.53586
0.1	0.55962	0.56356	0.56749	0.57142	0.57535
0.2	0.59871	0.60257	0.60642	0.61026	0.61409
0.3	0.63683	0.64058	0.64431	0.64803	0.65173
0.4	0.67364	0.67724	0.68082	0.68439	0.68793
0.5	0.70884	0.71226	0.71566	0.71904	0.72240
0.6	0.74215	0.74537	0.74857	0.75175	0.75490
0.7	0.77337	0.77637	0.77935	0.78230	0.78524
0.8	0.80234	0.80511	0.80785	0.81057	0.81327
0.9	0.82894	0.83147	0.83398	0.83646	0.83891
1.0	0.85314	0.85543	0.85769	0.85993	0.86214
1.1	0.87493	0.87698	0.87900	0.88100	0.88298
1.2	0.89435	0.89617	0.89796	0.89973	0.90147
1.3	0.91149	0.91309	0.91466	0.91621	0.91774
1.4	0.92647	0.92785	0.92922	0.93056	0.93189
1.5	0.93943	0.94062	0.94179	0.94295	0.94408
1.6	0.95053	0.95154	0.95254	0.95352	0.95449
1.7	0.95994	0.96080	0.96164	0.96246	0.96327
1.8	0.96784	0.96856	0.96926	0.96995	0.97062
1.9	0.97441	0.97500	0.97558	0.97615	0.97670
2.0	0.97982	0.98030	0.98077	0.98124	0.98169
2.1	0.98422	0.98461	0.98500	0.98537	0.98574
2.2	0.98778	0.98809	0.98840	0.98870	0.98899
2.3	0.99061	0.99086	0.99111	0.99134	0.99158
2.4	0.99286	0.99305	0.99324	0.99343	0.99361
2.5	0.99461	0.99477	0.99492	0.99506	0.99520
2.6	0.99598	0.99609	0.99621	0.99632	0.99643
2.7	0.99702	0.99711	0.99720	0.99728	0.99736
2.8	0.99781	0.99788	0.99795	0.99801	0.99807
2.9	0.99841	0.99846	0.99851	0.99856	0.99861
3.0	0.99886	0.99889	0.99893	0.99896	0.99900
3.1	0.99918	0.99921	0.99924	0.99926	0.99929
3.2	0.99942	0.99944	0.99946	0.99948	0.99950
3.3	0.99960	0.99961	0.99962	0.99964	0.99965
3.4	0.99972	0.99973	0.99974	0.99975	0.99976
3.5	0.99981	0.99981	0.99982	0.99983	0.99983

13.3.2 Percentage Points of the Chi-Square Distribution

Table 13.3.2 provides the lower $100\gamma\%$ point $\chi^2_{\nu,1-\gamma}$ for the χ^2_ν distribution for different values of ν and γ. See Figure 13.3.3.

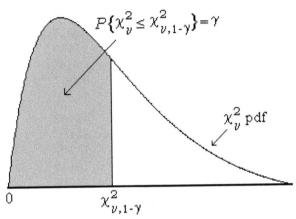

$$P\{\chi^2_\nu \le \chi^2_{\nu,1-\gamma}\} = \gamma$$

$$\chi^2_\nu \, \text{pdf}$$

$$\chi^2_{\nu,1-\gamma}$$

Figure 13.3.3. The shaded area is the probability γ.

Suppose that X has the χ^2_ν distribution with some appropriate ν. Now, consider the following examples:

With $\nu = 3, P\{X \le 0.1149\}$ $= 0.01 \Rightarrow \chi^2_{3,0.99} = 0.1149$

With $\nu = 10, P\{X \le 18.307\}$ $= 0.95 \Rightarrow \chi^2_{10,0.05} = 18.307$

With $\nu = 15, P\{6.2621 \le X \le a\}$ $= 0.925 \Rightarrow a = 24.996$

Remark 13.3.1 The percentage points of the χ^2_1 distribution can be easily found using the standard normal Table 13.3.1. Observe that:

$$\Phi\left(\sqrt{\chi^2_{1,1-\gamma}}\right) = \tfrac{1}{2}(1+\gamma)$$

For example, when $\gamma = 0.95$, one has $\Phi\left(\sqrt{\chi^2_{1,0.05}}\right) = 0.975$. Thus, from the Table 13.3.1b we find that $\sqrt{\chi^2_{1,.05}} = 1.96$, that is, $\chi^2_{1,.05} = (1.96)^2 = 3.8416$. In the Chi-square Table 13.3.2, we have $\chi^2_{1,0.05} = 3.8415$.

Remark 13.3.2 The percentage points of the χ^2_2 distribution can be easily found using direct integrals of the exponential pdf. Observe that:

$$\chi^2_{2,1-\gamma} = 2log(1/(1-\gamma))$$

For example, when $\gamma = 0.95$, one has $\chi^2_{2,0.05} = 2log(1/0.05) \approx 5.991465$. In the Chi-square Table 13.3.2, we have $\chi^2_{2,0.05} = 5.9915$.

Table 13.3.2. Lower $100\gamma\%$ Point $\chi^2_{\nu,1-\gamma}$ for the χ^2_ν

Distribution: $P(\chi^2_\nu \leq \chi^2_{\nu,1-\gamma}) = \gamma$

ν	$\gamma = 0.01$	$\gamma = 0.025$	$\gamma = 0.05$	$\gamma = 0.95$	$\gamma = 0.975$	$\gamma = 0.99$
1	0.00016	0.00098	0.00393	3.8415	5.0239	6.6349
2	0.02010	0.05064	0.10259	5.9915	7.3778	9.2103
3	0.11490	0.21570	0.35185	7.8147	9.3484	11.345
4	0.29711	0.48442	0.71072	9.4877	11.143	13.277
5	0.55430	0.83121	1.1455	11.070	12.833	15.086
6	0.87209	1.2373	1.6354	12.592	14.449	16.812
7	1.2390	1.5643	2.1673	14.067	16.013	18.475
8	1.6465	2.1797	2.7326	15.507	17.535	20.090
9	2.0879	2.7004	3.3251	16.919	19.023	21.666
10	2.5582	3.2470	3.9403	18.307	20.483	23.209
11	3.0535	3.8157	4.5748	19.675	21.920	24.725
12	3.5706	4.4038	5.2260	21.026	23.337	26.217
13	4.1069	5.0088	5.8919	22.362	24.736	27.688
14	4.6604	5.6287	6.5706	23.685	26.119	29.141
15	5.2293	6.2621	7.2609	24.996	27.488	30.578
16	5.8122	6.9077	7.9616	26.296	28.845	32.000
17	6.4078	7.5642	8.6718	27.587	30.191	33.409
18	7.0149	8.2307	9.3905	28.869	31.526	34.805
19	7.6327	8.9065	10.117	30.144	32.852	36.191
20	8.2604	9.5908	10.851	31.410	34.170	37.566
21	8.8972	10.283	11.591	32.671	35.479	38.932
22	9.5425	10.982	12.338	33.924	36.781	40.289
23	10.196	11.689	13.091	35.172	38.076	41.638
24	10.856	12.401	13.848	36.415	39.364	42.980
25	11.524	13.120	14.611	37.652	40.646	44.314
26	12.198	13.844	15.379	35.563	41.923	45.642
27	12.879	14.573	16.151	40.113	43.195	46.963
28	13.565	15.308	16.928	41.337	44.461	48.278
29	14.256	16.047	17.708	42.550	45.731	49.590
30	14.953	16.791	18.493	43.780	46.980	50.890

13.3.3 *Percentage Points of the Student's t Distribution*

Table 13.3.3 provides the lower $100\gamma\%$ point $t_{\nu,1-\gamma}$ for the Student's t_ν distribution for different values of ν and γ. See Figure 13.3.4.

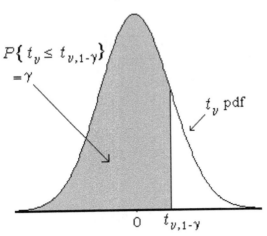

Figure 13.3.4. The shaded area is the probability γ.

Suppose that X has the Student's t_ν distribution with some appropriate ν. Now, consider the following examples:

With $\nu = 4, P\{X \le 2.1318\}$ $\quad\quad = 0.95 \Rightarrow t_{4,0.05} = 2.1318$

With $\nu = 10, P\{X \le -2.7638\}$ $\quad = 0.01 \Rightarrow t_{10,0.99} = -2.7638$

With $\nu = 15, P\{-1.7531 \le t_{15} \le a\}$ $\;= 0.94 \Rightarrow a = 2.6025$

Remark 13.3.3 The percentage points of the t_1 distribution can be easily found by directly integrating its *probability density function* (pdf), which happens to be the same as the Cauchy pdf. Observe that:

$$\gamma = \frac{1}{\pi} \int_{-\infty}^{t_{\nu,1-\gamma}} (1+x^2)^{-1} dx = \frac{1}{2} + \frac{1}{\pi} \arctan(t_{\nu,1-\gamma})$$

$$\Rightarrow t_{\nu,1-\gamma} = \tan\left(\pi\left(\gamma - \frac{1}{2}\right)\right)$$

For example, when $\gamma = 0.95$, one has $t_{1,0.05} = \tan\left(\pi(0.95 - \frac{1}{2})\right) \approx 6.3138$. In the t Table 13.3.3, we have $t_{1,0.05} = 6.314$.

Table 13.3.3. Lower $100\gamma\%$ Point $t_{\nu,1-\gamma}$ for the Student's t_ν

Distribution: $P(t_\nu \leq t_{\nu,1-\gamma}) = \gamma$

ν	$\gamma = 0.90$	$\gamma = 0.95$	$\gamma = 0.975$	$\gamma = 0.99$	$\gamma = 0.995$
1	3.0777	6.3140	12.706	31.821	63.657
2	1.8856	2.9200	4.3027	6.9646	9.9248
3	1.6377	2.3534	3.1824	4.5407	5.8409
4	1.5332	2.1318	2.7764	3.7469	4.6041
5	1.4759	2.0150	2.5706	3.3649	4.0321
6	1.4398	1.9432	2.4469	3.1427	3.7074
7	1.4149	1.8946	2.3646	2.9980	3.4995
8	1.3968	1.8595	2.3060	2.8965	3.3554
9	1.3830	1.8331	2.2622	2.8214	3.2498
10	1.3722	1.8125	2.2281	2.7638	3.1693
11	1.3634	1.7959	2.2010	2.7181	3.1058
12	1.3562	1.7823	2.1788	2.6810	3.0545
13	1.3502	1.7709	2.1604	2.6503	3.0123
14	1.3450	1.7613	2.1448	2.6245	2.9768
15	1.3406	1.7531	2.1314	2.6025	2.9467
16	1.3368	1.7459	2.1199	2.5835	2.9208
17	1.3334	1.7396	2.1098	2.5669	2.8982
18	1.3304	1.7341	2.1009	2.5524	2.8784
19	1.3277	1.7291	2.0930	2.5395	2.8609
20	1.3253	1.7247	2.0860	2.5280	2.8453
21	1.3232	1.7207	2.0796	2.5176	2.8314
22	1.3212	1.7171	2.0739	2.5083	2.8188
23	1.3195	1.7139	2.0687	2.4999	2.8073
24	1.3178	1.7109	2.0639	2.4922	2.7969
25	1.3163	1.7081	2.0595	2.4851	2.7874
26	1.3150	1.7056	2.0555	2.4786	2.7787
27	1.3137	1.7033	2.0518	2.4727	2.7707
28	1.3125	1.7011	2.0484	2.4671	2.7633
29	1.3114	1.6991	2.0452	2.4620	2.7564
30	1.3104	1.6973	2.0423	2.4573	2.7500
35	1.3062	1.6896	2.0301	2.4377	2.7238
40	1.3031	1.6839	2.0211	2.4233	2.7045
45	1.3006	1.6794	2.0141	2.4121	2.6896
50	1.2987	1.6759	2.0086	2.4033	2.6778
100	1.2901	1.6602	1.9840	2.3642	2.6259

13.3.4   Percentage Points of the F Distribution

Table 13.3.4 provides the lower $100\gamma\%$ point $F_{\nu_1,\nu_2,1-\gamma}$ for the F_{ν_1,ν_2} distribution for different values of ν_1, ν_2 and γ. See Figure 13.3.5.

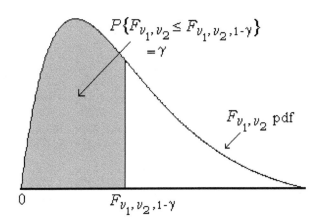

Figure 13.3.5. The shaded area is the probability γ.

Suppose that X has the F_{ν_1,ν_2} distribution with some appropriate ν_1, ν_2. Now, consider the following examples:

With $\nu_1 = 5, \nu_2 = 8, P\{X \le 3.6875\}\quad = 0.95 \Rightarrow F_{5,8,0.05} = 3.6875$

With $\nu_1 = 10, \nu_2 = 5, P\{X \le 10.051\}\quad = 0.99 \Rightarrow F_{10,5,0.01} = 10.051$

In some instances, one may need $F_{\nu_1,\nu_2,1-\gamma}$ where γ is "small." Recall that $Y = 1/X$ has a F_{ν_2,ν_1} distribution if X has a F_{ν_1,ν_2} distribution. Also, observe that $P\{Y \ge 1/b\} = P\{X \le b\}$. Suppose we wish to determine b such that with $\nu_1 = 5, \nu_2 = 8, P\{X \le b\} = 0.05$. In Table 13.3.4, for $F_{8,5}$ we find the number 4.8183, which corresponds to $\gamma = 0.95\ (= 1 - 0.05)$. That is, $b = \frac{1}{4.8183} \approx 0.20754$.

Remark 13.3.4 Observe that:

$$F_{2,\nu_2,1-\gamma} = \tfrac{1}{2}\nu_2\left\{(1-\gamma)^{-2/\nu_2} - 1\right\}$$

For example, when $\gamma = 0.90, 0.95, 0.975, 0.99$, and 0.995, this formula provides the values $F_{2,2,1-\gamma} = 9.0, 19.0, 39.0, 99.0$, and 199.0, respectively, which match with the corresponding entries in Table 13.3.4. Also, for example, this formula provides the values $F_{2,5,0.05} = 5.7863$ and $F_{2,8,0.025} = 6.0596$ whereas the corresponding entries in Table 13.3.4 are 5.7861 and 6.0595, respectively.

Table 13.3.4. Lower $100\gamma\%$ Point $F_{\nu_1,\nu_2,1-\gamma}$ for the F_{ν_1,ν_2} Distribution:
$P(F_{\nu_1,\nu_2} \leq F_{\nu_1,\nu_2,1-\gamma}) = \gamma$ Where ν_1, ν_2 Are Respectively the
Numerator and Denominator Degrees of Freedom

					ν_2			
ν_1	γ	1	2	5	8	10	15	20
	0.90	39.863	8.5263	4.0604	3.4579	3.2850	3.0732	2.9747
	0.95	161 45	18.513	6.6079	5.3177	4.9646	4.5431	4.3512
1	0.975	647.79	38.506	10.007	7.5709	6.9367	6.1995	6.1995
	0.99	4052.2	98.503	16.258	11.259	10.044	8.6831	8.0960
	0.995	16211	198.50	22.785	14.688	12.826	10.798	9.9439
	0.90	49.500	9.0000	3.7797	3.1131	2.9245	2.6952	2.5893
	0.95	199.50	19.000	5.7861	4.4590	4.1028	3.6823	3.4928
2	0.975	799.50	39.000	8.4336	6.0595	5.4564	4. 765	4.4613
	0.99	4999.5	99.000	13.274	8.6491	7.5594	6.3589	5.8489
	0.995	20000	199.00	18.314	11.042	9.4270	7.7008	6.9865
	0.90	57.240	9.2926	3.453	2.7264	2.5216	2.273	2.1582
	0.95	230.16	19.296	5.0503	3.6875	3.3258	2.9013	2.7109
5	0.975	921.85	39.298	7.1464	4.8173	4.2361	3.5764	3.2891
	0.99	5763.6	99.299	10.967	6.6318	5.6363	4.5556	4.1027
	0.995	23056	199.30	14.940	8.3018	6.8724	5.3721	4.7616
	0.90	59.439	9.3668	3.3393	2.5893	2.3772	2.1185	1.9985
	0.95	238.88	19.371	4.8183	3.4381	3.0717	2.6408	2.4471
8	0.975	956.66	39.373	6.7572	4.4333	3.8549	3.1987	2.9128
	0.99	5981.1	99.374	10. 89	6.0289	5.0567	4.0045	3.5644
	0.995	23925	199.37	13.961	7.4959	6.1159	4.6744	4.0900
	0.90	60.195	9.3916	3.2974	2.538	2.3226	2.0593	1.9367
	0.95	241.88	19.396	4.7351	3.3472	2.9782	2.5437	2.3479
10	0.975	968.63	39.398	6.6192	4.2951	3.7168	3.0602	2.7737
	0.99	6055.8	99.399	10.051	5.8143	4.8491	3.8049	3.3682
	0.995	24224	199.40	13.618	7.2106	5.8467	4.4235	3.8470
	0.90	61.220	9.4247	3.2380	2.4642	2.2435	1.9722	1.8449
	0.95	245.95	19.429	4.6188	3. 2184	2.8450	2.4034	2.2033
15	0.975	984.87	39.431	6.4277	4.1012	3.5217	2.8621	2.5731
	0.99	6157.3	99.433	9.7222	5.5151	4.5581	3.5222	3.0880
	0.995	24630	199.43	13.146	6.8143	5.4707	4.0698	3.5020
	0.90	61.740	9.4413	3.2067	2.4246	2.2007	1.9243	1.7938
	0.95	248.01	19.446	4.5581	3.1503	2.7740	2.3275	2.1242
20	0.975	993.10	39.448	6.3286	3.9995	3.4185	2.7559	2.4645
	0.99	6208.7	99.449	9.5526	5.3591	4.4054	3.3719	2.9377
	0.995	24836	199.45	12.903	6.6082	5.274	3.8826	3.3178

References

Abramowitz, M. and Stegun, I.A. (1972). *Handbook of Mathematical Functions* (edited volume). Dover Publications, Inc., New York.

Aitken, A.C. and Silverstone, H. (1942). On the estimation of statistical parameters. *Proc. Roy. Soc. Edinburgh, Ser. A*, **61**, 186-194.

Anscombe, F.J. (1949). The statistical analysis of insect counts based on the negative binomial distribution. *Biometrika*, **5**, 165-173.

Bahadur, R.R. (1954). Sufficiency and statistical decision functions. *Ann. Math. Statist.*, **25**, 423-462.

Bahadur, R.R. (1957). On unbiased estimates of uniformly minimum variance. *Sankhyā*, **18**, 211-224.

Bahadur, R.R. (1958). Examples of inconsistency of maximum likelihood estimates. *Sankhyā*, **20**, 207-210.

Bahadur, R.R. (1971). *Some Limit Theorems in Statistics*. NSF-CBMS Monograph No. 4. Society for Industrial and Applied Mathematics, Philadelphia.

Balakrishnan, N. and Basu, A.P. (1995). *The Exponential Distributions: Theory, Methods and Applications* (edited volume). Gordon and Breach, Amsterdam.

Barankin, E.W. and Maitra, A. (1963). Generalization of the Fisher-Darmois-Koopman-Pitman theorem on sufficient statistics. *Sankhyā, Ser. A*, **25**, 217-244.

Barndorff-Nielsen, G. (1978). *Information and Exponential Families in Statistical Theory*. John Wiley & Sons, Inc., New York.

Basu, D. (1955a). On statistics independent of a complete sufficient statistic. *Sankhyā*, **15**, 377-380.

Basu, D. (1955b). An inconsistency of the method of maximum likelihood. *Ann. Math. Statist.*, **26**, 144-145.

Basu, D. (1958). On statistics independent of sufficient statistics. *Sankhyā*, **20**, 223-226.

Basu, D. (1964). Recovery of ancillary information. *Contributions to Statistics, the 70th Birthday Festschrift Volume Presented to P. C. Mahalanobis*. Pergamon Press, Oxford.

Bayes, T. (1783). An essay towards solving a problem in the doctrine of chances. *Phil. Trans. Roy. Soc.*, **53**, 370-418.

Bellhouse, D.R. and Genest, C. (1999). A history of the Statistical Society of Canada: The formative years. *Statist. Sci.*, **14**, 80-125.

Bennett, J.H. (1990). *Statistical Inference and Analysis: Selected Correspondence of R. A. Fisher* (edited volume). Oxford University Press, Oxford.

Blackwell, D. (1947). Conditional expectation and unbiased sequential estimation. *Ann. Math. Statist.*, **18**, 105-110.

Box, J.F. (1978). *R. A. Fisher, The Life of a Scientist*. John Wiley & Sons, Inc., New York.

Brown, L.D. (1964). Sufficient statistics in the case of independent random variables. *Ann. Math. Statist.*, **35**, 1456-1474.

Buehler, R. (1980). Fiducial inference. *R. A. Fisher: An Appreciation* (S.E. Fienberg and D.V. Hinkley, eds.), pp. 109-118. Springer-Verlag, Inc., New York.

Cramér, H. (1946a). *Mathematical Methods of Statistics.* Princeton University Press, Princeton.

Cramér, H. (1946b). A contribution to the theory of statistical estimation. *Skand. Akt. Tidskr.*, **29**, 85-94.

Cressie, N., Davis, A.S., Folks, J.L., and Policello, G.E., II (1981). The moment-generating function and negative integer moments. *Amer. Statist.*, **35**, 148-150.

Darmois, G. (1945). Sur les lois limites de la dispersion de certaines estimations. *Rev. Int. Statist. Inst.*, **13**, 9-15.

DasGupta, S. (1980). Distributions of the correlation coefficient. *R. A. Fisher: An Appreciation* (S.E. Fienberg and D.V. Hinkley, eds.), pp. 9-16. Springer-Verlag, Inc., New York.

de Finetti, B. (1937). Foresight: Its logical laws, its subjective sources. Translated and reprinted, *Studies in Subjective Probability* (H. Kyberg and H. Smokler, eds.), 1964, pp. 93-158. John Wiley & Sons, Inc., New York. Reprinted in *Breakthroughs in Statistics Volume I* (S. Kotz and N.L. Johnson, eds.), 1992. Springer-Verlag, Inc., New York.

Efron, B.F. (1975). Defining the curvature of a statistical problem (with applications to second order efficiency). *Ann. Statist.*, **3**, 1189-1242.

Efron, B.F. (1978). The geometry of exponential families. *Ann. Statist.*, **6**, 367-376.

Efron, B.F. (1998). R. A. Fisher in the 21st century (with discussions by D.R. Cox, R. Kass, O. Barndorff-Nielsen, D.V. Hinkley, D.A.S. Fraser, and A.P. Dempster). *Statist. Sci.*, **13**, 95-122.

Feller, W. (1968). *An Introduction to Probability and Its Applications, Volume 1,* third edition. John Wiley & Sons, Inc., New York.

Fisher, R.A. (1912). On an absolute criterion for fitting frequency curves. *Messeng. Math.*, **42**, 155-160.

Fisher, R.A. (1915). Frequency distribution of the values of the correlation coefficients in samples from an indefinitely large population. *Biometrika*, **10**, 507-521.

Fisher, R.A. (1920). A mathematical examination of the methods of determining the accuracy of an observation by the mean error, and by the mean square error. *Monthly Notices Roy. Astronom. Soc.*, **80**, 758-770.

Fisher, R.A. (1922). On the mathematical foundations of theoretical statistics. *Phil. Trans. Roy. Soc.*, **A222**, 309-368. Reprinted in *Breakthroughs in Statistics Volume I* (S. Kotz and N.L. Johnson, eds.), 1992. Springer-Verlag, New York.

Fisher, R.A. (1925a). Theory of statistical estimation. *Proc. Camb. Phil. Soc.*, **22**, 700-725.

Fisher, R.A. (1925b). *Statistical Methods for Research Workers.* Oliver and Boyd, Edinburgh. Reprinted (1973) by Hafner, New York. Reprinted in *Breakthroughs in Statistics Volume II* (S. Kotz and N.L. Johnson, eds.),

1993. Springer-Verlag, Inc., New York.

Fisher, R.A. (1930). Inverse probability. *Proc. Camb. Phil. Soc.*, **26**, 528-535.

Fisher, R.A. (1934). Two new properties of mathematical likelihood. *Proc. Roy. Soc., Ser. A*, **144**, 285-307.

Fisher, R.A. (1956). *Statistical Methods and Scientific Inference.* Oliver and Boyd, Edinburgh and London.

Fréchet, M. (1943). Sur l'extension de certaines evaluations statistiques de petits echantillons. *Rev. Int. Inst. Statist.*, **11**, 182-205.

Gani, J. (1982). *The Making of Statisticians* (edited volume). Springer-Verlag, Inc., New York.

Gauss, C.F. (1821). *Theoria combinationis observationum erroribus minimis obnoxiae.* Its English translation is available in Gauss's publications (1803-1826).

Gayen, A.K. (1951). The frequency distribution of the product moment correlation coefficient in random samples of any size drawn from non-normal universes. *Biometrika*, **38**, 219-247.

Ghosh, J.K. (1988). *Statistical Information and Likelihood: A Collection of Critical Essays by Dr. D. Basu* (edited volume). Lecture Notes in Statistics No. 45. Springer-Verlag, Inc., New York.

Ghosh, J.K., Mitra, S.K., and Parthasarathy, K.R. (1992). *Glimpses of India's Statistical Heritage.* Wiley Eastern, Ltd., New Delhi.

Halmos, P.R. (1946). The theory of unbiased estimation. *Ann. Math. Statist.*, **17**, 34-43.

Halmos, P.R. and Savage, L.J. (1949). Application of the Radon-Nikodym theorem to the theory of sufficient statistics. *Ann. Math. Statist.*, **20**, 225-241.

Hinkley, D.V. (1980). Fisher's development of conditional inference. *R. A. Fisher: An Appreciation* (S.E. Fienberg and D.V. Hinkley, eds.), pp. 101-108. Springer-Verlag, Inc., New York.

Hipp, C. (1974). Sufficient statistics and exponential families. *Ann. Statist.*, **2**, 1283-1292.

Jeffreys, H. (1957). *Scientific Inference.* Cambridge University Press, London.

Johnson, N.L. and Kotz, S. (1969). *Distributions in Statistics: Discrete Distributions.* John Wiley & Sons, Inc., New York.

Johnson, N.L. and Kotz, S. (1972). *Distributions in Statistics: Continuous Multivariate Distributions.* John Wiley & Sons, Inc., New York.

Karlin, S. and Rubin, H. (1956). The theory of decision procedures for distributions with monotone likelihood ratio. *Ann. Math. Statist.*, **27**, 272-299.

Kolmogorov, A.N. (1933). On the empirical determination of a distribution. *G. 1st. Ital. Attuari*, **4**, 83-91. Reprinted in *Breakthroughs in Statistics Volume II* (S. Kotz and N.L. Johnson, eds.), 1993. Springer-Verlag, Inc., New York.

Kolmogorov, A.N. (1950a). Unbiased estimates (in Russian). *Izvestia Acad. Nauk. USSR,* **14**, 303-326. (Amer. Math. Soc. Translations No. 90).

Kolmogorov, A.N. (1950b). *Foundations of the Theory of Probability* (German edition, 1933). Chelsea, New York.

Lane, D.A. (1980). Fisher, Jeffreys, and the nature of inference. *R. A. Fisher: An Appreciation* (S.E. Fienberg and D.V. Hinkley, eds.), pp. 148-160. Springer-Verlag, Inc., New York.

Lehmann, E.L. (1983). *Theory of Point Estimation.* John Wiley & Sons, Inc., New York.

Lehmann, E.L. (1986). *Testing Statistical Hypotheses,* second edition. John Wiley & Sons, Inc., New York.

Lehmann, E.L. and Casella, G. (1998). *Theory of Point Estimation,* second edition. Springer-Verlag, Inc., New York.

Lehmann, E.L. and Scheffé, H. (1950). Completeness, similar regions and unbiased estimation–Part I. *Sankhyā,* **10**, 305-340.

Lehmann, E.L. and Scheffé, H. (1955). Completeness, similar regions and unbiased estimation–Part II. *Sankhyā,* **15**, 219-236.

Lehmann, E.L. and Scheffé, H. (1956). Corrigenda: Completeness, similar regions and unbiased estimation–Part I. *Sankhyā,* **17**, 250.

Lindley, D.V. (1980). L. J. Savage – His work in probability and statistics. *Ann. Statist.,* **8**, 1-24.

Lindley, D.V. and Scott, W.F. (1995). *New Cambridge Statistical Tables,* second edition. Cambridge University Press, Cambridge.

Lukacs, E. (1960). *Characteristic Functions.* Charles Griffin & Co., Ltd., London.

Mukhopadhyay, N. (2000). *Probability and Statistical Inference.* Marcel Dekker, Inc., New York.

Mukhopadhyay, N. (2005). Dependence or independence of the sample mean and variance in non-iid or non-normal cases and the role of some tests of independence. *Recent Advances in Applied Probability* (Ricardo Baeza-Yates, et al. eds.), 397-426. Springer-Verlag, Inc., New York.

Mukhopadhyay, N. (2006). Some comments on Hassairi et al.'s "Implicit Distributions and Estimation." *Commun. Statist. Theory & Meth.,* **35**, in press.

Neyman, J. (1935a). Sur un teorema concernente le cosidette statistiche sufficienti. *Giorn. 1st. Ital. Att.,* **6**, 320-334.

Neyman, J. (1935b). On the problem of confidence intervals. *Ann. Math. Statist.,* **6**, 111-116.

Neyman, J. (1937). Outline of a theory of statistical estimation based on the classical theory of probability. *Phil. Trans. Roy. Soc., Ser. A,* **236**, 333-380.

Neyman, J. (1961). Silver jubilee of my dispute with Fisher. *J. Oper. Res. Soc. (Japan),* **3**, 145-154.

Neyman, J. and Pearson, E.S. (1928a). On the use and interpretation of certain test criteria for purposes of statistical inference, Part I. *Biometrika,*

20, 175–240.

Neyman, J. and Pearson, E.S. (1928b). On the use and interpretation of certain test criteria for purposes of statistical inference, Part II. *Biometrika*, **20**, 263-294.

Neyman, J. and Pearson, E.S. (1933a). On the problem of the most efficient tests of statistical hypotheses. *Phil. Trans. Roy. Soc., Ser. A*, **231**, 289-337. Reprinted in *Breakthroughs in Statistics Volume I* (S. Kotz and N.L. Johnson, eds.), 1992. Springer-Verlag, Inc., New York.

Neyman, J. and Pearson, E.S. (1933b). The testing of statistical hypotheses in relation to probabilities *a priori*. *Proc. Camb. Phil. Soc.*, **24**, 492-510.

Neyman, J. and Scott, E.L. (1948). Consistent estimates based on partially consistent observations. *Econometrica*, **16**, 1-32.

Pearson, E.S. and Kendall, M.G. (1970). *Studies in the History of Statistics and Probability* (edited volume). Charles Griffin & Co., Ltd., London.

Ramachandran, B.R. (1967). *Advanced Theory of Characteristic Functions*. Statistical Publishing Society, Calcutta.

Rao, C.R. (1945). Information and accuracy attainable in the estimation of statistical parameters. *Bull. Calcutta Math. Soc.*, **37**, 81-91. Reprinted in *Breakthroughs in Statistics Volume I* (S. Kotz and N.L. Johnson, eds.), 1992. Springer-Verlag, Inc., New York.

Rao, C.R. (1947). Minimum variance and the estimation of several parameters. *Proc. Camb. Phil. Soc.*, **43**, 280-283.

Rao, C.R. (1973). *Linear Statistical Inference and Its Applications*, second edition. John Wiley & Sons, Inc., New York.

Reid, C. (1982). *Neyman from Life*. Springer-Verlag, Inc., New York.

Reid, N. (1995). The roles of conditioning in inference (with discussions by G. Casella, A.P. Dawid, T.J. DiCiccio, V.P. Godambe, C. Goutis, B. Li, B.C. Lindsay, P. McCullagh, L.A. Ryan, T.A. Severini, and M.T. Wells). *Statist. Sci.*, **10**, 138-157.

Savage, L.J. (1954). *The Foundations of Statistics*. John Wiley & Sons, Inc., New York. Reprinted (1972) by Dover Publications, Inc., New York.

Sen, P.K. and Singer, J.O. (1993). *Large Sample Methods in Statistics*. Chapman & Hall, Inc., New York.

Serfling, R.J. (1980). *Approximation Theorems of Mathematical Statistics*. John Wiley & Sons, Inc., New York.

Soper, H.E. (1913). On the probable error of the correlation coefficient to a second approximation. *Biometrika*, **9**, 91-115.

Soper, H.E., Young, A.W., Cave, B.M., Lee, A., and Pearson, K. (1917). On the distribution of the correlation coefficient in small samples. A cooperative study. *Biometrika*, **11**, 328-413.

Stigler, S.M. (1980). *American Contributions to Mathematical Statistics in the Nineteenth Century* (edited volume). Arno Press, New York.

Stigler, S.M. (1986). *The History of Statistics: The Measurement of Uncertainty Before 1900*. Harvard University Press, Cambridge.

Stigler, S.M. (1989). Francis Galton's account of the invention of correlation. *Statist. Sci.*, **4**, 77-86.

"Student" (W.S. Gosset) (1908). The probable error of a mean. *Biometrika*, **6**, 1-25. Reprinted in *Breakthroughs in Statistics Volume II* (S. Kotz and N.L. Johnson, eds.), 1993. Springer-Verlag, Inc., New York.

Tong, Y.L. (1990). *The Multivariate Normal Distribution.* Springer-Verlag, Inc., New York.

Wallace, D.L. (1980). The Behrens-Fisher and Fieller-Creasy problems. *R. A. Fisher: An Appreciation* (S.E. Fienberg and D.V. Hinkley, eds.), pp. 119-147. Springer-Verlag, Inc., New York.

Zehna, P.W. (1966). Invariance of maximum likelihood estimators. *Ann. Math. Statist.*, **37**, 744.

Zinger, A.A. (1958). The independence of quasi-polynomial statistics and analytical properties of distributions. *Theory Probab. Appl.*, **3**, 247-265.

Answers: Selected Exercises

Chapter 1

1.1.1 (i) $\frac{5}{18}$; (ii) $\frac{5}{12}$.

1.1.2 (i) $\binom{5}{2}$ elements in $S = \{(M1, M2), (M1, M3), (M1, M4), (M2, M3),$
$(M2, M4), (M3, M4), (M1, W), (M2, W), (M3, W), (M4, W)\}$; (ii) $\frac{1}{10}$; (iii)
$\frac{4}{10}$.

1.4.2 $\frac{126}{143}$.

1.4.3 (i) $\frac{16}{270725}$; (ii) $\frac{1}{270725}$; (iii) $\frac{4}{270725}$; (iv) $\frac{1}{270725}$.

1.4.4 E, F are dependent; E, G are dependent.

1.4.5 $\left\{1 + \dfrac{(1-p)}{p} \dfrac{\beta}{(1-\alpha)}\right\}^{-1}$.

1.4.6 $0.0000\,1; 0.99999$.

1.5.1 $p = 0.1$.

1.5.2 Any arbitrary $p \in (0, 0.3)$ works.

1.5.3 (i) $f(2) = 0.5, f(4) = 0.3, f(5) = 0.2$ and $f(x) = 0$ for other x;
(ii) $F(x) = 0, 0.5, 0.8$ and 1.0 respectively for $x < 2, 2 \leq x < 4, 4 \leq x < 5, x \geq 5$.

1.6.1 $c = \frac{2}{3}$; $F(x) = 0, 1$ for $x < 0$ and $x > 1$ respectively and for
$0 \leq x \leq 1, F(x) = \frac{4x - x^2}{3}$; $F(x)$ has no discontinuity points; $F(x)$ is non-differentiable at $x = 0, 1$; $P(-0.5 < X \leq 0.8) = 0.85333$.

1.6.2 $c = \frac{1}{2}$; $F(x) = 0, \frac{1}{2}(1 + x + \frac{1}{4}x^2), \frac{1}{2}(1 + x - \frac{1}{4}x^2), 1$ respectively
when $x < -2, -2 \leq x < 0, 0 \leq x \leq 2$, and $x > 2$; $F(x)$ is continuous and
differentiable everywhere; $P(-1.5 < X \leq 1.8) = 0.96375$.

1.7.1 (i) 0.057396; (ii) 0.99985; (iii) 0.070191.

1.7.3 (i) $0.\,265033$; (ii) 0.95957; (iii) 0.17547.

1.7.4 (i) Exponential with $\beta = \frac{1}{\pi}, c = \pi$; (ii) Normal with $\mu = 0, \sigma = 1/\sqrt{2\pi}, c = 1$; (iii) Normal with $\mu = -\frac{1}{8}, \sigma = 1/\sqrt{2}, c = \pi^{-1/2}e^{-1/64}$; (iv)
Gamma with $\alpha = 2$ or $3, \beta = \frac{1}{2}, c = 1$ or 2; (v) Gamma with $\alpha = 5, \beta = \frac{1}{4}, c = 4$; (vi) Beta with $\alpha = 5, \beta = 3, c = 2$.

1.7.6 $-\infty < x \leq \theta, F(x) = \frac{1}{2}e^{\frac{1}{\beta}(x-\theta)}$; $\theta < x < \infty, F(x) = 1 - \frac{1}{2}e^{-\frac{1}{\beta}(x-\theta)}$.

Chapter 2

2.2.1 \$22.

2.2.2 $\mu = \frac{1}{2}(n+1)$; $\sigma = \sqrt{(n^2 - 1)/12}$.

2.2.3 $\mu = P(A)$; $\sigma = \sqrt{P(A)(1 - P(A))}$.

2.2.4 $\mu = 1/p$; $\sigma = \sqrt{q}/p$.

2.2.5 (i) $c = \frac{2}{3}$; $\mu = \frac{4}{9}$; $\sigma = \sqrt{13/162}$; (ii) $\frac{16}{9}$; (iii) $\frac{128}{5005}$.

2.2.6 $c = \frac{1}{2}$; $\mu = 0$; $\sigma = 4/\sqrt{6}$.

2.2.7 (i) $\frac{524288}{969969}$; (ii) $\frac{3\pi}{2^8}$; (iii) $\frac{31188749}{669278610}$.

2.2.8 $\mu = \frac{2}{5}$; $\sigma = \frac{1}{5}$.

2.2.9 $\mu = \frac{1}{2}\sqrt{\pi\theta}$; $\sigma = \frac{1}{2}\sqrt{\theta(4 - \pi)}$.

2.2.10 $\mu = \frac{\beta}{\alpha}\Gamma(1/\alpha)$; $\sigma = \frac{\beta}{\alpha}\{2\alpha\Gamma(2/\alpha) - \Gamma^2(1/\alpha)\}^{1/2}$.

2.2.11 $\mu = e^{1/2}$; $\sigma = (e^2 - e)^{1/2}$.

2.3.9 $M_{|Z|}(t) = 2e^{t^2/2}\Phi(t)$.

2.4.1 (i) X is a degenerate random variable with $P(X = 5) = 1, \mu = 5, \sigma = 0$; (iii) X is Bernoulli(p) with $p = \frac{1}{2}, \mu = \frac{1}{2}, \sigma = \frac{1}{2}$; (iv) X takes values: $-2, 0, 1$ each with probability $\frac{1}{3}$, $\mu = -\frac{1}{3}, \sigma = \sqrt{14}/3$.

2.4.2 (i) X is $N(0, 50)$, $\mu = 0, \sigma = \sqrt{50}$; (ii) X is $N(0, 2)$, $\mu = 0, \sigma = \sqrt{2}$; (iii) X is Gamma$(2, 3)$, $\mu = 6, \sigma = \sqrt{18}$.

2.4.3 $\mu = -1.8$; $\sigma^2 = \frac{64}{25}$; X takes values: $-4, -2, 1$ with probabilities $\frac{1}{5}, \frac{3}{5}, \frac{1}{5}$.

2.4.6 $\dfrac{e^t(e^{tn} - 1)}{n(e^t - 1)}$.

2.4.7 (i) $d = -3, c = \frac{1}{\sqrt{\pi}}e^{-9/4}$; (ii) X is $N(\mu, \sigma^2)$, $\mu = -\frac{3}{2}, \sigma = 1/\sqrt{2}$; (iii) $\mu_3 = 0, \mu_4 = \frac{3}{4}$.

2.4.8 (i) $d = 2.5, c = \{3^{3.5}\Gamma(3.5)\}^{-1}$; (ii) Gamma$(3.5, 3)$.

2.4.9 Geometric$(p = \frac{1}{5})$; $\frac{576}{3125}$.

Chapter 3

3.2.1 (i) $E(X_i) = \frac{2}{3}, V(X_i) = \frac{1}{18}, i = 1, 2$; (ii) $\frac{1}{6}, \frac{17}{9}$; (iii) $f_{1|2}(x_1) = 2x_1 I(0 < x_1 < 1), f_{2|1}(x_2) = 2x_2 I(0 < x_2 < 1)$.

3.2.3 $c = 1$; $f_1(x_1) = f_{1|2}(x_1) = 2e^{-2x_1}I(0 < x_1 < \infty)$; $f_2(x_2) = \frac{1}{2}e^{-\frac{1}{2}x_2}I(0 < x_2 < \infty)$; Exponential.

3.2.4 $f_1(x_1) = 3x_1^2$ if $0 < x_1 < 1$; $f_2(x_2) = \frac{3}{2}(1 - x_2^2)$ if $0 < x_2 < 1$; $f_{2|1}(x_2) = 2$ for $0 < x_2 < 0.5$; 0.4.

3.2.5 30; 330.

3.2.6 $c = \frac{2}{\sqrt{\pi}}$; $f_1(x_1) = 2\exp(-2x_1)I(0 < x_1 < \infty)$, $f_2(x_2) = \frac{2}{\sqrt{\pi}} \times \exp(-4x_2^2)I(-\infty < x_2 < \infty)$, $f_3(x_3) = \frac{1}{2}\exp(-\frac{1}{2}x_3)I(0 < x_3 < \infty)$; $\frac{85}{8}$.

3.2.8 (i) $\text{Mult}_4(n, \frac{2}{16}, \frac{5}{16}, \frac{3}{16}, \frac{6}{16})$; (ii) $\frac{3}{16}n, \frac{39}{256}n$;
(iii) $\text{Mult}_4(10, \frac{2}{16}, \frac{5}{16}, \frac{3}{16}, \frac{6}{16})$; (iv) $\text{Mult}_3(10, \frac{2}{10}, \frac{5}{10}, \frac{3}{10})$.

3.2.10 (i) $f_1(x_1) = \frac{5}{42}, \frac{10}{21}, \frac{5}{14}$ and $\frac{1}{21}$ when $x_1 = 0, 1, 2, 3$; $f_2(x_2) = \frac{5}{21}, \frac{15}{28}$, $\frac{3}{14}$ and $\frac{1}{84}$ when $x_2 = 0, 1, 2, 3$;
(ii) $f_{1|X_2=0}(1) = \frac{1}{5}, f_{1|X_2=0}(2) = \frac{3}{5}, f_{1|X_2=0}(3) = \frac{1}{5}$;
(iii) $E[X_1 \mid X_2 = 0] = 2; E[X_2 \mid X_1 = 0] = \frac{9}{5}$.

3.2.11 $f_1(x_1) = pq^{x_1}, x_1 = 0, 1, 2, ...; f_2(x_2) = (x_2 + 1)p^2q^{x_2}, x_2 = 0, 1, 2, ...$

3.4.1 Independent.

3.4.2 Independent.

3.4.8 Independent.

3.4.10 Dependent.

3.5.2 $\{1 - 2\left(t\rho + \frac{1}{2}t^2(1 - \rho^2)\right)\}^{-1/2}$ with $t\rho + \frac{1}{2}t^2(1 - \rho^2) < \frac{1}{2}$.

3.5.3 $N_2; \mu_1 = \mu_2 = 0, \sigma_1^2 = 1, \sigma_2^2 = 4, \rho = \frac{1}{2}$.

3.5.4 $N_2; \mu_1 = 2, \mu_2 = -3, \sigma_1^2 = 9, \sigma_2^2 = 16, \rho = \frac{1}{2}$ and $k = \{12\sqrt{3}\pi\}^{-1}$.

3.5.5 0.43005; 0.64308.

3.7.3 (i) and (ii) one-parameter exponential family; (iii) two-parameter exponential family.

3.7.5 (i) Independent; (ii) No.

Chapter 4

4.2.4 (ii) $\frac{1}{2\beta}e^{-|v|/\beta}I(-\infty < v < \infty)$.

4.3.3 $g(y) = \begin{cases} \frac{3}{14}y^{\frac{1}{2}} & \text{if } 1 < y < 4 \\ 0 & \text{if } y \le 1 \text{ or } y \ge 4 \end{cases}$

4.3.4 $g(u) = \begin{cases} \frac{3}{2}(1 - u^2) & \text{for } 0 < u < 1 \\ 0 & \text{for } u \le 0 \text{ and } u \ge 1 \end{cases}$

4.3.9 (i) $f_{123}(y_1, y_2, y_3) = 6$ if $0 < y_1 \le y_2 \le y_3 < 1$, and 0 otherwise;
(ii) $f_2(y_2) = 6y_2(1 - y_2)I(0 < y_2 < 1)$; (iii) $g_1(u) = 6u(1 - u)I(0 < u < 1)$.

4.3.10 (i) $\frac{1}{4^n\Gamma(n)}\{log(\frac{4^n}{u})\}^{n-1}I(0 < u < 4^n)$; (ii) $E(U) = 2^n, V(U) = 4^n\{(4/3)^n - 1\}$.

4.4.1 $\text{Weibull}(\alpha = \frac{1}{2}, \beta^2)$.

4.4.2 $4w(1 - w^2)I(0 < w < 1)$.

4.4.3 $\frac{1}{\theta}exp(-u/\theta)I(u > 0)$.

4.4.6 $g_1(u) = \{\Gamma(2\alpha)\beta^{2\alpha}\}^{-1}e^{-u/\beta}u^{2\alpha-1}; g_2(v) = \Gamma(2\alpha)\{\Gamma(\alpha)\}^{-2}\times$

$\dfrac{v^{\alpha-1}}{(1+v)^{2\alpha}}$ for $0 < u, v < \infty$ and 0 otherwise.

4.4.7 $E\{(X_1 - X_2)^2 \mid X_1 + X_2 = x\} = 2\sigma^2$ and $E\{(X_1 + X_2)^2 \mid X_1 = X_2\} = 2(\sigma^2 + 2\mu^2)$.

4.5.1 (i) $\frac{1}{\pi}(1+v^2)^{-1}I(v \in \Re)$; (ii) $\frac{1}{\pi}(1+v^2)^{-1}I(v \in \Re)$; (iii) $f(v) = kv^{-\frac{1}{2}}\{1+v\}^{-1}I(0 < v < \infty)$; (iv) $(n-2)^{-1}F_{1,n-2}$ for $n > 2$; (v) $2(n-3)^{-1}F_{1,n-3}$ for $n > 3$.

4.5.2 $\sqrt{3}W \sim t_3$; $3W^2 \sim F_{1,3}$.

4.6.4 (i) $N_{n-1}(\mathbf{0}, \mathbf{\Sigma})$, $\mathbf{\Sigma} = \sigma^2[I_{(n-1)\times(n-1)} + \mathbf{11}']$; (ii) $N_{n-1}(\mathbf{0}, \mathbf{\Sigma} = (\sigma_{ij}))$ with $\sigma_{ii} = 2\sigma^2$ for $i = 1, ..., n-1$, $\sigma_{ij} = -1$ for $|i-j| = 1$, $\sigma_{ij} = 0$ for $|i-j| > 1$.

Chapter 5

5.2.1 $a = 2.15$; $b = 5.6275$.

5.2.4 (i) $a = e^\mu$; (ii) $a = \log(|\mu|)$ assuming that $\mu \neq 0$.

5.2.5 $b = 2$.

5.2.6 $0 < \gamma < 1$.

5.2.8 $a = \sqrt{2/\pi}$; $b = (2/\pi)^{1/4}$.

5.3.2 (i) $a_n = e^\mu$, $b_n = e^\mu/\sqrt{n}$;

(ii) $a_n = \log(|\mu|)$, $b_n = 1/(|\mu|\sqrt{n})$ if $\mu \neq 0$;

(iii) $a_n = e^{\mu^2+2\mu}$, $b_n = 2(\mu+1)e^{\mu^2+2\mu}/\sqrt{n}$;

(iv) $a_n = |\mu|^3 e^\mu$ and $b_n = |\mu|^2 e^\mu(1+|\mu|)/\sqrt{n}$ if $\mu \neq 0$.

5.3.4 (i) $c = e^\mu$; (ii) $a_n = e^\mu$, $b_n = \sigma e^\mu/\sqrt{n}$.

5.3.5 (i) $c = \theta e^{-1}$; (ii) $a_n = \theta e^{-1}$, $b_n = \theta e^{-1}/\sqrt{n}$.

Chapter 6

6.3.5 Nonexponential family.

6.3.6 Exponential family; $\Sigma_{i=1}^n X_i^2$.

6.3.7 $\max_{1 \leq i \leq n} |X_i|$.

6.4.1 $I_{\mathbf{X}}(p) = n/(p(1-p))$; $I_{\overline{X}}(p) = n/(p(1-p))$.

6.4.4 $I_{X_1}(\alpha) = 1/\alpha^2$ and $I_T(\alpha) = n/\alpha^2$.

6.5.1 $\mathbf{T}$ is ancillary; T_1, T_2 are independent.

6.5.3 Yes.

6.5.4 Yes.

6.6.1 No.

6.6.2 Yes.

6.6.3 No.

6.6.4 Yes.

6.6.5 Yes.

6.6.6 Yes.

6.6.7 Yes.

6.6.8 Yes.

6.6.9 Yes.

Chapter 7

7.2.1 $\hat{p}_{MLE} = (1 + \overline{X})^{-1}$ and sufficient.

7.2.2 (i) $\overline{X}^2$; (ii) $\overline{X}/(1 - \overline{X})$; (iii) $\overline{X}(1 - \overline{X})$.

7.2.3 (ii) Yes;

(iii) $1/\sqrt{n^{-1}\Sigma_{i=1}^n (X_i - \mu)^2}$;

(iv) $(\{n^{-1}\Sigma_{i=1}^n (X_i - \mu)^2\}^{1/2} + \{n^{-1}\Sigma_{i=1}^n (X_i - \mu)^2\}^{-1/2})^{1/2}$.

7.2.4 $n^{-1}\Sigma_{i=1}^n X_i^\beta$.

7.3.1 (i) T_i is an unbiased, $i = 1, ..., 4$;

(ii) The MSEs of $T_i, i = 1, ..., 4$, are respectively $2\sigma^4 + 2\sigma^2, 2\sigma^4/3, \sigma^4/2$ and $2\sigma^4/3$.

7.4.1 $\overline{X}(\overline{X} - \frac{1}{n})/(1 - \frac{1}{n})$.

7.4.3 $\overline{X}$.

7.4.5 $W = \frac{\Sigma_{i=1}^n X_i}{n}(1 - \frac{1}{n})^{\Sigma_{i=1}^n X_i - 1}$.

7.5.1 $U = \Sigma_{i=1}^n X_i^2$, (i) $\frac{1}{n}U$; (ii) $W = U^2/(n + n^2)$.

7.5.2 (i) Attained; (ii) Not attained.

7.5.4 $(n + k)n^{-1} \max_{1 \leq i \leq n} |X_i|$ if $n > -k$.

7.5.5 (iii) $\overline{X} + (1 - \overline{X})e^2$.

7.5.7 $e^{-2n^{-1}\sigma^2}e^{2\overline{X}}$; CRLB is not attained.

7.6.3 $\hat{\theta}_{MLE} = -(\frac{1}{n}\Sigma_{i=1}^n log(X_i))^{-1} - 1$.

7.6.4 $\hat{\mu}_{MLE} = \frac{1}{n}\Sigma_{i=1}^n log(X_i)$, $\widehat{\sigma^2}_{MLE} = \frac{1}{n}\Sigma_{i=1}^n (log(X_i) - \overline{U}_n)^2$.

Chapter 8

8.2.1 Test #1: $\alpha = 0.088508, \beta = 044038$;

Test #2: $\alpha = 0.046766, \beta = 036866$;

Test #3: $\alpha = 0.059897, \beta = 0.2858$;

Test #4: $\alpha = 0.017864, \beta = 0.18406$.

8.3.4 $k = 4, \gamma = 0.668$.

8.3.6 $k = 1 - exp\{-\tan(\pi\alpha/2)\}$.

8.3.9 $\psi(\mathbf{X}) = I\left(X_1 + \frac{3}{2}X_2 - \frac{11}{2}\mu_0\right)/\sqrt{11\sigma^2/2} < -z_\alpha\right)$;

Power $= \Phi\left(\sqrt{\frac{11}{2}}\frac{(\mu_0-\mu_1)}{\sigma} - z_\alpha\right)$.

8.5.2 $\alpha = \frac{1}{16}$; $Q(\theta) = 1.0, \frac{1}{16\theta^4}, 1 - \frac{15}{16\theta^4}$ when $\theta \leq \frac{1}{2}, \frac{1}{2} < \theta < 1$, and $\theta > 1$ respectively.

Chapter 9

9.1.1 Interval: $(X - c, X + c), c = -log(\alpha/2)$.

9.1.3 Interval: $(\frac{1}{d}|X|, \frac{1}{c}|X|), c = -log(1 - \frac{1}{2}\alpha), d = -log(\frac{1}{2}\alpha)$.

9.2.1 (i) 0.60303; (ii) $9\sqrt{\frac{2}{\pi}}\sigma$.

9.2.2 Interval: $\left(\frac{1}{b}X, \frac{1}{a}X\right), a = 1 - \sqrt{1 - (\alpha/2)}, b = 1 - \sqrt{\alpha/2}$.

9.2.5 $\left(\frac{2\Sigma_{i=1}^n(X_i - X_{n:1})}{\chi_{2n-2,\alpha/2}^2}, \frac{2\Sigma_{i=1}^n(X_i - X_{n:1})}{\chi_{2n-2,1-\alpha/2}^2}\right)$.

9.2.6 Interval: $\left(\frac{1}{b}\max_{1\leq i\leq n}|X_i|, \frac{1}{a}\max_{1\leq i\leq n}|X_i|\right)$ with $a = (\frac{1}{2}\alpha)^{1/n}$, $b = (1 - \frac{1}{2}\alpha)^{1/n}$.

9.2.7 (i) $\left(\overline{X} - a + \sqrt{\frac{(n-1)S^2}{\chi_{n-1,\gamma/2}^2}}, \overline{X} + a + \sqrt{\frac{(n-1)S^2}{\chi_{n-1,1-\gamma/2}^2}}\right)$ with $a = t_{n-1,\gamma/2}\frac{S}{\sqrt{n}}$.

9.3.1 $\left((\overline{X}_1 - \overline{X}_2) - z_{\alpha/2}b, (\overline{X}_1 - \overline{X}_2) + z_{\alpha/2}b\right)$ with $b = \{\sigma_1^2 n_1^{-1} + \sigma_2^2 n_2^{-1}\}^{1/2}$.

Chapter 10

10.2.1 $k(\theta; t) = \frac{1}{\Gamma(n+1)}(t + \alpha)^{n+1}exp\{-(t + \alpha)\theta\}\theta^n I(\theta > 0), t > 0$.

10.3.1 $N(a, b^2)$ with $a = (t + \frac{5}{\tau^2})(c + \frac{1}{\tau^2})^{-1}, b^2 = (c + \frac{1}{\tau^2})^{-1}, c = \frac{7}{3}$.

10.4.1 $T \equiv \Sigma_{i=1}^n X_i, \widehat{\vartheta}_B = (t + \alpha)\beta(n\beta + 1)^{-1}$.

10.4.2 $T = X_1 + \frac{2}{3}X_2, \widehat{\vartheta}_B = (t + \frac{5}{\tau^2})(c + \frac{1}{\tau^2})^{-1}$.

Chapter 11

11.2.2 Reject H_0 if and only if $\Sigma_{i=1}^n X_i^2/\sigma_0^2 < \chi_{n,1-\alpha/2}^2$ or $\Sigma_{i=1}^n X_i^2/\sigma_0^2 > \chi_{n,\alpha/2}^2$.

11.2.3 Reject H_0 if and only if $\Sigma_{i=1}^n (X_i-\mu)^2/\sigma_0^2 < \chi_{n,1-\alpha/2}^2$ or $\Sigma_{i=1}^n (X_i-\mu)^2/\sigma_0^2 > \chi_{n,\alpha/2}^2$.

11.2.7 Reject H_0 if and only if $|\sqrt{n}(T-\mu_0)/\sigma| > z_{\alpha/2}$ where $T = \frac{1}{n}\Sigma_{i=1}^n log(X_i)$.

11.3.5 $T_i = \frac{1}{n_i}\Sigma_{j=1}^{n_i} X_{ij}^2, i = 1, 2$; Reject H_0 if and only if $T_1/T_2 < F_{n_1,n_2,1-\alpha/2}$ or $T_1/T_2 > F_{n_1,n_2,\alpha/2}$.

11.4.1 $t_{calc} = 2.9567, \nu = 7$. Reject H_0 at $\alpha = 0.05$.

11.4.4 $t_{calc} = 2.8849, \nu = 8$. Reject H_0 at $\alpha = 0.05$.

Chapter 12

12.4.1 $z_{calc} = -14.79$; Reject H_0 at *approximate* 1% level.

12.4.2 (3.9773 fl. ounces, 3.9827 fl. ounces).

12.4.3 (18.743 MPG, 19.817 MPG).

12.4.4 Reject H_0 at *approximate* 5% level.

12.4.5 $z_{calc} = -15.704$; Reject H_0 at *approximate* 1% level.

12.4.6 $\widehat{p}_{Lower} = (1+c)^{-1}\left[\widehat{p}_n + \frac{1}{2}c - \sqrt{c\widehat{p}_n + \frac{1}{4}c^2 - c\widehat{p}_n^2}\right]$,

$\widehat{p}_{Upper} = (1+c)^{-1}\left[\widehat{p}_n + \frac{1}{2}c + \sqrt{c\widehat{p}_n + \frac{1}{4}c^2 - c\widehat{p}_n^2}\right], c = n^{-1}z_{\alpha/2}^2$.

12.4.7 $z_{calc} = -9.759$; Reject H_0 at *approximate* 1% level.

12.4.8 (iii) A randomized upper-sided test with $Z_{calc} = \sqrt{n}(2\overline{Y}_n - 1)$.

12.4.9 (i)

Power of the Test
$n = 30$

	μ		
	0.03	0.02	0.01
Test #1	0.06935	0.06234	0.05589
Test #2	0.06497	0.05964	0.05465

$n = 50$

	μ		
	0.03	0.02	0.01
Test #1	0.07595	0.06635	0.05771
Test #2	0.06995	0.06272	0.05607

12.5.1 $log(x), x \in \Re^+$; $\sqrt{n}[log(S_n^2) - log(\sigma^2)] \xrightarrow{\pounds} N(0, \eta^2)$ with $\eta^2 = 2$.

12.5.2 $log(x), x \in \Re^+$; $\sqrt{n}[log(\overline{X}_n) - log(\beta)] \xrightarrow{\pounds} N(0, \eta^2)$ with $\eta^2 = 1$.

12.5.3 $log(\frac{1}{2}x), x \in \Re^+$.

Author Index

Subject Index

A

B

Milton Keynes UK
Ingram Content Group UK Ltd.
UKHW040447071024
449327UK00020B/1070